The Invention of Medicine

ROBIN LANE FOX

The Invention of Medicine

From Homer to Hippocrates

BASIC BOOKS

New York

Basic Books
Hachette Book Group
1290 Avenue of the Americas, New York, NY 10104
www.basicbooks.com

Printed in the United States of America

Originally published in 2020 by Allen Lane, Penguin Random House UK

First US Edition: December 2020

Published by Basic Books, an imprint of Perseus Books, LLC, a subsidiary of Hachette Book Group, Inc. The Basic Books name and logo is a trademark of the Hachette Book Group.

The Hachette Speakers Bureau provides a wide range of authors for speaking events. To find out more, go to www.hachettespeakersbureau.com or call (866) 376-6591.

The publisher is not responsible for websites (or their content) that are not owned by the publisher.

Print book interior design by Jouve (UK), Milton Keynes. Set in Sabon LT Std.

Library of Congress Control Number: 2020942472

ISBNs: 978-0-465-09344-1 (hardcover), 978-0-465-09345-8 (ebook)

LSC-C

Printing 1, 2020

FOR LEO LANE FOX

ἐν τοῖς παθήμασιν ἀνδρειοτάτῳ *Medicine is not only a science and an art: it is also a mode of looking at man with a compassionate objectivity. Why turn elsewhere to contemplate man's moral nature?*

Owsei Temkin, *The Double Face of Janus* (1977), 37

The doctor sees terrible things, touches unpleasant ones and reaps a harvest of personal distress from other people's misfortunes, whereas the sick are turned from the greatest hardships, sicknesses, distress, pain and death thanks to the [medical] craft.

Hippocratic *On Breaths*, c. 400 BC, 1.1

Contents

PART THREE

The Doctor's Mind

List of Illustrations

Book jacket: From an Attic red-figure shallow cup, or *kylix*, *c.* 470 BC, in which it is the tondo, or image inside on the bottom, which the drinker would see as he drank the last of his wine mixed with water. Achilles is bandaging the pained Patroclus, and an arrow, probably the source of the wound, is in the ground before him. The episode is not known in Greek poetry: it is discussed on my pages 37–8. An inscription states that Sosias made the cup. The painter is widely known as the 'Kleophrades painter', while his actual name is still uncertain. Found in the cemetery at Vulci in 1828. Now in the Staatliche Museen zu Berlin, Antikensammlung, inv. no. F 2278.

1. Doctor Iapyx fails to heal Aeneas' arrow wound, while Aeneas' mother, Venus, appears behind with her cocktail of healing drugs, including Cretan dittany. Aeneas' son Ascanius weeps beside him, as in Virgil Aeneid XII.384–424; my page 22 discusses it. Fresco-painting, *c.* AD 45–60, from the House of P. Vedius Siricus, near the Stabian Baths in Pompeii, Ins.VII.1.25, excavated in 1851. Naples Archaeological Museum, inv. no. 9009.
2. Votive relief in the shape of a small temple, dedicated by Archinos to the healing hero Amphiaraos, at Oropos in north-east Attica, *c.* 360 BC. From left to right it shows three episodes of divine 'healing', a widespread resort for the sick from at least the later sixth century BC onwards, but the very opposite of the Hippocratic doctors' medicine. The kindly hero (*left*) leans on his stick and attends a young patient, presumably representing Archinos, and treats his right arm, possibly with a bandage, not a knife. Behind, the right shoulder of the boy, now asleep on a

couch, is being licked by a divine snake, related to the healing hero: snakes were also connected with the healing god Asclepius. Behind stands a young worshipper with his right hand raised, surely young Archinos now healed. The plaque, once painted, may have shown the item he dedicated in thanks. Amphiaraos appeared in the dreams of clients who slept in his sanctuary at night and was considered to heal them. Athens, National Archaeological Museum, inv. no. Γ 3369.

3. Recently published fragment on papyrus of an unnamed medical text, found in the town of Oxyrhynchus in Egypt, first to second century AD. It is the first surviving reference by a Greek author to the Hippocratic Oath and to its use in the early training of medical students, but the author implies that the Oath was not used in that way by all medical teachers: 'for those young men who are being introduced to medicine in a reasoned way ... it is proper, as I see it at least, in the first place to make the beginning of learning from the Hippocratic Oath [the words are in the third line of the text] since it was established as a most just law and one which is extremely useful for life. For to those who have been initiated through it ...' The words may belong at the beginning of a medical text. They testify to the Oath's continuing fame and its appreciation as 'most just'. They also refer to pupils being 'initiated', presumably into medicine, a term used in the separate Hippocratic Law, discussed on my pages 84–5. P.Oxy 74.4970.

4. Sculpture of a young male, or *kouros*, *c.* 550 BC, inscribed on its right leg with the words 'Of Som[b]rotidas the doctor son of Mandrocles'. It marked his grave. Made of marble from Naxos and imported when already cut into shape, but only partly, as its left arm is not detached from the body. The inscription is in lettering which is probably Samian. Found in the south cemetery of Megara Hyblaea, in Sicily, and discussed on my pages 44–5. Archaeological Museum Paolo Orsi, Syracuse, inv. no. 49401.

5. Silver tetradrachm of Abdera, *c.* 500–475 BC. Obverse depicting a winged griffin and signature of the moneyer in the form of the first four letters of his name: PERI-. My pages 155–6 suggest a possible link with the patient Pericles in the city. Paris, Bibliothèque nationale de France, inv. no. M1694.

on one of the gateways in the south-west section of the city's wall. As that sculpture is now very worn, this coin's fine image is evidence of its original details which the Epidemic doctor will have seen on his rounds. From the Pixodarus Hoard, discussed by A. Meadows in *Coin Hoards* 9 (2002) pls. 21, 26d.

11. Thasian marble showing a naked woman, legs crossed, playing the double pipes, or aulos, *c.* 470–460 BC. Found in 1887 in the grounds of the Villa Ludovisi in Rome, part of the ancient gardens of Sallust. It is one of three sculpted sides on the marble block, the back being hollowed out, but as it is far the biggest naked female in Greek art up to its apparent date, it has been widely argued to be a late fake. However, a companion piece to it, also of Thasian marble, in the Museum of Fine Arts, Boston, has now been established to be authentic. The two pieces probably stood at either end of a big altar, I suspect on Thasos itself, although claims were made for its location at Locri in southern Italy before the marble's source was fully recognized. If it is original, it is far superior to other marble sculptures of that date found on Thasos. It was removed by a Roman to Rome in antiquity and is a reminder that much in the island's artistic history has been lost. See my page 180. Bought for the Italian state in 1894 and now in Palazzo Altemps, Rome, inv. no. 85702.

12. Sculpted panel of the three Graces, *c.* 480 BC, set in the wall on the east side of the so-called 'Passage of the Theoroi' in Thasos city, the passage in which the big list of theoric magistrates was later inscribed. The Graces process toward the god Hermes, who holds out his right hand, standing on their right across an intervening niche and small altar. Beneath him were inscribed the words 'For the Graces a goat forbidden and also a pig [as sacrifices]'. The figures exemplify the beginnings of the change from the stiff archaic style to the newer severe style. The sculptures and the little ash altar overlooked a main axial street, where the doctor will have seen them on his travels. See my pages 161 and 258. Louvre, Paris, inv. no. Ma 676.

13. Marble banquet relief dedicated in the centre of Thasos city, found in 1907 and removed to Istanbul Museum in 1908. About thirty such reliefs are known on Thasos across the centuries, but

this one is especially big and a fine example of the severe style. The reclining male is usually interpreted as a heroic ancestor, and indeed Thasos had a festival for hosting heroes, the Heroxenia, not attested in any other Geek city. However, even at this early date I incline to see him as a commemoration of the dedicator. His wife is depicted in a separate woman's room. She is dipping what seems to be a pin into her alabaster vase, perhaps in order to apply scent from it to her hair and person. Parts of the relief were painted. See my pages 204–5. Istanbul Archaeological Museum, inv. no. 1947.

14. Upper part of the marble Thasian funerary *stele* of Philis daughter of Cleomenes, a fine work *c.* 450–440 BC in the classical, not early severe, style, as her head well shows. Below it she is shown handling a piece of cloth above a small chest, an item whose meaning is uncertain. She was probably very young or new-born during the doctor's time on Thasos. Funerary inscriptions there name a respectable woman by her own name, although polite outsiders, including the doctor, did not do so in everyday life. Louvre, Paris, inv. no. Ma 766.

PICTURE CREDITS

1. H. J. Rix. 2. E. Nikolaou. 3. Egypt Exploration Society and Oxyrhynchus Papyri Project (Oxford). 4. Costello Images. 5. Bibliothèque nationale de France, M1694. 6. Ashmolean Museum, Heberden Coin Room, Oxford. 7. Antikenmuseum und Sammlung Ludwig, Basel. 8. R. Fleischmann. 9. E. McFadden CNG Coins. 10. Nomos Auction 16 (2018), lot 56. 11. J. E. Skinner. 12. Alamy Images. 13. E. Yildiz. 14. D. Mercier. 15. Map adapted and updated from wikimedia.org/wiki/File: Plan_der_antiken_Stadt_Thasos; drawing on Y. Grandjean, F. Salviat, 'Plan générale de la ville antique', in *Guide de Thasos*, fig. 12.

List of Maps

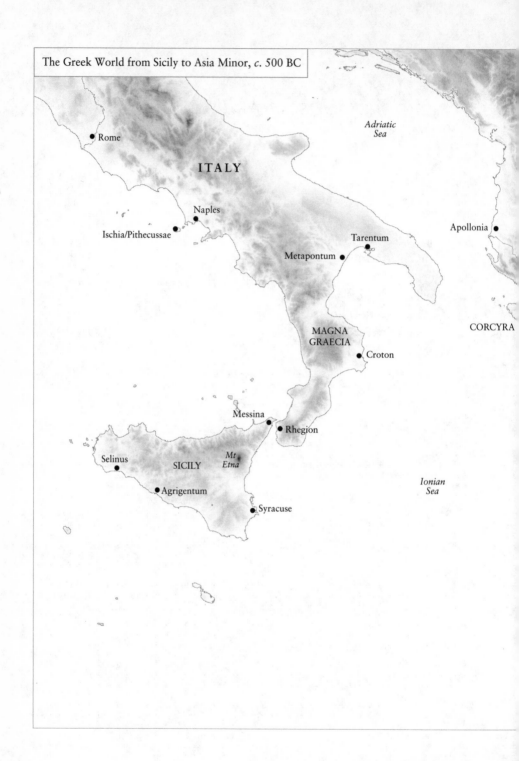

The Greek World from Sicily to Asia Minor, *c.* 500 BC

Rome

*Adriatic
Sea*

ITALY

Naples

Ischia/Pithecussae

Apollonia

Tarentum

Metapontum

MAGNA
GRAECIA

CORCYRA

Croton

Messina

Rhegion

Selinus

*Mt
Etna*

SICILY

*Ionian
Sea*

Agrigentum

Syracuse

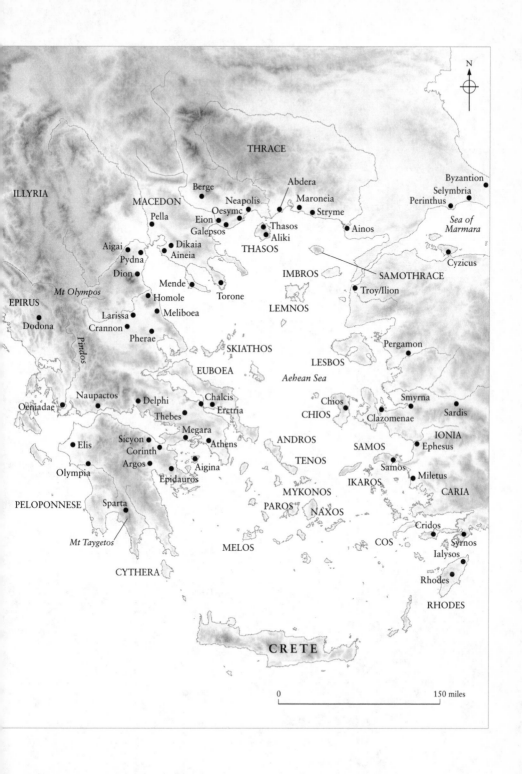

Northern Greece and the travels of the Epidemic doctors, *c.* 470–355 BC

Preface

'Medicine' and 'doctor' are words based on Latin, but they have Greek practical, ethical and intellectual roots which are among the most remarkable legacies of the ancient world. I approach them as a historian. Excellent studies by specialized scholars of ancient medicine have continued to clarify almost every text or topic which this book discusses. Historians, meanwhile, make most use of medical texts which they can set directly beside the great historians Herodotus and Thucydides. The two branches of scholarship, historical and medical, have moved in parallel, not always making contact. I hope to show how both can gain, even now, when they are closely interrelated.

My title has a specific reference. The terms doctor, healer and medicine have modern overtones, not always apt for the ancient world, but there were skilled practitioners in many of its cultures, including Babylonia and Egypt, both within reach of the Greeks. There had also been widespread doctoring and use of remedies based on plants long before Homer or any surviving Greek medical text. Scholars of them rebel, as I do, against the notion of a single invention of the skill in the fifth century BC. However, only among Greeks were there authors at that time who proclaimed a new method and, therefore, a new craft of medicine. It is in that specific sense that this book's title applies.

In this long-studied field, I have new things to say. I have worked throughout with the underlying ancient evidence, enabling me to select, and at times reject, positions advanced in modern studies on my chosen topics. I have also selected what to adduce from the ancient evidence itself. Selection and rejection give my book a distinctive texture and emphasis. I also present a new date for the texts on which its second and third parts focus, and I give them a new social, cultural and intellectual context as a result. The consequences are far-reaching.

My main subjects are Epidemic doctors, people who have acquired a haunting relevance as I finish this book. My interest in them grew from quite other roots, some of which go far back in time. Hippocratic medicine was one of the first subjects my Oxford tutor-to-be, Geoffrey de Sainte Croix, discussed with me during our first pre-tutorial meeting in 1967, addressing it even then with his exceptional range and energy. I am pleased to find that my chapter on Thucydides and medical authors coincides at a crucial point with his own penetrating pages on the subject, published in 1972. I had forgotten this overlap when I first wrote mine. It is to his inspiring grasp of Greek science, a side of his scholarship not always remembered, that I owe my continuing interest in the subject.

Ten years later I found myself appointed as his tutorial successor. While preparing new lectures on the age of King Philip and Demosthenes, I discovered that the first references to Philip's military catapults were lurking, relatively ignored, in one of the medical Epidemic books. In perpetual rebellion against a prescribed syllabus I diverted to reread the other Epidemic books and then asked myself why modern scholars gave them dates with such confidence. I soon found that they related the earliest two books to the big inscriptions of public magistrates on Thasos, but when I looked at the prevailing reasons for the dating of names in those lists, I was struck by the first of the two serious difficulties I discuss in chapter 13 of this book. The fragmentary lists, I hastily concluded, needed to be restored in a different order. With an amateur rearrangement of the main inscription in one hand, I went to see the great epigraphic expert, David Lewis, in Christ Church, Oxford. He heard me out and then remarked that he had been travelling in a similar direction since 1955 and that I needed to be bolder and that evidently he should speak on the topic one day. I prudently withdrew from it meanwhile, but in autumn 1993 he did so, in a single Oxford class attended by two others whose identity I cannot recall and have so far failed to establish. After a slow start he presented us with his own compressed, but detailed, rearrangement of the two main lists of Thasian magistrates and concluded generously that it would not surprise one member of his audience. Any notes he made to accompany his lecture are not among those preserved after his early death in 1994. His literary executor, Peter Rhodes, has confirmed to me that he had no instruction to preserve anything relating to this topic, one he had not known Lewis to have studied in depth. Since 1994, study of the Thasian inscriptions has continued to progress and new pieces have been published. My recall of the general shape of Lewis's reconstruction suggests it is one he would undoubtedly have revised. However, I believe that while writing this book I too have seen and developed another obstacle to the prevailing dating, one which, I think, he must have adduced in 1993. I have also been sustained by his agreement that the date of the medical texts, generally related to the date of the magistrates in the inscriptions, is far earlier than scholars have assumed.

Anyone who ventures into the challenging epigraphy of Thasos owes major debts to previous interpreters of the finds. My awareness of its relevance for medical studies was soon transformed by Karl Deichgraeber's two brilliant studies, made in 1933 and 1982, bold though some of their conjectures are. Another debt is more fundamental. There would be nothing to study without the decades of devoted work on Thasos by members of the French School in Athens. Their publications underlie my chapters on the archaeology, topography and epigraphy of the island and remain the vital resource for any scholar concerned with them. The work of J. Pouilloux, F. Salviat and P. Hamon, among others, is fundamental for the continuing restoration and interpretation of the major inscriptions. I would understand none of the issues involved without it.

Patrice Hamon has brought penetrating lucidity to them, not only in two brilliant articles published in 2015/16 and 2018, but in the discussion and comments which he kindly shared with me in 2019. I am so grateful to him for his time and generosity. As I finished, his volume III of the Corpus of Thasian inscriptions was published, a work of exemplary judgement, range and scholarship. I have had time to re-check my text with its help, and in an Endnote I have touched on some of the connections it proposes which bear on my chapters 12 and 13. They will be clearer when the lists of Thasian magistrates are published in future volumes of the Corpus. His grasp of the prosopography of the lists, the letter-forms of the inscriptions and the likely shape of the lists as a whole are crucial contributions to questions of their underlying chronology, but answers still rest on probabilities, not on certainties. I must emphasize that my re-dating of the medical texts does not depend on my suggested re-dating of the columns of these lists. Decisive reasons for it remain, even if a re-dating of the columns is eventually ruled out by other conclusive evidence.

I gained first-hand detailed knowledge of Thasos with Alexandra Kasseri, an exemplary companion even when confounded by forest fires and night-time lightning as if from Zeus. Her acute grasp of the topography and archaeology of the mainland transformed what I saw in it. Subsequently, Yangos Chalazonitis shared his expertise too, sending me relevant extracts of his Oxford doctoral thesis which I acknowledge gratefully. On Thasos, Konstantina Panoussi was an invaluable guide to the Museum and its contents and to the island's setting and antiquities, of which she shared her profound local knowledge. She then kindly read a draft of my Chapter 14 and commented on it helpfully. I thank her for her time, hospitality and attention, and Tony and Manuela Wurch Kőzelj for theirs too. Leah Lazar followed up with yet more help.

As a foot follower of the big battalions who study ancient medicine I owe so much to their commanding scholars old and new. The works of Émile Littré remain awesome for their grasp and range and I continue to learn from them. Those by Ludwig Edelstein and Owsei Temkin are ever-rewarding, as are the English translations and notes by W. H. S. Jones, who would, I hope, appreciate the role given to malaria in my text. Geoffrey Lloyd and Helen King sharpen the thinking and concepts of everyone who works in their high-powered slipstreams. Ann E. Hanson's many writings have raised my awareness of women's role in the subject and also of medical papyri. I am especially grateful to two speed-readers of the first draft of my text, John Wilkins and Vivian Nutton, the great expert in ancient medicine, who deftly probed its weak spots and made me rethink. Catherine Darbo Peschanski kindly shared a text she delivered in Oxford on the Hippocratic doctors' implicit thinking about heat and fever and followed up with helpful comments on related topics. Robert Arnott invited me to the conference at which she and others spoke most helpfully on ancient and modern diagnoses of disease. Elizabeth Craik has been a kind guide, both in print and in person, and organized a conference panel at St Andrews in July 2018 which became an immensely helpful gathering of experts in the field. Tomas Alusik and Matthias Witt have helped me with their offprints. I particularly thank Laurence Totelin for her work on medicinal plants, a field in which I too have long laboured. Alain Touwaide then transformed my understanding of this subject by sharing the years of research he has conducted in his Institute for the Preservation of Medical Traditions, an essential contribution to knowledge on this subject from Hippocrates to Dioscorides and beyond. My discussion of retrospective diagnosis in chapter 18 owes much to my friend Annelieke Oerlemans in Leiden, to Lutz Graumann in Giessen, who kindly read the chapter in advance, and especially to Penelope Frith in New College, who transformed my hazy medical understanding and responded with chastening rapidity.

On the Epidemic books, I am exceptionally grateful to Robert Alessi, expert especially in the difficult books 2, 4 and 6 and in the use of them and the other Epidemic books in the Islamic world. He kindly gave a close and penetrating reading to my pages on both these subjects, sending me his comments and insights, including corrections to my Arabic, on the very eve of Brexit, a memorable last gift to a fellow EU member. My greatest debt is to the work of Jacques Jouanna, who has transformed international study of the Hippocratics in the past fifty years. In spring 2017 I was two years into this book and was battling with the many problems of the Greek text of the *Epidemics*, which were causing my translations to look like minefields of square brackets and question marks. I then acquired in Paris his magisterial new edition and commentary on books 1 and 3, the essential work on the subject. Where I already held views which he too expounded I let mine stand, but I specifically note those which I owe to him or which he has expressed with greater range and clarity. I have not always agreed with him, especially on questions of dating and topography, but I need hardly say how much I and every other worker in this area owe to his edition, a philological masterpiece, and to his many essays on related Hippocratic questions.

In Oxford it is a pleasure and instruction to talk daily with my colleagues in New College, with William Poole, especially on early modern scientific writing, Andy Meadows, especially on coinage, Paolo Fait on Plato, Jane Lightfoot on Greek literature and so much else, Robert

Parker on Greek religion and the perils of being too sure of my guesses, Stephen Anderson on finer points of Greek language, and David Raeburn on Greek tragedy, of which he, the maestro, and his College productions have transformed my understanding. I am also grateful to Yana Sistovari and her Thiasos company for their mesmerizing performance of the *Bacchae* in the College gardens in summer 2017, followed by an invitation to Poland in 2018 for a rerun which caused me to address the relation between Euripides and medical knowledge. I owe particular thanks to Anna Blomley for sharing her acute topographical scholarship, especially on Thessaly, and for much helpful guidance and encouragement on related subjects. Elsewhere in my wider Oxford orbit, I have been kindly helped at various points by Peter Rhodes, Peter Thonemann, Aneurin Ellis-Evans, Alexy Karenowksa, Danuta Shanzer, now in Vienna, David Potter, now in Ann Arbor, Peter Wilson, now in Sydney, and one of my earliest ex-pupils, Andrew Erskine, now in Edinburgh.

My publishers Stuart Proffitt and Lara Heimert each commented in great detail on the shape, obscurities and details of my text and made me think more widely and exactly. I am also grateful to Alice Skinner for reading and assisting, to Mark Handsley for expert copy-editing, and to Claudia Wagner for help with finding pictures. Estella Kessler, Daniel Etches and Charlie Baker took over typing and co-ordinating my Bibliography with great skill.

Just as I finished work on this book, an epidemic broke out in China and began to travel west, giving contemporary force to the issues of transmission, observation and prognosis which are prominent among my topics. It has made me reflect on how the doctor who is my main subject would have addressed it. The diffusion of an epidemic by travellers or by people taking flight elsewhere is not one which his text considers. Meanwhile I would like to thank the doctors who have saved members of my family in other circumstances, John Sichel and John Ledingham for their help, Tom Cadoux Hudson, Richard Keys and his Oxford team for saving my daughter, James Ng for saving half my eyesight, and Vesna Pavosevic and her team in Great Ormond Street for saving my grandson, to whose exemplary bravery this book is dedicated.

Robin Lane Fox

Introduction

In the fifth century BC, there were individual Greeks, all men, who started to think and write in ways which no one had attempted before. Not all Greeks did so, but the thoughts and texts of this minority still shape ways in which we think and write too. They include writers and thinkers on medicine, the subject of this book.

Their 'invention of medicine' was one item in a broader wave of new thinking. In the fifth century BC Greeks began to write history by using the first-person pronoun, 'I', and expressing their own views. In the Near East, written narratives about the past had always been anonymous and impersonal, as if they were 'the' narrative, not one person's subjective account.[1] In Babylon and Egypt there were many texts which used numbers and calculations, but it was a Greek, Hippocrates of Chios (c. 440 BC), who wrote theoretically about mathematics for the first time.[2] Philosophers, another Greek invention, began to explore differences between knowledge and belief and to contrast what is natural with what is conventional. Mid-century, some of them invented political philosophy, a major conceptual change to political argument ever since.[3] By the end of the century there were even some who argued that slaves and free men did not differ by nature and that being the master of a slave was actually contrary to nature and therefore unjust: it is the first attested challenge to this basic element in ancient Greek societies.[4] Others, meanwhile, related differences in human communities to differences of climate and environment. They wrote about the very nature of man for the first time.[5]

There were changes, too, to styles of speaking and writing. Orators began to be taught by experts in a new skill, rhetoric, which influenced ways of writing prose and verse ever afterwards. In Athens, from the

430s onwards, dramatists, including Aristophanes, composed the world's first political comedies: some of their plays are the first, but not last, word on populism and on demagogues, a term their comic authors invented.[6] New subjects were addressed in prose too. Near the end of the century, Simon the Athenian wrote the first text about horsemanship. It is a defining moment in man's 'Centauric pact', his bargain with the horse as if the rider and his steed are, in theory, one.[7] Meanwhile, in Greek Sicily, Mithaecus wrote the world's first cookbook, paining the puritanical Plato, but not paining posterity, as his recipes have all been lost.[8]

These changes are unmatched in neighbouring civilizations, whether Egyptian, Babylonian, Hebrew or Persian: for a start, they had no philosophers. Yet, Greeks did not stop there. During the fifth century BC, they wrote texts on painting and on the making of sculptures, arts which Greek practitioners changed too, influencing the entire history of Western art and marking it off decisively from cultures to the east and, when finally discovered, the west.[9] The scenes painted on decorative Greek pottery also took on a new style and favoured a new range of subjects. By the last quarter of the century, painters learned to mix and shade colours so as to suggest contours and volume for their figures. Apollodoros was identified later as the 'first to depict objects as they appear', using 'shadow painting' to enhance a lifelike effect.[10] We know nothing else about him, a forgotten master.

Free-standing sculptures were revolutionized too. Before the fifth century BC, Greek sculptors had already shown the male body naked, but in the first decades of the century they showed it in new poses, presenting an idealized beauty and a facial expression of inner contemplation which was quite different from the set smile of previous Greek statues. The fifth-century masters stylized the lines of the male body in new ways and changed the distribution of their figures' weight. They may have been helped by a new technique, preliminary modelling, whereby they could build up the body from within 'adding muscle and flesh in clay . . .', in the view of their lifelong expert, John Boardman, and then translating the life-size model into stone: 'Archaic sculpture was essentially carved; Classical sculpture is essentially modelled.'[11] One of the classical masters, Polycleitus of Argos, *c.* 450 BC, wrote a text called the *Canon*, which defined the measurements and proportions of the ideal male body for the first time. It has not survived; nor has his bronze masterpiece, the Spear Bearer, which is known only through later Roman marble versions. In them, the male body has a broad-shouldered chunkiness which is not everyone's current ideal of male beauty.[12]

These changes of thought and technique were contemporary with changes in political practice. As the fifth century BC advanced there were ever more Greeks who practised democracy, the system invented by the Athenians in 508 BC. It was not that pale modern shadow, 'representative' democracy with political parties and 'deputies'. It was government by the majority decision of all citizens, all of them males, expressed by voting in person at public meetings on every major issue. In the second half of the fifth century, its cultural impact was evident, not least when big theatre audiences enjoyed the hilarious outspokenness of Athenian political comedy, inconceivable without a democratic context.

The Greek theatre's other wonder, tragedy, owed democracy a debt which was more oblique. There was no necessary link between democracy and the genre of tragic drama itself. Some of the fifth century's greatest tragic plays were composed and first performed for monarchies outside Athens.[13] In due course, however, there was a new note to the scenes of dispute and debate which had become central to many of the plots. Their protagonists disputed the rights and wrongs of decisions and actions, arguing at length over guilt and responsibility, the voluntary and the involuntary. Since 462 BC, in Athens's lawcourts, speeches on these topics had begun to be delivered before big democratically chosen juries, newly instituted as courts of first and last instance for most of the city's criminal cases. Many of their jurors, male citizens, were also members of the huge audiences for whom dramatists explored similar issues in the centre of their plays.[14]

The Athenian democracy commissioned some of the century's finest buildings, including the iconic Parthenon, begun in the 440s, but here democracy did not foster a new style. The basic architectural orders had all been invented earlier. As for political philosophy, democracy did not monopolize discussion there, either. Theorists considered the origins of political society, even proposing that a social contract underlay it, but democracy's main theoretical impact was to provoke contrary theorizing by those who loathed it.[15] Pamphleteers expressed views which could never have been presented in a democratic assembly. They wrote on oligarchy, not democracy.

This amazing range of achievement did not result from a technological revolution. There was no such thing. The revolution behind it was conceptual. In his fine tragedy, the *Antigone*, composed very probably in the 440s, Sophocles gave his chorus stirring words on man's ingenuity and progress. His crowning example was not art or politics: it was medicine. 'From Death alone man will not procure a refuge, but from diseases hard to resist he has devised escapes'.[16]

His chorus's words continue to resonate, never more so than when man's devising of 'escapes' has become critical in a global pandemic, hard indeed to resist. Unlike our terms for most of the sciences, 'medicine' is not based on a Greek word. Its root is Latin, but as the Romans adopted Greek medical skills, its historical roots are Greek nonetheless. Unlike philosophy, it was not a Greek monopoly. Doctors had long been famous in other older societies, especially in Egypt, India and Babylon, but in the fifth century BC there were Greeks who were conscious of having a new method: they looked on medicine as a craft they had invented. The 'invention of medicine' is, in this sense, theirs. They are headed by the great name of Hippocrates, born on the island of Cos and still admired throughout the world. He is widely regarded as the founding father of rationally based medicine and medical ethics, but his exact dates and writings, if any, have remained matters of scholarly dispute.

These new Greek thinkers addressed the human body, both male and female, as a natural entity and discussed the environment and lifestyles which could best maintain or restore health. Here too there was no technological revolution, no wonder drug or breakthrough in treating a disease. The change, again, was conceptual. Not everyone who propounded medical theories was necessarily even a doctor. As man was a part of the natural world, study of him and his constitution was a part of natural philosophy, to be discussed in that context too. Philosophers, therefore, pronounced on the subject, whereas doctors worked with, or against, ideas which philosophers first formulated.[17] In Egypt or Babylon there was no such interplay.

As the new Greek medicine was not confined to doctors' practice rooms, effects of its theorizing have been looked for in other fifth-century BC modes of expression. It has been related to the new style of sculpture: 'balance, rhythm, proportion, harmony and symmetry', on one rather optimistic view, 'are the language of Greek medicine, but also of representational art', visible in male statues, especially, from the mid-fifth century onwards.[18] Improved medical understanding of veins and muscles has also been seen, controversially, behind fifth-century sculptors' representations of the male nude.[19] In drama, 'Hippocratic medicine' has been upheld, also controversially, as a formative element in the making of Euripides' tragedies.[20] Its presentation of madness and mental stress has been considered to underlie the dramatist's presentation of similar conditions in his plays from the 430s onwards, whether the madness of Heracles or the trance from which poor, crazed Agave regains sanity in the *Bacchae*, his late masterpiece (composed by 406 BC).

The new medical texts have been related to a democratic context, as if they too owed a debt to the culture of open and competitive debate in public meetings. They have even been linked to the new field of history-writing. On a first encounter, medical awareness may not seem to typify history's founding text, Herodotus' Histories, complete in the 420s, but many examples and stories of doctoring can be assembled from his nine long books. In Thucydides' great Histories, which were still being composed two decades later, medicine's presence is unmissable, both in his description of the deadly plague which afflicted the Athenians from 430 BC onwards and as an implicit parallel for his notion of the value to posterity of studying history at all. He also presents a leading political figure as endorsing medicine in a public speech. Fifteen years after the start of the plague, the Athenians decided in 415 BC to send a military expedition against Sicily, but the decision was then reopened in their democratic assembly. Thucydides tells how the general Nicias, wanting the idea to be abandoned, told the presiding member of the council to put the matter to a public vote for a second time: if he was afraid to do so, let him consider that if he did so, he would be acting as the 'doctor of the city when it has deliberated badly'. This advice, still pertinent, is then amplified. Good governance, Nicias says, is when 'someone benefits his country as much as possible or does it no willing harm.' This same precept, to 'do good or at least no harm', had already been commended to doctors in at least one of the new-style medical texts.[21] It was to have a long life in medical ethics, but less of a life in political practice.

One difficulty in assessing the new medicine's influence is the difficulty of knowing where and when the surviving Greek medical texts were written. Eventually they were all gathered into a collection which is now known as the 'Hippocratic Corpus', but they carry no dates and the famous name of Hippocrates is no help in fixing them. It is not just that scholars contest which of these texts, if any, were written by Hippocrates himself. The first surviving mention of him is in a philosophical dialogue by Plato set in c. 433 BC, but is minimally informative.[22] Already in antiquity, biographers and storytellers tried to fill the gap, but they wrote long after Hippocrates' death.

While reconsidering these problems, I will focus on two famous medical books in the Hippocratic Corpus, originally united as one. I will explain why they are rightly regarded as high points of the new medicine, but I will argue that they originated in a period which nobody has yet considered. They have been universally dated late in the fifth century BC, in the age of Socrates and other great thinkers and writers whose rational methods of argument seem to be the natural context for their author's thinking. I wish to overturn that consensus and, with it, the prevailing view of their relationship to fifth-century thought.

I will proceed in three stages. First, I will consider these texts' predecessors in order to see how far they themselves were something new: for this reason I will survey what we know of doctors and healing from the Homeric poems to the mid-fifth century BC. Then, from the big collection of Greek medical writings now known as the Hippocratic Corpus, I will sketch the invention of medicine to which these two books relate. I will then narrow the focus and engage in some concentrated local history in order to place and date them. Most of their contents connect to a particular Greek island-city which is also known through archaeology and the many texts inscribed there, especially on blocks of stone. Historians of the fifth century still concentrate on Athens and Sparta, the dominant powers in Greece after 480 BC. In my account the 'doctors' island' turns out to be the Greek place which we can know best at the very start of this period, one in which Athens is only dimly attested.

I will then consider the thinking which the two texts contain. I will examine its relations, if any, to democracy and philosophy and to drama and history-writing: in the fifth century BC innovations did not all move simultaneously in one direction. I will also consider how these medical texts differ from those known in adjoining cultures, whether Etruscan, Egyptian or Babylonian. Finally I will compare them with their most evident heirs, one of whom, a Greek, is another neglected genius of the later fifth century. He acknowledges them as predecessors but his approach is strikingly different from theirs. Another of their heirs is not Greek at all. He is one of the geniuses of the Islamic world, some 1400 years later, aware of these books by an indirect route. His use of them, too, is tellingly different, in ways which accentuate their own qualities.

This three-part study is also, implicitly, an exploration of what we can know about the distant Greek world, how and within what limits. The first part, from Homer to the Hippocratic Corpus, rests on beguiling but randomly preserved scraps of evidence, texts, especially poems, objects and inscriptions with a medical reference. They are memorable shafts of light, but beyond them so much remains hidden from our view. The second part rests on apparently more promising survivals, not only the medical texts themselves, but the results of archaeology in one Greek city for more than 150 years and especially its inscriptions, more detailed in scope and time span than those found in other Greek cities of this period. Yet here too there are limits to what apparently rock-solid evidence can establish. The third part rests on well-preserved texts, but they need interpretation as much as any newly found object from the ground.

In the interplay between these types of evidence and the reality which once surrounded them, part of the fascination of ancient history resides. We should not simply take this distant past for granted, as if it is as well attested as our own. There is, however, another part to it, the recognition that even after 2500 years, new places and dates can be cogently assigned to very old items, hitherto misplaced. They change our sense of the Greek past, its passage from what we call an archaic to a classical age and the thinking and expression it involved. They also bear on how we regard that enigmatic figure, Hippocrates, the father of medicine himself.

PART ONE

Heroes to Hippocrates

To be sure, none of the things which have been found [by medicine] can be known by someone looking only with the eyes: that is why I have called them 'not evident' and also why they are adjudged so by the craft [of medicine]. It is not at all that what is 'not evident' has predominated, but rather it has [so far] been dominated: for it to predominate is possible, in so far as the natures of the sick allow examination and those of the researchers are naturally suited to research.

On the Art, 11 (c. 430–400 BC)

They [the Greeks] have taken an oath among themselves to kill all barbarians by medicine, and they do this very thing for a fee, so that trust will be put in them and they may destroy with ease.

Cato the Elder, *To His Son*, in the 170s BC,
cited by the elder Pliny, *Natural History*, 29.1–28:
an implicit reference to Greek doctors'
'Hippocratic Oath' has been detected here

I

Homeric Healing

I

Long before texts about medicine were composed in the fifth century BC, healers and doctors had been practising in the Greek world. They existed in the palace society of the early Mycenaean period (c. 1400–1200 BC), but the first doctors known to us by their personal names are doctors active in Homer's epic poems some 500 years later (c. 760–730 BC). Doctors are then mentioned in Herodotus' Histories (composed c. 440–425 BC). In these classic works they are crucial to two pivotal moments. Without a doctor's role Homer's Achilles would never have returned to the fighting round Troy. Nor, Herodotus says, would the Persian king, Darius, have become interested in invading the Greek world. Before texts of the new medicine survive, doctors in Greek literature have much to answer for.[1]

In Homer's Iliad, the Trojans are pressing hard on the Greeks, and one of the two top doctors on the Greek side, Machaon, is wounded in the fighting and carried off in elderly Nestor's chariot. The Greeks' supreme warrior, Achilles, is absent, refusing to fight, but is watching from his beached ship and thinks he can identify the wounded man as he is driven past. If such a great healer is himself in need of a 'blameless healer', the Greeks' predicament, Achilles sees, is indeed dire. So he calls to his beloved companion Patroclus, who comes from the tent, 'the beginning of evil for him', words very close to those which Herodotus uses 300 years later for the Athenians' first, and similarly fatal, despatch of help to Greek rebels against the Persian king. Achilles sends Patroclus to find out more, the start of his own re-engagement with the Greek army. It leads on to Patroclus' compassionate attention to the wounded and then to his fateful plea to Achilles to be allowed to enter the fighting.

It is not only that a doctor is pivotal to the Iliad's plot. In the view of Galen, the greatest doctor of the second century AD, Homer was the founder and patron of medicine.[2] This unusual tribute to an epic poet is based on a distinctive aspect of his Iliad: its description of about 300 wounds. The majority are briefly described as fatal wounds through a hero's chest or skull, but about thirty are followed in inner and outer detail through the body. Homer's range of bodily detail far surpasses the range in other epic traditions. Unlike most modern readers, his Greek audience was able to relate to poetry on the passage of spears through entrails, the lower stomach or, most gruesomely, the face: 'Hector hit him under the jaw and the ear, and the spear thrust out his teeth by the roots and cut through the middle of his tongue . . .'[3]

The emphasis of these detailed descriptions is on killing, not dying.[4] The poem's first audience surely included warriors who knew very well what a wound in close combat could involve. The details were not given for amusement. They are dispassionate and precise, and if there is an implicit emphasis, it tends to fall less on the victim's poignant suffering than on the striker's might.[5] It is never moralized as violence, though some of the extreme wounds are surely meant to seem horrific, at times when the violence of the fighting in the poem is increasing. Three of the most gruesome are inflicted by one and the same warrior, Meriones, himself a grisly character. Even so, most of them befall victims whose past or present behaviour is relevant to what they receive.[6]

These scenes of wounding show doctors cutting out weapons and applying medicaments. Long before any medical texts about wounds and surgery, they seem to be drawing on close observation of parts of the male body. They also relate to two critical questions for medicine's future: the degree to which traumas and diseases were ascribed to interventions by the gods and the social status of doctors themselves.

II

Readers who have a medical training still admire what they regard as Homer's 'anatomical topography': it impels them to analyse it clinically. This type of study began in Italy in the early seventeenth century and by 1879 Hermann Frölich, himself a military doctor, concluded that Homer must have been one too, not the top doctor in king Agamemnon's camp but perhaps the second-in-command who could take a general view of the action.[7] 'Doctor Homer' continues to be discovered by surgeons and pathologists. They count and tabulate Homeric wounds as data (53 in heads and necks or 54 thoracic, of which 70.37 per cent are fatal . . .) and continue to claim Homer as a surgeon like themselves.[8] Their wound counts vary but the premise behind such studies is unsound. Homer's descriptions of wounds owe much to phrasing inherited from his poetic predecessors. They need not owe anything to his own witnessing or surgical skill.

His sense of the body's workings, inside and out, is certainly not ours, even if a modern contention that he had no idea of a unified body and no word for it has collapsed before scholarly ripostes.[9] Nonetheless his naming of parts is specific. He refers to vertebrae and to a membrane that covers the liver. Some of his body language is so precise that it still poses problems for translators. What exactly is the *inion* (probably the occiput, the little lump on the back of our heads above the neck)? Several such words are used only once or twice in Homeric poetry and remain rare thereafter, but there is no reason to regard them as Homer's own coinages. He refers once to 'what men call the *kotyle*', using their graphic term, 'cup', for the hip joint and revealing it to be current outside his poem.[10] When later doctors use one of these rare words they too may know it from general Greek vocabulary. In fact, Homeric words for exact bodily details occur seldom in Greek medical texts, especially in those written in the fifth century BC, and some of the most striking words are not picked up at all. Homer uses the word *laukanie* for Hector's gullet, pierced by Achilles' fatal spear.[11] No Homeric context was more famous, but the word never recurs in a medical text. Even when a medical writer uses the same word as Homer, he may use it with a different sense: '*inion*' no longer means 'occiput' to a medical author. Similarly, *phlegma*, in Homer, appears only once and means 'fire'.[12] To the classical doctors it means phlegm, a cold and wet element inside the body. Homer's legacy to later doctors was not a ready-made lexicon.

How much did he really understand about the body's inner organs and interconnections? He knows that fat is around the kidneys and that a heavy blow on the collarbone will cause a hand to be numbed at the wrist (a 'brachial plexus lesion').[13] The climactic killing of Hector by Achilles contains accurate detail too. As a superhero, Achilles first throws his massive ashen spear and then uses the same weapon for thrusting, a double use which is beyond ordinary mortals and spears nowadays. Nonetheless, he makes contact with bodily reality when he thrusts at Hector's gullet, his unarmoured spot, described as just above the collarbone. He drives his spearpoint right through it, but the windpipe is expressly said not to be severed. As a result, Hector can utter his last poignant plea for mercy in the brief moments before he dies.[14] The windpipe would indeed have remained untouched by such a wound, one which modern doctors credit with cutting the carotid artery and jugular vein, and so Hector could have made one last utterance.

When Hector is dead, Achilles drives leather thongs through his legs in order to drag him behind his chariot. He pierces him from 'heel to ankle' through his tendons (the origin of our phrase Achilles tendons).[15] Thereby, he pierces the one point above the heel where such penetration is readily possible. In general when Homer's wounded heroes fall, they fall forwards or backwards, but the direction is always 'compatible with what a modern physician would predict given the localization of the wound'.[16] The vivid detail that a spear continued quivering throughout its length when driven into Alcithous' heart may seem incredible, but the heart would briefly continue throbbing and could cause a spear to throb too.[17] Somebody had seen such vibrations and so they passed into oral epic's repertoire, where, not being commonplace, they survived. So did the spray of blood from a man's nose and mouth when he receives a severe wound there and, like Homer's Erymas, exhales.[18] By the eighth century BC Greeks had noticed some dramatic bodily effects in battles.

Nonetheless, oral poets exaggerated, as their heir, Homer, exemplifies. Bone marrow exists in the spine, but it would have oozed out, not 'leapt out' in a Homeric spurt.[19] Eyeballs might have been dislodged by a spear blow which smashed the surrounding bone but they would not have fallen dramatically onto the ground.[20] The general context of Homeric body wounds is riddled with ignorance. Like all Greeks, Homer had no idea of the circulation of the blood or the heart's role in it, truths never discovered in antiquity. Nor did he know anything of the brain's role in our thinking and emotions, both of which he attributes to the *phrenes*. These *phrenes*, much discussed by modern scholars, are the lungs, not the midriff or the diaphragm.[21] Homer does not distinguish arteries from veins and has no idea of the nerves. When he describes the death of Thoon, he refers to the big vein which 'runs all up the back to reach the neck'. As a recent medical critic crisply remarks, 'there is no such vein'.[22]

As poetic phrases passed down through the generations from one oral poet to the next, they became extended to situations which they did not fit exactly. This development shows clearly in the impact of weaponry. The gruesome effects of sword thrusts became transferred to those of spear wounds and the angle of spear wounds became transferred to arrow wounds. In the Iliad, when the carpenter Phereclos tries to run away, Meriones catches up with him and thrusts his spear through Phereclos' right buttock with such force that it passes on, piercing the bladder and emerging through the front of his body. A spear could only take this route by passing through the narrow gap which is nowadays known as the greater sciatic notch, but there is no need to credit Homer with exact understanding of this intimate detail.[23] Epic poets needed only to have seen or heard about spears going in at the back and coming out at the front, causing their victims to urinate as their bladder split. These details became a sequence of poetic phrases and, tellingly, Homer uses them again. An arrow, again from grim Meriones, follows the same passage through its victim's right buttock, but in this case an arrow shot horizontally is being poetically, and impossibly, credited with the course of a spear thrust from above.[24] In neither case had the skeleton's detail been carefully examined. In the tangle of a shattered corpse it is not surprising if Greeks did not understand the inner complexities of the bodily mess before them. They also feared pollution from corpses, a powerful deterrent to research on the dead, although Homer never refers to it.

In general the ignoble is omitted in Greek epic, including the stench of the dead after a battle, which Homer also never mentions. Often he singles out the whiteness of the heroes' skin, but usually only in order to contrast it with the blood and cutting which it will undergo from weapons.[25] Because of such omissions and stylized emphases his wound-ings have been called cinematic. Unlike an everyday camera, lines of poetry can go below the skin, but when Homer's go there, they are not accurate. Nonetheless they co-exist with assumptions about healing and disease which, for historians of medicine, are important.

III

One of these assumptions is the role of the gods. In the Iliad, a poem of heroic warriors, the doctors are described attending war wounds, not everyday illnesses. The origins of these wounds were manifest: spear- and swordblades or large hunks of rock, like the rock which dazed Hector and caused his companions to throw cold water over him in order to revive him. There was no need for a doctor to reason from the visible to the invisible in order to diagnose a wounded patient's condition. The cause was evident, and so attendants of the wounded never pray first to the gods in perplexity or ask for their help. Once, a wounded hero appeals to a god, but, even then, he is in no doubt about the cause of his pain: it had resulted from an arrow wound in his shoul-der, narrated about 2600 lines earlier in the poem and now causing his shoulder to be 'heavy'.[26] What he wants is quick relief in the absence of a nearby healer. He invokes the god wherever he may be, an open-ended location which is conventional in Greek prayers: 'Hear me, lord (Apollo), who are somewhere in the rich land of Lycia or in Troy; you can hear everywhere someone who is in distress, as distress has come on me.' Ever-hearing Apollo was on the peaks of Mount Ida. By long-distance healing he stopped the pains and the dark blood's flow and put might into Glaucus' heart beside Troy.

Like heroes, gods heal one another.[27] First, they wash or staunch the blood; then they apply soothing 'drugs' (*pharmaka*), acquired from plants. The similarity between their practice and that of mortal healers is not intended to be humorous: it is one more instance of the poet and his audience envisaging their gods like glorified aristocrats and project-ing onto them what was known to a lesser degree among mortals. When human healers act, they too wash the blood from a wound or suck it away. They too apply 'kindly' drugs, but unlike gods they sometimes cut round a wound with a knife. They cut away, pull out or sometimes push through the obvious offender, the weapon.[28] A wound may also be bandaged, but it is never left deliberately to ooze what classical doctors valued, and medieval authors also knew, laudable pus.

Homer's reservoir of phrases for healing omits the pain which these actions would have caused. Patroclus cuts round a wound, but, like his herbal remedies, he himself is 'kindly'.[29] He shows kindness by staying with a wounded Greek hero, talking to him after treatment. The agonies of slash and burn by a real doctor are passed over. The kindliness of Greek doctors was to become a recurring literary theme, but it is one which begins with Homer.

In the Iliad kindly drugs have been taught to great heroes by superhuman teachers, Chiron the centaur being repeatedly mentioned, not least as Achilles' instructor. However, most of these drugs were not divine items. For centuries before Homer, people had found healing properties in plants, their roots and leaves. The very phrase 'kindly drugs' implies an awareness of others which were not kindly at all. Homer does not identify the plants which his warrior-heroes use, but in the Odyssey he dwells on a very special one which Helen had acquired in Egypt, a powerful tear-stopper which we cannot now identify. In real life, drugs used by Greeks included opium, which was being traded in the east Aegean in small poppy-shaped flasks about a hundred years before the Iliad was composed. Homer mentions poppy heads, but never as a source of pain control.[30]

As the Iliadic wound is typically a wound treated by humans on human terms, Homeric wound-healing has been admired for existing in a 'non-theological space' into which subsequent surgery and wound care could expand.[31] However, such a space sits oddly with the omnipresent divine interventions in Homeric poetry: even for wounds it was not consistently maintained. The Odyssey describes how Odysseus was once tusked by a wild boar, whereupon his hunting companions treated the leg wound in two ways. They were not specialized doctors but they bound it 'expertly' and also uttered 'incantations' over it.[32] Human expertise and appeals to divine intervention were not mutually exclusive. In a crisis, doctors, like hunting companions, surely resorted to both at once.

What about cases of sickness? The great cultural historian Jacob

Burckhardt was struck by the hardiness of Homer's heroes: 'Nestor (the old man) with Machaon (the field surgeon), coming back heated from battle, at once expose themselves to the wind on the shore, to the horror of all present-day victims of rheumatism'. He ascribed it to a 'wonderful vigour' of the Greeks: 'we may well ask whether the ancients ever noticed a draught'.[33] However, an epic about heroic warriors had no reason to dwell on mundane aches. Such diseases as they mentioned needed, they thought, a different expertise.

The Greek word we translate as 'disease' derived from an Anatolian non-Greek root, meaning 'without well being, (divine) favour'. At first it had a scope wider than sickness, like our word dis-ease, but even so the gods were integral to it: they still are in the Iliad.[34] The poem begins with the first epidemic known in western literature. It started by affecting animals, killing mules and dogs first, and it then went on to kill humans, during ten days in which 'the pyres of the dead were burning densely'. Nobody regarded it as an infection or a virus. A seer, not a doctor, was called in to discern its origin; he diagnosed the god Apollo, angry at the dishonouring of his priest. The plague duly stopped, not through any medicine or social distancing, but when Apollo heeded his priest's prayer for it to do so and when the Greek army was purified and the god was placated by their offerings. The plague's origins were explicitly related to Apollo, who came down, the sun-god, 'like the night'. This simile, the first in the Iliad, is one of the shortest, but its reversal, night for day, makes it the most terrifying. Apollo's invisible arrows struck without being seen, just as bacteria or viruses still strike in a pandemic.

In the Odyssey, Homer mentions another exceptional affliction, one which besets an individual. The Cyclops is wounded by Odysseus and howls to the others in the surrounding hills, asking for their help with what 'No-man' (Odysseus' artful pseudonym) has inflicted on him. They interpret his agonized howls as those of an intensely sick man, as he seems to be shouting that no mortal has pained him, and so they assume that his sickness is divinely sent: they reply that a disease sent from Zeus cannot be avoided and tell him to pray to his father, Poseidon, another god.[35] Both their diagnosis and their remedy appeal to divine intervention.

The Cyclops's agony and the plague with which the Iliad opens are

exceptional afflictions, but Homer also refers to people whose suffering is not so unusual. Significantly, they occur in two of his similes,
passages which tend to present what is familiar to their listeners so as
to help them to envisage abstract qualities. In the Iliad the acute pain
of a spear wound in Agamemnon's arm is compared to the pains of a
woman when the goddesses of birth pangs, the Eilythuiai, send 'a
sharp piercing shaft'.[36] The human dimension of labour pains was of
course known, but goddesses were assumed to be active too because
the pains were so extreme and the outcome for the mother's life so
uncertain. Prolonged acute pain might also have a divine power
behind it. In the Odyssey, a magnificent simile presents a 'hateful'
divinity, a *daimon*, inflicting disease on a long-suffering father and
causing him mighty pains until the gods free him from his distress, to
the delight of his children.[37] His recovery from a chronic disease is
the first in Western literature, but both its beginning and its end have
a supernatural dimension.

Once, Odysseus may seem to take a different view. During his visit
to the Underworld, he asks his dead mother if a 'long disease' or
'Artemis with her gentle arrows' killed her.[38] Artemis' usual killing
field was childbirth, which Odysseus evidently considers to have been
still possible for his mother. She replies that it was not Artemis nor
Apollo's shafts nor 'a disease which very often takes away the spirit
from the limbs by grievous wasting-away'. Touchingly, it was her
'missing of Odysseus and his counsels and gentle-heartedness', no
less, 'which took away her honey-sweet life'. Here, she and Odysseus
are sometimes taken to be contrasting a death from a natural disease
with a death from the arrows of a goddess, but, in context, that contrast was not their concern: they were considering whether death had
been slow or swift. The swift death was due to a god but the slow
death might have been due to a god too, although neither speaker
needed to say so.

A third simile, in the Iliad, might seem to differ. When old Priam
finally sees Achilles bearing down on his son Hector, Homer compares
the glitter from Achilles' bronze armour to the bright star of the harvest season, which men call the Dog of Orion, the one which we still
call the Dog Star. It is the brightest star in the 'gloom of night' and 'has
been fashioned as an evil sign and brings much fiery heat to wretched

mortals'.[39] The word for this 'heat', *puretos*, was to be the regular word for the heat of a fever in medical texts of the fifth century BC, and, as a result, a later grammarian at the turn of our era understood it in that sense here. He was wrong. For Homer, as for other archaic poets, *puretos* meant seasonal heat, the extreme heat from mid-July onwards. Homer is not implying here that fevers are caused by a star or a season, nor that they can be natural. He has no idea of an abstract 'Nature'.

Historians of Greek thought tend to present rational medicine as emerging after a time, attested in Homer, when all illness was ascribed to random divine intervention and when healers did not think in terms of natural effects inside the body. In Homer, pains or dreams or sleep are said to be sent from outside the body, the latter two from a god, rather than arising as processes inside it.[40] That way of thinking is indeed different from ours. However, what Homeric poetry presents is stylized and selective. In real life, contemporaries sometimes ascribed a sickness to pollution, or *miasma*, that invisible agent which Homer never mentioned. He said nothing about everyday sicknesses, either, the small change of Greek winters and unwashed food. These common-place conditions were outside the range of heroic epic, but nonetheless they occurred in Homer's own world, the sickness, perhaps, of a travel-ling Euboean oarsman on his way to the Bay of Naples or the upset stomach of a Boeotian slave kept on very poor food. Like us, Homer's contemporaries coughed up unwanted matter or sneezed or excreted it. Even if they considered that the whole of life was ultimately in the gods' capricious care, they would surely treat these visible signs of sickness as if something was out of order inside themselves. Epic ignores these lowly facts of life, but they pointed in a direction in which principled medicine could later expand, leaving the gods to one side.

IV

Not all societies value healers and their art, but Greek medicine sets a notably high value on its ablest exponents. Was this value already present in Homer's poems? Respected healers were not confined to the Greeks (the Trojans also had them), but the social status of the Greeks' top two is significant. Podaleirios and Machaon were brothers, but there is no hint that their skill was a secret one or that it was related to temple service of a god: among Greeks, 'it would appear not that the doctors grew out of the priests but that the priests were parasitic on the doctors'.[41] These two healers were also fine fighters.

When Machaon is wounded in his shoulder by a three-barbed arrow, Idomeneus comments to old Nestor that he must be taken to safety because 'a healer is worth many other men, both for cutting out arrows and applying soothing remedies'.[42] However, Machaon and his brother are never described as tending soldiers who were not nobles or their companions. Away from the heroes at war, healers were a different class of person. In a famous passage in the Odyssey, the swineherd Eumaeus reminds an insolent suitor that nobody goes off and invites a stranger from abroad unless he is one of the 'public workers', a 'prophet, a healer of evils, a craftsman of wooden planks' (for ships as well as houses) 'or a divine singer' of poems.[43] In Homer's world skilled labour could migrate freely, and then too, doctors were welcomed abroad for their talents. As 'healers of evils' they were not just attending wounds and fractures: they were also attending sicknesses. They were 'public workers' because they did not confine their skills to the noble class or their own friends and families. The word also implies that they were publicly rewarded. These migrant healers were certainly not aristocrats or people living off inherited assets. They were professionals, grouped with singers and carpenters, and they were willing travellers. Travelling doctors were to remain crucial assets throughout antiquity, from the first surviving Greek medical texts in the fifth century BC to the Greek cities in the Roman empire where they received citizenship and honours away from the cities of their birth, including inscriptions which praised their unselfish offers of skills.[44] The Odyssey is the first witness to them.

In Homer, but not in cities of the Roman empire, all the healers are men. Even when a woman, Hecamede, washes a wounded hero, she is only a captive slave girl, old Nestor's concubine, whose jobs include serving at table. She is not a specialized nurse. She offers a wounded hero cheese grated in wine, not as a remedy but as a consolation. It sounds repulsive, but cheese graters found with drinking cups in contemporary burials show that it was much to the taste of Homer's contemporaries.[45]

Virgil presents a telling contrast here. Near the end of our *Aeneid*, Aeneas is wounded by an arrow and the healer Iapyx struggles to extract the weapon.[46] Now elderly, he owes his skill not to a centaur but to a predatory god. Apollo had lusted for him as a young man and offered him one of his arts, music or lyre playing or archery, if he would grant him sexual favours. As a pious Virgilian son, Iapyx asked for the gift of healing, because he had an elderly father who had long been lying ill near the point of death. Sex assisted the young man's career and, according to Virgil, he was enabled to practise 'mute arts, ingloriously' as a result. The comment shows a typical Roman disdain for medicine's status among the arts.[47]

Iapyx is not only inglorious. He fails to extract the arrow, even by using Apollo's potent herbs and a pair of tongs. Aeneas, 'fretting bitterly', leans on his spear meanwhile, unmoved by the tears of his son and the gathering of young men around him. Columns of dust rise into the air as enemy horsemen draw near and their shots begin to find targets. Aeneas is healed only when his goddess mother, Venus, plucks some Cretan dittany with purple flowers, our woolly-leaved *Origanum dictamnus*, the modern *erontas*, which is still collected and cultivated on Crete and exported for herbal medicine and for adding a distinctive tang to drinks.[48] Following Aristotle and Cicero, Virgil chose it for a different reason, its reported effect on goats, who seek it out when they are struck by an arrow and need to be healed.[49] Veiled in mist, Venus steeps it in pure water and mixes it with two divine agents, ambrosia and a healing panacea. Unseen, she brings her heavenly cocktail for Iapyx to apply, whereupon the blood stops even in the bottom of Aeneas' wound and the arrow falls out, to Iapyx's astonishment. He rouses the warriors, recognizing a divinity's work, and Aeneas promptly puts on his armour, brandishes his spear and prepares to return to the fray.

Unlike the top Homeric healers, Iapyx received his gifts from a harassing god in return for sex. Even so, they did not always work: Iapyx succeeds only after an intervention by his patient's divine mother. Failure and the inglorious practice of a mute art were never ascribed by Homer to his healers. The healing of Virgil's Aeneas is vividly staged and scripted: Virgil's wound, not Homer's, is the one which is cinematic.

2

Poetic Sickness

I

The Homeric poems are our best evidence for pre-classical Greeks' words for parts of the male body, for the status of those who attended the sick and wounded, and for their assumption that gods were active in sending epidemics and severe illness. The skills of wound care and elementary surgery which the poems present were to have a long history. From the mid- to later fifth century BC onwards, Greek medical texts survive on the topics of head wounds, fractures and damaged joints, but much of what they set out is likely to have been known and applied long before it was written down, probably already in Homer's own eighth-century BC world: in this branch of doctoring there was no major fifth-century invention.[1] There was also to be a long further history to the age-old use of kindly drugs. In the Homeric poems, they were applied to the wounded part only after surgery. During it patients continued to be treated without sedation or painkillers, presumably because it was too dangerous and difficult to calibrate a dose correctly for someone who was being cut. In the first century AD, the doctor Celsus still gave revealing advice: a surgeon should be filled with pity, he wrote, but 'not moved by his patient's cries to go too fast or cut less than necessary; he must do everything as if the cries of pain cause him no emotion'.[2]

Homeric doctors attend people wounded in war, but curiously no Greek war-doctors are known to have accompanied armies, those battlefields of pain, in all the years of fighting between *c.* 700 and 400 BC. Even Xenophon, the most military of Greek historians, gives only a few details, none in a typical Greek city-state. He assumes that doctors would accompany a Spartan army on campaign and he also constructs a fictitious little discussion between Cyrus the Great and his father which assumes that doctors accompany a Persian army, as indeed they sometimes did.[3] Doctors were also considered by Greeks to accompany Egyptian armies and give treatment for free. The only factual reference to Greek army-doctoring occurs during Xenophon's long march into Asia with the Ten Thousand, on which, near Nineveh, he remarks that they appointed eight 'doctors' to attend to the wounded after some hard local fighting.[4] In 401 BC they seem to have been Greeks, not locals, but Xenophon says nothing about their origins or training and there is no hint that they belonged to a defined medical corps. A medical text, *On the Doctor*, probably written just after Xenophon's exploits, advises aspiring doctors to join military expeditions in order to gain experience in treating the wounded: perhaps Xenophon's eight were present for that reason.[5] The silence about doctors on other Greek campaigns is presumably fortuitous, but the silence about leaders who could heal, like Homer's Achilles or Patroclus, may not be. None is attested until Alexander the Great, who recommended treatments to his soldiers and was even said to have dreamed a remedy which turned out to be a cure. Plutarch, his later biographer, considered that Alexander seemed 'especially fond of medicine' and that he might have learned it from Aristotle, his tutor. An even more potent model was Homer's Achilles, whom Alexander greatly admired and rivalled.[6] Previous Greek generals had no such tutor and no such Homeric role model to inspire them.

Off the heroic battlefield real people in Homer's lifetime suffered lives which were quite different from those of his noble heroes, the 'fine and fair'. What would a practising healer of evils confront in them? For a start there was life's great killer, birth. The epics had no reason to dwell on details of childbirth or the related phases of a woman's life, but these ordeals surely attracted much lore, especially between women, long before general theories were advanced to explain them. As for deformity and disease, both were widespread, even without a ten-year siege of Troy.

The best hope of recovering evidence of them is to study excavated skeletons and bones. A recent analysis of the bones of forty Dark Age individuals at the northerly edge of the Homeric world, at Pydna just beyond the gods' Mount Olympus, has shown quite frequent evidence of anaemia, bad teeth and the emergence of 'degenerative joint diseases early in life from 13 to 24' in people alive during the epics' formative years. Nearly a quarter of them were suffering from 'arthritic changes, mainly in the spinal column'.[7] On their nearby mountain, Homer's gods regarded deformity and disability as a disgrace. In disgust, the goddess Hera had thrown her disabled child Hephaestus out of heaven. Even when he returned there, his limping struck the gods as comic.[8] Homer's Olympians would find our paralympics hilarious.

As cultural life increased from *c.* 700 BC onwards, self-inflicted damage increased too. Athletic sports, as ever, multiplied it. Boxing was already ferocious in Homeric epic, as was wrestling, in which, said an ancient observer, men 'trip each other up, while others are choking and twisting each other and grovelling together in the mud, rolling like swine'.[9] Once, Homer mentions riding, though only in a simile: in real life, without stirrups, it greatly increased risks of injury.[10] So did hunting, especially when conducted on foot against wild boar. As there were no rules for health and safety, accidents of everyday work and play proliferated. Parents, too, could add to them. In a remarkable comment, the author of *On Joints*, *c.* 450–420 BC, states that 'tellers of myths' say that Amazon women 'dislocate the joints of their male children at once when they are infants, some at the knees, some at the hips, so that they would become lame, presumably, and the male sex would not plot against the female'. After this pre-emptive strike in the sex war, 'the women use these males as manual workers for whatever leather work or bronze work needs doing or any other sedentary job.'[11] The author concludes honestly, 'whether this is true, I do not know, but I do know that such things would happen if anyone were to dislocate little infants as soon as they are born.' Amazons were mythical beings, but in the second century AD, the doctor Soranus remarked that infants in Macedon and Thrace were indeed fixed down on flat boards or planks in order to correct the softness of their necks and heads. The result was that their heads became elongated.[12] This condition, macrocephaly, may already have been visible among Alexander the Great's all-conquering

Macedonian and Thracian soldiers, though nobody remarked on it at the time: modern scholars have not made anything of Soranus' remark, either.

For physical defects, material objects, too, are suggestive evidence. A painted clay figurine, found at Gortyn on Crete, shows a woman, c. 600 BC, with a lopsided face and an oddly positioned hand, details which have been explained as a craftsman's attempt to represent the effects of a stroke.[13] A curious small vase, found in a funerary urn at Selinus in west Sicily, shows a male adult with legs ending like stumps at the knee and only one arm, perhaps a realistic model of deformity since birth. Texts also mention a baby born with no skeleton and another with an arm attached to its flank.[14] Not until the early fifth century BC did painters represent dwarfs, but dwarfism had existed long before it appeared in art.[15] Partly as a result of it, hunchbacks were to be seen in cities' populations. Just as slim modern Americans stand out in a rising tide of obesity, so Greek nobles and the beautiful boys whom they courted stood out in sharpened contrast against many of the physiques around them.

So far as these deformities were treatable, they could be addressed by skills akin to those of Homer's heroes: cutting, bandaging and applying kindly drugs. Diseases raised quite different problems, especially those attested by the bones of archaic skeletons. They include conditions which Homeric epic had no reason to describe. Bio-archaeology has found traces of the effects of cancer, whether in ancient Mycenae or in even more ancient Egypt, while the bones of two faraway Scythians in separate parts of the Altai region in modern Siberia, one a lady in her twenties, the other a prince in his forties, show evidence of death from breast cancer and prostate cancer respectively in the fourth century BC.[16] Neither of these cancers, therefore, is a new modern disease. Nor is malaria, even in its most acute form: evidence of it has been extracted recently from the pharaoh Tutankhamun's mummy.[17] DNA analysis has indeed opened new vistas here: samples of it from ancient Greek teeth have confirmed the presence of tuberculosis. However, despite its seductions, DNA will never give a full record of ancient pathology as it does not attest diseases of the soft tissues.[18] Some well-known soft-tissue conditions may have been absent in antiquity, including measles and scurvy and most of the sexually transmitted diseases, but others were certainly at large, as descriptions in medical texts show. Diseases of the eyes were also very widespread, though DNA from bones never reveals them.

How would a 'healer of evils' set about treating these extreme conditions, whose true nature was quite unknown to him? As in Homer, he would use kindly drugs, age-old recipes which could be swallowed, inserted or strapped onto a patient. He might try to cut swellings away, as if they were akin to wounds. He would also use the supposed skill of divination, to intuit the role of the gods behind the disease and then try to placate or avert it. This link to divine aid is especially clear for those hardy perennials, disorders of the mind. In the Iliad, Machaon's brother-healer Podaleirios was detained by fighting in the front line and so nothing was said about his skills. However, an early post-Homeric epic, the Sack of Troy (c. 680 BC), credits him with 'knowing everything in the breast of men accurately, both realizing what cannot be seen and curing the incurable'. This gift enabled him to be the 'first to recognize the eyes flashing like lightning and the heaviness of mind of Ajax in his anger', implying that he realized from these signs not just that Ajax was angry, a rather obvious inference, but that there was enough to make him go on to kill himself, as indeed he did.[19] This exceptional insight and therapeutic skill make Podaleirios the first psychiatrist attested in the Greek world.

The poet's word for the objects of Podaleirios' knowledge here, *akribea* or accurate things, is an exacting one. It survives nowhere else for more than two centuries, but it was then to have an important role for exponents of the new medicine. Podaleirios is also presented as 'curing the incurable', a miraculous achievement, and is said to owe his talents to the special gift of Poseidon, a god. His knowledge of *akribea* is god-given and is not based on any natural craft.

This same stress on divinely given skill recurs in other tales of psychological affliction, a category which was already known to post-Homeric poets and is widespread in Greek mythology. At Tiryns, in the Peloponnese, a fine tale, variously told but already known to Hesiod, described how King Proetus' daughters once mocked the goddess Hera by claiming to be more beautiful.[20] She punished them by turning them into cows, an animal with which she was associated, and left them to wander, mooing, while the hair of their heads fell out and their skins were affected with a whitening disease. They were only cured by the seer Melampus, who demanded a share of the kingdom in return. His divine insight had realized that Hera was responsible and needed to be placated and that the water at which these moo-girls were drinking had to be purified.

These just-so stories gave no effective role to medicine or human healing. Mythical madness originated from the gods, and if a healer was involved, he had to be divinely gifted, a healer-seer. There is an exception, however, at Miletus in a story of uncertain date. There was a time, it said, when the city's girls went mad and were afflicted with what Plutarch (c. AD 100) calls a 'longing for death and an impulse to hang themselves'.[21] Details vary between various tellings of the story, but a divine power was involved, Plutarch assumed, in the girls' collective frenzy. They were only stopped when the city passed a law that the bodies of girls who killed themselves would be processed naked in public through the city. For once, seers and divination played no part in curing heaven-sent mania. Female 'shame' was strong enough to do so, at least in a story invented and enjoyed by men. Men would never have been deterred by the threat of being seen naked. It was the state in which they exercised and competed for athletic prizes.

II

About 250 years divide the Homeric epics and their immediate successors from the first surviving Greek medical texts. These years are ones of headlong change and innovation in many parts of Greek life, in warfare, politics, poetry, architecture, and from c. 580 BC onwards the beginnings of philosophical thought: what, if anything, did that long interval contribute to medical thinking and treatment meanwhile? The self-styled inventors of medicine in the fifth century BC have to be seen in this longer context in order to appreciate how far they were indeed new. Between c. 650 and 480 BC, medical evidence is scarce but at least it has a wide location, extending from Sicily and south Italy through Athens, Cyprus and the coast of south-west Asia. It even extends to the great palace-city of the Persian kings, Susa. In the fifth century, exponents of the new medicine were often to be travelling doctors; study of their predecessors will show that mobility was not their novelty.

As prose texts did not begin to be written by Greek authors until the mid-sixth century BC, the only texts available earlier are yet more poems, stylized, therefore, and selective. They do not describe the symptoms of any disease, except, on one view of it, love. Nonetheless, they repay a closer look, not as 'literary reflections of pathological reality', but because they present ways of imagining medicine before its self-styled inventors claimed to revolutionize it. Six texts stand out, two by Solon, the great Athenian lawgiver, and two by Hesiod, Homer's successor in the epic tradition. A fifth is a poem by Sappho, the first Western female known in her own words, and the sixth is by Pindar, the master of poetic flattery.

In what survives of his poems, Solon uses two arresting medical analogies. One, implicitly, is the body politic, the first known use of an image which was to have a very long life. In 594 BC, he presented a scathing picture of the Athenian 'leaders of the people', history's first known demagogues. They pursue unjust, rapacious and luxurious ends, he writes, and thereby inflict, not a disease, but an 'inescapable wound' on their entire city.[22] That wound is sent by Justice herself and affects all citizens because of the greed of a few. Financial crises still do. Solon moralizes and generalizes this affliction as if it applies not merely to his own Athenians but to all city-states: he considers them all to be woundable bodies. His later biographer, Plutarch, also describes him as refraining from applying 'healing' to what was satisfactory in the city as he found it, but he is not necessarily using Solon's own words here.[23] It is uncertain, therefore, if Solon used another image which is still all too popular, that strong medicine, applied by a ruler or lawgiver, can heal a sick or broken society.

This eloquent medical metaphor is not Solon's only reference to the subject. In verses on the mutability of human fortune, he makes the first surviving reference to healers who try to heal, but fail. He mentions them among 'those who attend to the business of the god of healing, Paion, and his many remedies'.[24] Sometimes, he says, a small pain may become a great distress 'which nobody would resolve by giving gentle drugs', whereas at other times a man who is 'being thrown into turmoil by evil and grievous diseases is quickly made healthy by someone who has touched him with his hands'. As an example of life's uncertainties, Solon contrasts gentle drugs' failure to cure what began as a little pain with the instant success of a hand put on someone whose disease has already become acute. He is not confining healing to the laying-on of hands: to make his point he does not need to mention healers whose hands succeed because they wield a knife. He merely wishes to illustrate the unpredictability of human endeavours, but it is significant that he chooses doctoring as his main example. Tellingly, he represents all healers as attending to the work of a god, Paion, often regarded as Apollo himself.

About a century earlier, Hesiod had explained why diseases exist at large. They are part of Zeus' purpose, he says, but were first let into the world by that recurrent culprit in male literature, a woman. Pandora took the lid off her jar and out flew 'countless diseases' which fly 'spontaneously', some by night, some by day, bringing evils to mortals who had previously lived without diseases at all.[25] The force of Hesiod's word spontaneous (*automatai*) here has divided scholarly opinion. Some read it as evidence that Hesiod regarded diseases as natural afflictions, or even that 'from Hesiod a straight line leads to the theorizing and empirical approach' of the great fifth-century BC doctors. Others insist that the 'natural explanation of illness is rejected altogether'.[26] The latter view is correct, as a comparison with Homer's usage shows. In Homer, the god Hephaestus is presented as building wondrous tripods on wheels which are intended to move 'automatically', the world's first robots, whenever they go into a gathering of the gods.[27] However, Hephaestus was their initiator and director. When the goddess Hera drives up to the gates of Heaven, they open 'automatically' on their hinges, like modern airport doors, although they were usually opened and shut, Homer says, by those lesser goddesses, the Hours.[28] Hera's presence gives the gates the impulse to open without the Hours' prompting. In Homer, therefore, what is automatic moves by itself, but at the instigation of a divinity. So too do Hesiod's diseases. They move in silence because Zeus has taken away their voices, but they are still at his beck and call. Later in the poem, they are said to befall whole cities as punishment sent by Zeus, not for the greed of grasping leaders, but for the injustices of a single man.[29] There is no line from Hesiod's thinking here to fifth-century doctors' theories about natural epidemics.

For Hesiod, as for later poets, there are myriad diseases, not a limited number of conditions which recurrently attack mortals. Like Homer, he also holds inaccurate views of essential physiology and has no idea of contagion or infection. However, he is not demonstrably wrong in his assumption that sex with a god always makes a woman pregnant or even that those nine sisters, the Muses, began life because Zeus made love to Memory on nine consecutive days and impregnated her each time.[30] The nine Muses took a year, not nine months, to grow on and be ready for birth, but, as they were a superhuman brood, they did not need to obey normal mortal rules. Hesiod also mentions the Dog Star, but, unlike Homer, he describes how it 'dries' mortals, making men most 'feeble' and women 'most lustful'. He is not referring to male workers' exhaustion from heat. The Dog Star, he says, dries out the head and knees, specific parts of the body which were linked in archaic thinking to the inner bone marrow, a substance which was thought to be packed in the knee joints and also in the brain and to be the crucial ingredient in male seed.[31] By drying it out, the star enfeebles the male fluid. As women's bodies were often considered to be colder and wetter than men's, the Dog Star's heat has the opposite effect on them. It hots them up and increases their desire.

Feelings linked to sexual desire were more likely to be described accurately than its physiology. They were, after all, a matter of subjective experience. 'Falling in love again' is a recurring theme in Greek personal poetry of the seventh and sixth centuries BC, and a brilliant poem by Solon's contemporary, Sappho, sets out the effects.[32] She presents herself in a triangular setting, with one too many in the relationship. Whenever she sees a man sitting beside a particular girl who chats intimately with him and laughs sweetly, Sappho describes how her own voice fails her, how she comes out in a cold sweat, how a fire runs beneath her skin, how buzzing sounds in her ears and how she turns pale, 'greener than grass'. Some of these symptoms recur in later medical case histories of sickness, but Sappho was not using specialized medical language here; nor would she necessarily see her own condition as an illness. In other poems she relates her love life to the goddess Aphrodite's intervention, but in what survives of this one she describes it without any reference to a god or a divine power.[33] The only cause of her condition is the girl, the object of her desire. The first accurate description of symptoms in Western literature is a Lesbian's description of erotic jealousy.

In Sappho's poem the girl is anonymous. Not for another hundred years or so does the name of a real patient survive, himself afflicted with disease. In the 470s BC, the poet Pindar sent a commissioned ode to his patron Hieron, forcible ruler of Syracuse and much of eastern Sicily. The poem was not a conventional poem of praise for a recent victory in athletic games.[34] It began with Pindar's wish that he could summon back Asclepius the healer from the dead. Nearer the end Pindar expressed the wish that he himself could come to Hieron in Sicily, personally bearing 'twin favours', golden health and a poetic revel, so as to bring glory to the crowns which the champion, Pherenicus, had won 'formerly' in games at Delphi.[35]

Pherenicus was Hieron's great racehorse, the chestnut stallion who won in Delphi's Pythian games in 478 BC, perhaps also in 482, and certainly in the Olympics in 476 in which he ran 'like the storm-wind' and was never headed throughout the race.[36] He was the ancients' equivalent of the modern wonder horse Frankel. His Olympic victory was even honoured, in a still fashionable way, by a commemorative statue of himself standing in a bronze group with Hieron's other four-legged winners.[37] However, in 474, Pherenicus did not win again at Delphi. Pindar's poem probably belongs in the aftermath of that defeat. It is tactfully offering to help Hieron to turn his thoughts to his horse's former triumphs.

Before expressing the wish that he himself could bring Hieron 'golden health', Pindar recalls how mortal Asclepius learned healing from the centaur Chiron, a claim that was already present in Homer.

Thanks to his lessons, Asclepius was able to heal three distinct groups: those who were beset with sores, those wounded by bronze or by rocks thrown from afar, or those 'ravaged in their body by summer fire or winter'. Some he tended with 'soft incantations'; others were those who drank soothing remedies or those on whose limbs he fastened 'remedies from all places', presumably herbal ones. Others 'by cutting, he set upright'.[38]

Pindar has been presented in recent scholarship as an amateur of medicine or even as someone medically alert to an exceptional and well-informed degree. In fact he shows no detailed grasp of medicine at all. The skills which he ascribes to Asclepius are mostly no different from those ascribed by Homer to his hero-healers. The most interesting novelty is the word he applies to the sores which Asclepius healed. They are 'self-grown', *autophuea*, a word which Homer never used.[39] However, Hesiod had applied it to the bronze threshold of Tartarus beneath the earth whose metal was wondrously self-grown, he tells us, not a man-made amalgam. Later authors applied it to harbours or unquarried mountains, items which were natural, not artificial.[40] In Pindar the items are sores, not diseases. The word *autophuea* distinguishes them from man-made wounds caused by metal or rock, the next items in Pindar's list, and from the ravages of 'summer fire and winter', presumably fevers induced by the seasons. Sores, Pindar means, break out spontaneously. Where Hesiod would have seen the punitive action of a god, Pindar saw a natural outbreak.

The idea behind this particular word was not a breakthrough. It was not even novel. Most of Pindar's patrons were Greek aristocrats for whom excellence, as he often stated, was a matter of nature, not learning. This nature was the creed of superior people and it was obvious, at least to them, that it was best transmitted by breeding with nobly born partners.[41] Pindar's natural sores were one more example of a natural-born attribute. They were not a new bit of medical knowledge related to Greek philosophers' important new idea of an underlying abstract Nature in the world.

Instead, Pindar plays with a literary conceit, that song has curative properties. It is a gift, he states, which he can give to his suffering patron Hieron, but, unlike a healer, a poet, he points out, can confer immortality. Hieron was by then well over fifty, but he was still the powerful ruler of Syracuse. He had recently founded an enlarged city on the site of Catania at the foot of Mount Etna and later in 474 BC he was to win a crucial naval victory against the Etruscans off the west coast of Italy.[42] He was to live for at least seven years after Pindar's consolatory poem, but Aristotle states that he suffered from difficulty in urinating, a complaint which later authors assumed to be due to gallstones.[43] As Aristotle was drawing on Syracusan informants, his comment was at least based on local tradition in Hieron's own city. Pindar's healing song encouraged Hieron to think, not of his acute pain, but of his magnificent horse, 'formerly' storming to victory.

In this period nothing survives from a doctor in person, illuminating his view of disease and the body and the remedies he might apply. Sappho at least describes her own physical symptoms and Solon uses doctoring as a metaphor or prominent analogy. Hesiod still thought of diseases as individual entities, not as processes, a notion which was also missing from Pindar. Nonetheless Pindar assumed something absent from Homeric epic: a direct relation between seasons and disease, linked to the effects of 'summer fire and winter'. This interrelation was an easy inference for anyone who noticed the debilitating effect of intense heat or the unhealthiness of cold or wet winters, but even Hesiod was assumed rather more, a direct impact of the Dog Star on the body's make-up, both on male seed and on female sexual desire. There was scope here for closer study of the effects of particular types of season and for much more thought about why these effects occurred. Such study becomes apparent in the fifth-century BC doctors who are conscious of inventing medicine as a genuine 'craft'.

3

Travelling Doctors

On one negative point the poets and Greek medical posterity agree. They never imply that a healer or doctor should labour, as nowadays, to prolong the lives of the elderly. In early Greek poetry, old age was execrated, never more so than by the gods themselves. In the superb post-Homeric hymn in her honour (perhaps as early as *c.* 680 BC), Aphrodite, goddess of love and sex, states a brutal truth to the mortal Anchises, whom she has just seduced.[1] When Dawn seduced Tithonus, Aphrodite tells him, Dawn made him immortal, but forgot to make him unageing. She stopped going to bed with him 'as soon as the very first grey hairs began to stream down from his fair head and noble chin'. When old age 'hateful in every way, began to press on him', Dawn locked Tithonus away in a separate bedroom. Aphrodite states bluntly that she too is repelled by the realization that old age will soon befall Anchises, her new mortal conquest, because it is 'relentless, ruinous, wearying, and the gods detest it'. Nonetheless, true to the rules of sex between a mortal and a Greek divinity, she was impregnated by her one-night stand with Anchises. The result was to be little Aeneas.

From a shrine, probably Aphrodite's, in the Greek settlement of Naucratis in Egypt comes near-contemporary evidence of a male condition which would have distressed the goddess even more than grey hair. Around 600 BC, someone dedicated a representation of an erect penis with a hole in its base, not its tip, apparently part of a painted vase: it has been understood as a representation of peno-scrotal hypospadia.[2] It is one of the earliest surviving dedications by a Greek with an anatomical reference and was offered presumably in the hope of healing rather than in thanks for a cure.

It is only one little part of a much broader body of evidence. All across the Greek world, as Homer merely exemplifies, divine assistance was sought in prayers and vows by human sufferers. They were not hindered by any conceptual line between medical science and divine healing. Importantly, the gods, since Homer, had even gained a specialized healer in their company. In the Iliad, Asclepius appeared only as a mortal, noble hero, but in the post-Homeric world he began to be worshipped as a healing divinity. By the later sixth century BC there was a cult of him at Epidaurus in southern Greece and another, probably older, can be inferred at Tricca in Thessaly.[3] Like other gods and heroes, Asclepius seemed to appear to his worshippers in dreams which he sometimes sent, and to prescribe treatments for their afflictions. This type of divine healing was supported by the personnel at Asclepius' temples and in due course by doctors who spent time there. By origin it owed nothing to medical science and did nothing to inspire it, but, as time passed, Asclepius' prescriptions became more like the prescriptions of mortal doctors. He reflected the ambience to which he owed his existence. His ever-spreading presence kept healing, miracles and divine intervention closely interlinked in almost every Greek's mind. The exceptions, therefore, were to be even more remarkable.

How did doctors treat patients in real life? As in Homer and for many centuries before him, they continued to use plants, the value of whose roots, juices and leaves had been discovered by people who lived and worked among them in the natural world and also used them for food and cooking.[4] One such remedy even found a place in a poetic hymn, one to the goddess Demeter (c. 600 BC). In it the goddess asks her human hosts to console her for the loss of her daughter with 'barley and water, mixed with pennyroyal' (a herb in the mint family).[5] In this context the mention of the drink related to its ritual use by worshippers of Demeter, but pennyroyal is indeed a herb with strong side-effects, especially on the stomach and in early pregnancy. In later Greek medical texts, barley gruel with pennyroyal was still being prescribed for women, including women who had just had a miscarriage.

For further glimpses of pre-classical doctoring in action, the main evidence is archaeological. Three finds are particularly revealing: a finely painted cup, a skull and a collection of items in bronze. The cup was found at Vulci in Etruscan Italy and dates from *c.* 480 BC.[6] Its painted interior is the work of a master artist, and an inscription identifies the maker of the pot itself as Sosias. It shows, on the right, a beardless Achilles, intently bandaging the left arm of bearded Patroclus, his older companion, as indeed Homer had represented him. Patroclus sits on his shield and looks away, baring his teeth and stretching his left leg full out and upwards as if braced against a wall, the frame of the scene. Emotions are notoriously hard to read in many vase paintings but there is no doubt that the painter wished to show Patroclus in pain. Beside him lies an arrow, perhaps the source of his wound. Like Patroclus' teeth, the bandage is coloured white.

This episode of heroic care is not known in any text, but is a neat illustration of the skills involved. Bandaging was widely practised by Greeks, but also by Egyptians, by Persians and as far as the central Asian steppes of Scythia and western India. During their invasion of Greece in 480 BC, Persian warriors were greatly impressed by the bravery of Pytheas from the Greek island of Aegina. Even when his ship was captured he continued to fight on until they 'butchered' him, Herodotus' very word.[7] He fell, though still alive and breathing, and 'because of his valour' Persians set the highest value on keeping him alive. They 'healed his wounds with ointments and wrapped him round with straps of linen cloth' and then, back in camp, showed him off to their entire army, presumably as a tribute to his heroism rather than to their capture of him or their bandaging.

In India, too, experts applied bandages, wrapping them tightly on snakebites in order to prevent the poison from spreading through the body: it was one of the techniques, surely, which made Alexander use Indian healers as a first resort for bites from snakes.[8] Bandaging was also a skill in Scythia. A finely engraved gold vase shows, among other incidents, a Scythian warrior bandaging the leg of a wounded companion. The scenes on it may be based on a lost Scythian epic poem, a local substitute for Homer, in which heroes were challenged to string a bow, but failed. The wounded man may be one of these heroes, damaged by the bow while attempting the challenge.[9]

The skill of bandaging was indeed widespread in the world, but Greeks became exceptionally adept at it, as their medical texts attest. When the great doctor Galen reviews some of the most conspicuous styles, he includes such virtuoso bandaging as the 'royal' and 'the hare with ears'.[10] Greeks also used the tourniquet, bandaging a limb tightly to stem the flow of blood, but as they had no proper understanding of the veins and blood vessels, they were less clear what to do next. They were at least aware that if they did nothing, a tight bandage would probably promote gangrene in a wound.[11] On the cup from Vulci, Achilles is shown bandaging Patroclus' arm by criss-crossing each end of the bandage which he is applying. This type of bandaging is known in Greek texts, but Achilles' bandage is about to end up with both ends on the same side of Patroclus' arm. The presentation is an artistic convenience, not a realistic image, as the artist cannot have wished to imply that Achilles, so intent on his work, was making a bad mistake.

The second discovery, a skull, shows evidence of something more drastic: trepanning. From China to Peru, archaeologists continue to find early examples of this practice, the opening of a hole in the head by boring or scraping. In the Greek world, evidence of trepanning goes very far back in time, even earlier than the Mycenaean age. It was still being practised in the time of Homer, as the bones of a skull happen to show, found in fragments in an eighth-century BC grave in eastern Crete. The skull had been wounded and then trepanned, but the patient appears to have survived the treatment.[12] In 2005, much clearer evidence was discovered on a well-preserved skull which had been excavated in the mid-1980s from a grave at Abdera, a Greek settlement on the coast of Thrace.[13] The back of it, on its right side, showed a wound from a missile, probably a sling-stone, which had been mitigated by scraping and rasping the shattered surrounding bones. A circular hole was then left open in the cleft which the missile had caused. The victim was a female and an analysis of her bones suggests she was aged about twenty and in reasonably good health when treated. As the trepanner had proceeded with great skill, the wound and the hole in her head had healed and hardened some years before her death. Other items in the grave date her burial to about 620 BC: the trepanning had not killed her.

On the north-west coast of the island of Rhodes, a grave in a cemetery at Ialysos has produced more varied evidence.[14] In the 1920s, its excavators found a range of bronze items and identified them at first as an innkeeper's equipment for preparing hot food and drinks. In fact they are medical tools, presumably the dead man's own. The find is unique in its period, *c.* 500–460 BC, and includes bronze needles, a dilator to widen passages into the body, a scalpel, some hook-shaped instruments, among which is a probe, two funnel-shaped tubes and six bell-shaped bronze cups with narrow necks.

Some of the bronze knives and the scalpel may have been used for cutting up compounds of carefully chosen plant material, pounded and chopped into recipes for healing. The other instruments were used directly on patients. The two tubes are enemas, or *clysteres*, for funnelling a prepared liquid into the body, up the backside or through the vagina. Later Greek medical texts confirm that honey and olive oil were frequently mixed into a fluid which was poured first into a pouch of animal membrane and then squeezed out down the funnel so as to address disorders of the bladder or to encourage the intestinal tract to move and excrete.[15] Enemas with short tubes were especially suited to applications through the vagina: two of the Rhodian examples are of this type. These well-crafted survivals testify to an entire field of expertise which, naturally, went unmentioned in elegant lyric poetry.

The bell-shaped bronze cups had a different use: they were cupping instruments. They were applied in two distinct ways, both of which are still in use.[16] Dry cupping required a flame to be held inside the cup to force out air and make a vacuum. Then, the cup was pressed on dry skin so as to draw blood up to the surface. It left circular bruise marks on the skin, like love bites. For wet cupping, the skin was wetted, often with olive oil, and only then was the cup pressed on it. Again, it brought blood to the skin's surface, whereupon the doctor pricked the skin in the bruised areas, causing blood to flow. Cupping was not a Greek invention, nor did it vanish with the ancients. It is still practised by self-styled specialists and its effects continue to be visible on bodies in the public eye, whether an Olympic gold-medal swimmer's or the bare backs of filmstars on, and off, a Hollywood red carpet.

Trepanning, cupping and enemas involved varying levels of skill, but they also implied general beliefs about the body. Greek trepanning was not aiming to release an evil spirit from the head or to recalibrate the 'blood–brain volume', the delusion of modern self-trepanners who seek a perpetual high from their boring.[17] Each of the archaic Greek examples addressed a head wound. The one on the skull at Abdera shows the exceptional skill of its practitioner. As the missile wound had fallen on a join in the woman's underlying skull, the trepanner had not used a drill. He had used a scraping rasp to take the bone away layer by layer and avoid sundering the skull by boring into it at a vulnerable join. He acted just as the author of a classic fifthcentury BC text on head wounds was later to recommend.[18] His aim was no doubt the aim specified there, to release the blood which concentrated below the surrounding surface of bone before it could go bad and cause problems.

Cupping, too, was used to remove blood from a patient's body, surely because excessive blood was considered to be an element in fevers and sickness. It was not extracted simply to enhance wellbeing, let alone 'organic' health, a notion which the ancients never formulated. Like trepanning, it was a means to an end. So were enemas. They provoked the excretion of waste matter, not only in cases of constipation but at other times when this waste matter was considered to be provoking illness. Nonetheless, the owner of the tools at Ialysos shared a significant area of ignorance with Abdera's trepanner. Neither of them had any idea that they needed to sterilize the items which they inserted into their clients' bodies. For all their dexterity, they risked causing far more harm than they could begin to understand.

II

Between Abdera's trepanner, c. 620 BC, and Rhodes's doctor with his tools, c. 480 BC, very important changes occurred in Greek perceptions of the body and the natural world. In the mid-sixth century, a prose text was written with the title, for the first time, of *On Nature*. Its author, Anaximander of Miletus, treated the world as a natural, explicable entity, not just as a field for random interventions by gods and divine powers. As man is a part of nature, thinkers now began to consider him as part of this natural universe, one whose composition could be defined and understood in terms of regularities. Such grand thoughts were probably far beyond the Rhodian doctor and his cupping glasses, but they helped to relate healing and health to a general type of theorizing which had not been fully considered before. It enabled medicine, not merely healing, to leave the Homeric world of random gods and demons to one side. In eighteenth-century Britain, new theories about Nature were to lead, after some sixty years, to a new craft, landscape gardening. In the Greek world, after a similar time lag, they were to lead to what exponents also claimed to be a new craft, a new sort of medicine conducted with a new method.[19]

The new idea of 'nature' put the human body, too, in a new light. It was, in principle, intelligible through its own processes without the old assumptions that dreams or sleep or wandering diseases at Zeus' direction invade it at random from outside. In the last two decades of the sixth century, there was also a change in art, compatible with these new ideas but not necessarily dependent on them. Artists in Athens began to represent the male figure with a new attention to its structure and interconnections. Particular painters and sculptors indicated lines of muscles and sinews on the naked male body where earlier artists had left a blank surface.[20] The most conspicuous of these innovators is the great Euphronius, active both as a painter and as a sculptor in Attica between c. 530 and 490 BC. By shading and by showing lines on a naked body's limbs he conveyed a more complex impression of its tension and composition than any expressed in Homeric poetry.[21]

There was an obvious social context for this new attention and expression. Naked male bodies were to be seen all around these artists in a post-Homeric Greek invention, the gymnasium. No early groundplan of a gym has been discovered, and several literary references to early ones are not necessarily sound, but they existed in Athens at least by c. 520 BC. So did wrestling rings there and elsewhere, promoting fitness for those who could attend.[22] In athletic contests, as in these training grounds, naked male bodies were shown and seen at their peak. In the early sixth century the sites for grand athletic contests had multiplied, adding to the games at Olympia. Statues, meanwhile, honoured naked victors in prominent public settings: it is convincing to relate the exceptional prominence of the male nude in Greek art to the existence of naked male athletes in the artists' and patrons' real world.[23]

The masterworks of Euphronius and his fellow-pioneers include fine scenes of wrestling and athletics. A fragment of an amphora, decorated probably by Euphronius, shows on one side the throwing of a discus and on the other the throwing of a javelin by a naked young man whose leg, thigh and arm are patterned with such apt lines that their most attentive modern scholar has concluded that the artist used an anatomical model.[24] Euphronius' own intention is, as ever, arguable. Our fullest Greek text on exercise and the body is a rhetorical one, composed in the early third century AD, and does not give much attention to big muscles in its account of the desirable marks of an athlete's fitness. Excessive muscles are even said to be 'bonds' inhibiting running.[25] It is difficult to think that anyone in the real world ever thought the same about the muscles of boxers or wrestlers. Euphronius and his pioneers may not have wished to represent the ideal Mister Muscles of modern body-trainers, but they certainly wished to suggest the fit lines of their heroes and athletes as never before.

The milieu of these male athletes was linked with medical treatment too. In the big public games, upper-class Greeks, Pindar's natural patrons, competed freely as contestants, boxing, wrestling and running in person. There was no cult of the amateur athlete. Anyone who could pay, or was helped to pay, used an expert trainer to enhance their skill and fitness. No record of these trainers' teachings survives, but they were aware of the need to enhance their clients' strength and stamina. One obvious item to address was diet. It mattered greatly what their trainees ate, how they exercised and even whether they continued to have sex while in training: in antiquity, too, sexual abstinence was recommended for competing athletes.[26]

Fifth-century medical texts have surprisingly little to say about muscles, their role and problems. Nobody is described as pulling or tearing a muscle, though spasms and swellings are noted in a general way. Perhaps the workings of the muscles were still little understood.

The interrelations of food, exercise and health were another matter. For doctors too they became an important part of their craft. They began to use the word *'diaita'* for a regulated way of living, giving it a new scope. Prescriptions for a good *diaita* became known as 'dietetics' in medically related texts: they emerged as one of the distinctive emphases of Greek medicine. When did they begin? Despite later polemics against athletic routines, it was probably in response to athletic trainers' results that doctors began to stress the effects of a well-regulated regime on health and to apply it to patients. As late as *c.* 400 BC an author implies that a regulated *diaita* is his own innovation, while later surveys of medicine credit the subject to doctors who were at work no earlier than *c.* 450 BC.[27] Nonetheless, dietetics probably began to be used by doctors during the sixth century BC, before anyone wrote a text on the skill. They saw from trainers that it worked. It also fitted well with the new notion that the body and its processes were part of nature, with regularities and their own dynamism.

III

Thanks mainly to archaeology, we can at last give names to individual doctors in this fertile period between *c.* 550 and 470 BC. One comes from Athens, one from east Sicily and others from Cyprus, neatly illustrating the distribution of talent across the Greek world. As ever, evidence for them raises questions of context and social status, latent in our evidence for doctoring since the Homeric poems.

From Athens comes a disc of marble painted *c.* 500 BC with a seated figure who is wearing a red robe and stretching out his right arm.[28] As it was found in eleven fragmentary pieces when it was about to be shipped out of Athens in the 1880s, its place of discovery was never recorded. Since its first publication, its condition, especially its paint, has deteriorated, but the inscription on it, a hexameter verse, is clearly legible. The disc, it says, was the memorial of the 'wise skill of Aineios, best of doctors'. Interpretations of its purpose have varied, but the likeliest is that it was the marble lid which once covered Aineios' funerary urn. The praise of his wisdom suggests that he was a doctor admired for more than the routine application of traditional healing. The name Aineios is known to have been borne by other Athenians, and the inscription is in Attic lettering, but this Aineios need not have been an Athenian himself. He might have been a travelling doctor, one of those skilled migrants whom Homer had already presented as valued facts of life.

About fifty years earlier, a marble statue honoured another doctor, certainly a medical migrant, this time to the east coast of Sicily. It was inscribed 'Of Som[b]rotidas the doctor son of Mandrocles' and was found in 1940 after landslides just to the south of the Greek settlement of Megara Hyblaea, twelve miles north of Syracuse. It had stood in the city's cemetery.[29] The inscription ran down the statue's right leg and was deeply cut into its marble. The statue itself is now headless but it was certainly not a portrait; nor did it show that it was representing a doctor. It was one more example of a typical sculpture of a young male, in this case a somewhat elongated one, but with an elegantly sculpted backside, a notably narrow waist and a highly stylized V-shaped marking to define its upper torso. Its thighs are closely joined at their tops; parts of its surviving arm, the left one, are also joined to the side of the body. According to the most recent analysis, its marble is likely to have come from a quarry on the Aegean island of Naxos, far from Sicily. If so, it was cut in outline there before being detached from its marble block and was then transported in a partly finished state so as to avoid damage to its extremities.[30] On arrival it was completed by a local cutter in Sicily. Some of its details look rather provincial, but it is uncertain how many of them are due to a local sculptor's chisel.

Imported marble of this quality was not cheap, and a statue for a doctor is a unique survival from this period. It is more likely to have been dedicated by Sombrotidas' son or a family member than by grateful patients as the latter would have mentioned themselves in the inscription in return for meeting its expense. The cutting of his name down one leg is a practice matched on statues from the east Aegean island of Samos: the lettering, too, has been assigned to Samos by an expert in archaic Greek letter forms because of the shape of its letter 'r'.[31] The father's name, Mandrocles, points in the same direction. It is related to the famous Maeander river, which flows out on the south-west coast of Asia Minor with the island of Samos nearby. Bearers of it tend to come ultimately from this region and in this very period a prominent Mandrocles, a structural engineer, is attested on Samos, born in the late sixth century BC.[32] The inscription and the father's name suggest that Sombrotidas was a man of Samos who migrated west to Sicily and won fame and no doubt fortune by his medical skills. Family members then had his name inscribed when this nearly finished statue of high-class marble arrived: they used the style of lettering which was current in their own place of origin. One interpretation of his name, Sombrotidas, is 'saver of mortals', implying that his father, if he gave him it, was already a doctor himself.

More detailed evidence comes from an even more telling survival, a tablet made of fine local copper which was found in 1850 at Idalion in central Cyprus. It is inscribed with a long bilingual text in Cypriote and Greek which honours a doctoring family, Onesilas and his brothers.[33] They had been summoned to Idalion and had attended the wounded when the city was being besieged by Persians and people from Citium, modern Larnaca. The tablet is dated partly by reference to the reign of King Stasikypros, but it is still disputed exactly when in the fifth century this siege belonged. Of the various options, either 499–498 or c. 477–475 are the most likely, and in my view the former is preferable.[34] If so, Onesilas and his brothers are near-contemporaries of the practising doctor and his tools at Ialysos and of Aineios 'best of doctors' in Athens.

The text is not only the first surviving mention of a Cypriote doctor. It is of crucial importance for our understanding of the Cypriote language and its syllabic script, because it gives a well-preserved translation of the Cypriote into Greek. It also contains invaluable evidence for relations between a Cypriote city and its king.[35] Above all, it concerns doctors whose skill ran in the family, all of them being brothers, presumably taught by their father. It reveals the very high value placed on their abilities.

The text begins very politely by citing the brothers' services to the wounded in the siege: they had treated them without charge. It goes on to specify terms already agreed for their continuing reward and to prescribe severe penalties for anyone who ever dares to overturn them. By depositing a sworn copy of the text for safety in the local temple of Athena, the king and the city show that they are committed to upholding the doctors' privileges in the long term. Whereas the king and the city swear to the terms recorded by the tablet, the doctors are not required to swear too. They are being wooed as part of a job offer.

Their privileges are twofold. Onesilas and his brothers had already been voted a talent of silver, a very big sum indeed, but they are now to receive land to the same value, whose income they and their descendants are to enjoy free of tax. The land's location is given in careful detail, allowing us to place some of it about three miles southwest of Idalion's main settlement.[36] Onesilas himself is to receive even more. Its value is to be equivalent to the sum which has already been voted to him personally in silver 'instead of fees and extras'.[37] As the senior doctor among the brothers, it seems that on arrival in the city he had been allowed at first to charge for his services and receive 'extras'. Like his brothers', his reward was to be commuted into land. Like them, he was to be given all the young plants growing on land which was to be his and his family's thenceforward: the gift to him personally was to include a garden, the 'one which is in the farmland of Simmis, the one which Diweithemis the Armanaian used to hold as an orchard, the one which extends to the property of Pasagoras son of Onasagoras'.[38] Here too the young plants and the future revenues were to belong to Onesilas and his descendants for ever, free of tax.

Onesilas' name shows he was a Cypriote, but as he had been 'bidden' by Idalion and its king to come and help them in the crisis, he and his brothers were presumably migrants from elsewhere on the island. He had evidently been appointed as a public doctor and rewarded by a very big down payment. The payment to him of a silver talent was not defined by a fixed time span, but the substitution of land to the same value had a clear intention. It would tie Onesilas and his brothers to Idalion's territory and keep them there to treat patients in the longer term. They and their descendants were presumably being made citizens of Idalion: the king and the city were encouraging them to remain and put down roots.

Like Aineios 'best of doctors', Onesilas and his brothers were masters of a highly valued skill. Like Sombrotidas, they had probably learned it from their father, another example of medical skills being transmitted within families. Like Sombrotidas and, possibly, Aineios, they were migrants, practising away from their home city. Even so, it was considered possible that others might resent the new immigrants and try to drive them or their heirs off their properties. Onesilas has a unique distinction. He is the first doctor known to have owned a garden, the haven of so many practising doctors since. It is the first garden whose delimitation is known in Greek history, but it was not just for relaxation and ornament. Doctors required plants for their medical prescriptions and so, beneath those fruit trees, Onesilas was being given an essential resource for his craft.

4

From Italy to Susa

I

Sombrotidas and Onesilas' family are doctors in action in the Greek world, far from the Homeric battlefield. They are given a wider context by Herodotus' inexhaustible Histories, mostly composed *c.* 440–425 BC but ranging back into earlier times. In them Libyan, Egyptian, Persian and even Scythian healers all feature, especially Egyptian ones who, Herodotus remarks, are specialized doctors, one for each disease.[1] Presumably he makes this point because it was so different from the general practice of Greek doctors. Were the likes of Sombrotidas and Onesilas heavily indebted to non-Greek neighbours, especially those in the most ancient cities of Asia? Was 'Greek' medicine until *c.* 500 BC basically Near Eastern medicine, but less detailed and without kings, usually, to fund it?

Babylonian medicine had been practised long before the sixth century BC, or indeed Homer, but Herodotus states that Babylonians have no doctors at all. It is a massive error, although he had visited Babylon himself.[2] It is also an indicative one: even if an inquiring Greek reached Babylon, he could remain completely ignorant of its doctoring and medical skills. The main Babylonian medical texts which survive are usually in fragments and mostly belong to what modern scholars call the Old Diagnostic Handbook: they contain a few notions similar to Greek ones before *c.* 400 BC, but claims to have detected a 'Babylonian Hippocrates' are unconvincing. There are many fundamental differences between the two types of medicine and similarities are probably only parallels. Greeks' real chance to study Babylonian medicine came after the city's conquest by

Alexander the Great in 331 BC, and even then it is arguable that they never made much of it.

The one type of borrowing traceable earlier is cultic, not textual. A cult and festival on the small Cycladic island of Anaphe honoured Apollo Asgelatas, a uniquely odd name which derives from the name of a major Babylonian goddess of healing, Azugellatu. On Samos, too, the sanctuary of the goddess Hera contained three bronze statuettes of a man praying with a big dog beside him, imagery which relates to similar statuettes in the cult of this same Babylonian healing-goddess. Dogs also feature in the cult of the Greek healing-god Asclepius, perhaps because they were adopted from a similar source. These transfers from east to west were due either to a visitor from Asia or to a Greek who had encountered the Babylonian cult directly in the Near East. In a local crisis a foreign god of healing might have been worshipped by desperate Greeks as a potential helper, but the transfer of an adapted name and some imagery is much simpler than the transfer of medical techniques and practice.[3]

Greeks' contact with Egypt and awareness of its medical culture were another matter. In the Odyssey, Homer already presented 'each doctor' in Egypt as 'expert beyond all men, for indeed they are of the family of Paieon'.[4] As Paieon was doctor to the gods, there could be no more honourable pedigree. Everything in Egypt, Herodotus remarks, 'is full of doctors'. He could not speak their language, but his Histories show that techniques could indeed be explained to a Greek, either by bilingual doctors or by interpreters. His long description of embalming and mummification by Egypt's experts is a proof of such learning, as it is an accurate masterpiece which is still applauded by Egyptian masters of this difficult craft.[5] A modern fantasy that he never actually visited Egypt has now been refuted. An Egyptian tale of a wicked pharaoh, punished by the gods with blindness, has been discovered in a fragmentary papyrus which came from an Egyptian temple library in the village of Tebtunis.[6] Dating to the second century AD its story manifestly goes back to an earlier Egyptian version. In it, the pharaoh is told in a dream that he will only be healed of his blindnesss by the tears of a virtuous woman. After the tears of noble women in the palace and forty women in the harem have been tried and found wanting, he hears of one other candidate who is living virtuously outside. Her tears are brought to him, it seems, in a golden cup and he is duly healed. Herodotus tells a very similar tale of an Egyptian ruler, whom he names, significantly, Pheros, as if 'Pharaoh' was a personal name: he is healed eventually, despite minor differences from the Egyptian story, including the need for a virtuous woman's urine, not her teardrops.[7] Herodotus had certainly heard the Egyptian story in Egypt, presumably through interpreters. Medically it was nonsense.

Herodotus remarks, tellingly, that the Egyptians are the healthiest of all men, except for the Libyans, because their seasons do not vary, 'diseases coming about in changes, especially changes in the seasons'. They also purge themselves, he believes, for three consecutive days in every month with emetics and enemas, for 'they think it is from the foods that nourish us that all diseases come upon men.'[8] Herodotus does not commit himself to that view, but he is correct about Egyptians' interest in emetics and so forth. At least since the 630s BC, there had been scope for exchanges between Egyptian doctors and Greeks resident in Egypt, but despite the Egyptians' long-standing concern for personal beauty, doctoring and hygiene, actual borrowing of techniques from them by Greeks living among them is not proven. Greeks did not purge themselves daily, let alone three times. In later Greek texts, a small proportion of the medical recipes, especially those for women, recommend items from Egypt, but they may be using them in a different way to their use in Egypt and may be recommending them partly because they are impressive and exotic items.[9] A specialist in Greek medicine in Egypt, Heinrich von Staden, has remarked on the 'pathological preoccupation with the anus that seems to characterize Pharaonic medicine' and observes that 'neither [Greek] pathophysiological theories nor their versatile practice reflects this Pharaonic emphasis.'[10] The clearest proofs of cross-cultural borrowings are loanwords, but there are no loanwords from Egyptian for medical items or conditions in classical Greek.

Parallels are another matter. The invasive treatments which Greeks are known to have been using by c. 500 BC were all current in non-Greek societies: cutting, trepanning, inserting enemas and cupping. So, too, was cauterizing, to which Greeks also resorted, albeit as a last resort. Its effects were especially evident among Scythians and

Libyans, according to considered Greek opinion, peoples who bordered two extremities of the Greek world.[11] However, parallels are not necessarily proofs of borrowings. In China, cupping, trepanning and bloodletting were practised by Chinese doctors, but no Greek ever went to China, nor Chinaman to Greece.[12]

II

Once, in his Histories, Herodotus interweaves medicine in the Greek and non-Greek world in much more detail: he tells the story of a Greek doctor who spanned both Greek and non-Greek company, from south Italy to the heart of the Persian empire.[13] He was younger than Sombrotidas, older than Onesilas, and was active from the 520s BC onwards. For the first time, a wandering Greek doctor's career can be followed: its implications are revealing.

Herodotus tells its ups and downs with a clear sense that, being Greek, its protagonist was superior to foreign practitioners. Democedes came from the Greek city of Croton in southern Italy, 'a doctor', Herodotus says, 'and the best of his contemporaries at practising the craft'.[14] His tales of him are not explicitly dated, but events with which they interrelate show that, from 520 BC onwards, Democedes was held in the highest honour at the court of Darius, the usurper who had just established himself as the Persian king. There, he had occasion to watch other doctors, especially Egyptians, in action. Herodotus never says that he learned from them.

By a remarkable chance, from these very same years the inscribed statue of Darius' chief physician survives, one Udjahorresnet. It recounts his great honour as a 'friend' of the king at court and as an 'administrator of the palace'. It even states that he received gold ornaments of honour: they are duly illustrated by the precious bracelets which the statue shows on his arm.[15] Herodotus presents Democedes too as receiving a 'very big house' at Susa in return for curing Darius and as becoming a 'table companion of the king', a status which other Greek texts attest for people in high royal favour at court.[16] Eunuchs at Susa are even said to have escorted him to the king's wives, each of whom gave him gold coins as a reward. One after another, they used

51

a cup to scoop them out of an overflowing chest. Persian riches always impressed Greeks, and in Democedes' case they pictured them as coined money, a telling detail in the story, as Darius was indeed the first Persian king to have coins minted, his new daric coins, though Herodotus calls them staters. Thanks to his skill, Democedes became a match for Udjahorresnet and his finery.

Memorably, Herodotus presents Democedes as leaving his home city in Italy because he was unable to bear the harsh temper of his father. He says that he went to Aegina instead and after a year on the island was hired there 'publicly' for the huge sum of a talent.[17] The Athenians, presumably their ruling tyrant-family, then headhunted him away with an even bigger offer. The tyrant of Samos, Polycrates, capped that offer a year later and so Democedes moved to Samos. After the fall of Polycrates, Democedes eventually travelled as a slave-prisoner, Herodotus says, from Sardis to Susa, a seat of the Persian king. There, Darius damaged his foot while jumping down from his horse out hunting, and Democedes is said to have healed him, whereas other non-Greek doctors failed. As a reward Darius is said to have given him a pair of fetters made of gold. Forerunners of our golden handcuffs, they were a typically regal gift to an inferior: Democedes is said to have replied with a joke, questioning why the king was rewarding him with a double evil when he had just made him whole and healthy, that is, 'one'. Darius was amused, and that was when he sent him to visit his many women.

Soon afterwards Democedes is said to have healed one of them, Darius' wife, the famous Queen Atossa. She had what Herodotus calls a 'growth' on her breast and at first, he says, she had been ashamed to show it to others.[18] When it burst and began to spread she showed it to Democedes, who treated it and healed it. She promised him a reward, and he is said to have instructed her to direct Darius' attention to a conquest of Greece, for which the king would need to send a reconnaissance party westwards. Who better to lead it, he taught her to say, than Democedes? She persuaded the king, and so Democedes left for the Greek world, setting out from Sidon, we are told, with an escort of fifteen Persians and a rounded ship in attendance. It is said to have been full of treasures for his family, including some for the father whose bad temper had caused him to leave home in the first place.[19]

When the reconnaissance party reached Italy, Democedes deserted and continued to elude his Persian escorts. Eventually, they found him in the marketplace of his native Croton, but fellow-citizens refused to surrender him and even used their staffs to beat off his pursuers. The Persians returned to Asia without Democedes but with his parting message to Darius: he was betrothed, he announced, to none other than the daughter of Milo the wrestler, a man, says Herodotus, whose 'fame was great' with the Persian king. Later Greek sources credit Milo with marriage to the philosopher Pythagoras' daughter.[20] If Democedes' bride was Pythagoras' granddaughter, she was probably aged 12–14 when he married her, a very young bride to us, but not unusually young for an ancient Greek girl.

Milo, her father, is the most famous athletic victor on the circuit of Greek games. He became credited with great feats of strength and meat-eating, including the tale that he once carried an ox into Olympia and ate it himself.[21] He won victories in six consecutive Olympic games, spread over twenty years, a number which far exceeds the successes of his herbivorous rival, four-legged Pherenicus. Suggestively, Herodotus considered that the Persian king in faraway Susa would know about this megastar of Greek athletics. He did not regard Democedes' final riposte to Darius as discreditable. The doctor's artfulness continued to appeal to him despite his supposed role in promoting a Persian invasion of Greece.

Democedes' beguiling story has earned him a novel, in German, and ought to earn him a film (Rosario Dawson as Atossa and a final reconciliation with that father, brought about by Milo's mediating daughter . . .).[22] A few other details about him are given in later authors but none is likely to be true.[23] Herodotus' invaluable account is the fullest presentation of a doctor and his career before any of the new medical texts existed and claimed a real invention of medicine. Democedes is the best example yet of a wandering doctor at large. Like Sombrotidas before him, he travelled between Samos and the Greek West, albeit at first in a different direction. Like Onesilas on Cyprus, he is said to have served as a public doctor, even in the 520s BC before the age of democratic assemblies which vetted candidates for such jobs and voted on them. Successors to those public workers in Homer, public doctors served a city-state, but did not offer services to one and all without charge. Democedes began by receiving a talent from Aegina, the very sum which Onesilas later received at Idalion, but it was then increased by competing offers from Greek tyrants. Like Homer's 'public workers', he was not of noble birth. The proof is not his willingness to accept a paid job: he was in voluntary exile and the job was exceptionally well rewarded. It lies in his final message to Darius, as retold by Herodotus. He wanted the king to know that he was now of high standing in his home city.[24] The implication is that, originally, he had not been.

Though socially mobile, Democedes was not remembered as politically progressive. Later philosophic sources claimed that he was a follower of the great Pythagoras, who was indeed in residence at that time in Croton, and that like others in Pythagoras' group he was a political reactionary who took part in what became a local civil war.[25] If he had indeed married Pythagoras' granddaughter, he had family ties to the Pythagorean group, but Herodotus gives no hint of such a sequel. It would surely have prejudiced his view of the man.

There are problems, nonetheless. The details, dialogue and tone of Herodotus' story raise important questions of method for historians. In Herodotus' view, Democedes' request to Atossa, repeated by her in the king's bedroom, accounted for the beginning of Darius' interest in conquering Greece. In fact, Democedes' actions, *c.* 520 BC, preceded Darius' despatch of an army against Athens, 490 BC, by some thirty years. The entire story is a web of personal encounters and actions, leading to momentous results, a pattern Herodotus regularly deploys, but how did he know any of what he tells? He was writing eighty years or more after Democedes' Persian adventures. The framework of his story has been adorned with details which are typical of a good tale, orally transmitted, and the very existence of its protagonists has been queried.[26] As yet, no Atossa is known in non-Greek evidence, let alone as a wife of King Darius: might she be a Greek invention? Democedes' name implies 'care for the people': was he really a historical person? On an extreme recent view, he was simply 'a folk-tale hero who has taken over the identity of a historical doctor'.[27] However, he may have come from a medical family and perhaps, like Sombrotidas, received his apt name from a father who was already a doctor himself.

Despite the interval of some eighty years, stories about Democedes were accessible to Herodotus through Greeks on Samos, where he had important informants, and through families in the West where he himself settled. Their forebears had known the doctor in person: Croton was an easy stopping point for Herodotus during his own years in south Italy. When he heard the stories, they had been embellished with 'rags to riches' incidents and tales of Democedes' exchanges with Persian royalty, but their protagonist was a real person.

Elsewhere in his Histories, typical story patterns elaborate a narrative, but historians can confidently identify them and detach them from a core which is historically grounded. In Democedes' story, the exact point where invention begins and fact ends is not always clear, but it exists nonetheless. Before healing Atossa's breast, Democedes is said to have made her promise to give him whatever he might ask as a reward. This sort of promise is a pattern in Herodotean tales about Persian royalty, and in this case he makes Democedes add, 'except for anything improper'.[28] However, Herodotus has not added all the fictitious extras himself: he comments on Darius' motive in one of his conversations with Democedes and thereby implies that the gist of it is not his own invention.[29] Some of the details may have been added over time to the historical core of the story in order to give it spurious precision, but the general sequence of Democedes' career is not a folk tale. Unlike the biblical Joseph, he is a historical person in historically coherent settings, open to Herodotus' verification.

Even the rags-to-riches details and the invented dialogues are excellent evidence for Herodotus' views of a Greek doctor, Greek medicine and its non-Greek rivals. He assumes that medicine can simply be called 'the craft' (*technē*) by participants in his story.[30] In the real world around Herodotus, texts by Greek doctors from *c.* 440 BC onwards were indeed struggling to assert this very status for their skill. He also contrasts the success of this lone imprisoned Greek with the failures of those 'who seemed to Darius to be the best Egyptian doctors'. Democedes' 'Greek cures and gentle remedies' for Darius' foot succeed, whereas the forceful methods of the Egyptians fail. Pride in Greek medicine shines through Herodotus' account.[31]

In it he presents two medical conditions. The 'growth' or swelling on Atossa's breast has been hailed by modern medical readers as the world's first attested case of breast cancer. However, Democedes is not said to have used surgery to cure it, as a cancer would probably have required. Other proposed diagnoses are mastitis with complications or, more plausibly, an abscess, but Herodotus is not being clinically exact and the queen's condition cannot be classified for sure in modern terms.[32] Darius' injury is another matter. He 'twisted his foot', says Herodotus, and it was twisted in some violent manner 'because one of his bones', in Greek the *astragalos*, 'came out of the joints'. The bone in question is the talus bone, part of the ankle's complexity.[33] It is not held in place by any muscles and so it can indeed be dislocated without fracture, just as Herodotus assumes. Dislocation is said to have caused Darius so much pain that he could not sleep, an effect which fellow-sufferers still recognize. Long after Darius' lifetime, and probably after Herodotus, fifth- and fourth-century Greek medical texts describe damage to this same ankle bone and advise that it should be bathed with warm water and treated by bandages and dressings.[34] Modern doctors prescribe much the same.

Herodotus was most unlikely to have detailed knowledge of the types and causes of swellings in the female breast. Injuries to a horseman were another matter. Greeks already had a good idea of the damage that can be done to particular bones, not least by observing the effects on their valued companions, horses. A horse's talus bone is just above its navicular bone, a recurrent point of weakness in hard-worked legs. Injuries to a horse's navicular are still best treated by soothing drugs, stronger, however, than Democedes' 'Greek cures'. By the time of Herodotus, Greeks understood much more about a horse's bones than about a woman's anatomy, let alone about mysterious swellings on her bosom.

III

Croton was not only famous for Democedes. Between the 580s and 480 BC, it was the home of at least thirteen victors in running races at the Olympics as well as of Milo, victor in wrestling contests at six consecutive sets of Olympic games. One of Democedes' contemporaries, Philip of Croton, was an Olympic victor and was known to

Herodotus as the most beautiful man of his time. None of these victors had been imported to Croton from another city-state.[35]

Strikingly, this winning phase did not continue. In the fifth and fourth centuries BC Croton's athletes ceased to be prominent. The city was still known as a very healthy one – '[even] healthier than Croton' became a proverb – but its earlier sequence of sporting successes cannot be explained by its healthy site alone: otherwise, they would have continued.[36] Good local trainers had surely helped its cluster of winners and it is attractive to relate their unusual skill to a cluster of famous thinkers in the city, known to have been interested in the nature of health.

From c. 530 till c. 510 BC Croton was the seat of Pythagoras, another westward émigré from the island of Samos. Almost every act or statement ascribed to him is disputable, but he is highly likely to have expressed views on health and healing. Short dogmatic sayings were later ascribed to him, one of which is clear enough. 'What is the wisest thing in the human realm? The art of medicine.'[37] He was later described as 'not having neglected medicine, either', and his famous views on the merits of vegetarianism and his notion of harmony were probably extended to apply to the general goal of health.[38] After about twenty years in Croton, he moved to Metapontum, another Greek city in south Italy, apparently c. 510 BC. His pronouncements did not shape the main course of the medical texts which survive from the fifth century, but his image and legend retained respect.

After Pythagoras had gone, an exceptionally interesting thinker emerged in Croton, one whose views on parts of the body, health and sickness were expressed in a theoretical framework of cardinal importance for medical writers of the next generation. He addressed three of Pythagoras' pupils outside Croton in a text 'to' them.[39] The author, Alcmaeon, is known only from a very few later comments and citations, a problem which compounds historians' attempts to date and understand him. The line between his own words and the paraphrasing of these later texts is not always clear. In some places, even the later paraphrases have been inaccurately transmitted by scribes. Alcmaeon's dates have also been much disputed. However, one of those he addressed was called Bro[n]tinus, a very unusual Greek name but known to have been borne by one of Pythagoras' early pupils.[40] Brontinus was a man of Metapontum, as was another recipient, Leon, and so Alcmaeon's text is most likely to belong after *c.* 510 BC when Pythagoras left for that city. Alcmaeon was certainly aware of topics which interested Pythagoreans, but he was not a Pythagorean himself. By the very act of writing a text, he implied as much. Pythagoreans mostly preferred oral teaching, orally transmitted.[41]

Alcmaeon exemplifies the relation, evident since the mid-sixth century BC, between what we now separate as medicine and philosophy. However, the two subjects should not be seen as wholly interchangeable fields of inquiry. The many healers and humdrum doctors in Alcmaeon's lifetime made no contribution whatsoever to philosophy and were uninfluenced by it. Likewise, Alcmaeon was not a doctor himself. He was one of a newer type of thinker, a natural philosopher, and so he was interested in questions about the nature of man. In Galen's view, and others', his text was even called 'On Nature', like Anaximander's text before it.[42]

In Alcmaeon's lifetime, questions about the functioning of the inner body and man's health were part of that important new idea of 'nature' which Greek intellectuals had quite recently postulated. They much interested Alcmaeon, who was considered by a later classifier to have written 'mostly' on medical topics.[43] In 1956, another bit of his teaching was identified in a ninth-century Arabic source, translating a Greek original which cited his view of plants and their relation to the sun and moisture. It is a reminder that Alcmaeon indeed ranged widely and, like medical writers after him, considered plants and their growth as analogies for the workings of the human body.[44] None of the very sparse evidence about him associates him with treatments of wounds or sicknesses. He matters, rather, for his general theories about the body and health.

Semen, the senses and sleep are topics on which he is known to have made important pronouncements. In sleep, Alcmaeon believed, blood withdrew from the 'blood-bearing veins', a theory which was to have a long life, but no link to reality.[45] His views on semen were also misguided. Pythagoreans are said to have considered male seed to be 'a drop of the brain', and Alcmaeon specified that it was part of the white marrow which was considered to be in the brain's grey squashy mass. On the likeliest understanding of later reports of his views, he accepted that semen also existed in the spinal cord, another place where marrow was known to be present. During sex, therefore, semen passed from the male brain, ran down the spine, travelled round the testicles and rushed out through the penis.[46] Elements of this roundabout route to ejaculation were not Alcmaeon's invention, but he also had views on reproduction. He thought that both the male and the female contributed 'seed'. He believed that mules were sterile because the male's semen was too thin and the female's womb 'remained wide open', whereas it closed in women who became pregnant. He also believed that the head is the first part of an embryo to develop.[47]

His emphasis on the role of the brain had an even more important dimension. He argued that sense-perceptions are relayed to it through 'channels' and that they contribute to humans' distinctive ability, 'understanding'.[48] Whereas Homeric poetry had located thinking and feeling in the heart and chest and only a vague life force in the head, Alcmaeon is the first writer known to have amalgamated both feeling and understanding in the brain. His view did not triumph. It was later rejected firmly by Aristotle, one theme, no doubt, of the text which he is said to have written *Against the Views of Alcmaeon*.[49]

Alcmaeon is also credited with 'daring to engage in excision' or cutting-out. He is said to have applied it to the eye, though only by a Latin commentator on Plato about 900 years after Alcmaeon's lifetime.[50] The scope, aim and object of this cutting-out are all uncertain. In antiquity, animals were cut up without scruple, not least as offerings to the gods, but dissection of the human body is not attested in any evidence about previous Greek healers: it remains the most significant absentee in Greek medical inquiry hitherto. As Alcmaeon's cutting required 'daring', it clearly involved a human, not an animal, item. The important point is that it was restricted to an eye. As eye wounds are well attested in Greek warfare from Homer onwards, Alcmaeon need only have cut up an eye which had been dislodged or severed from an unfortunate combatant. He engaged in an inquiry, not a controlled experiment, perhaps with the eyeless victim to help him. He decided that 'channels' run back from the eye to the brain, but he did not understand them as nerves: nerves were only identified in the early third century BC. He probably observed what we now know as the big optic nerve, but believed, wrongly, that the eye contains fire and is surrounded by water. Although he correctly linked perceptions back to the brain, his notions of the underlying physiology were erratic.[51]

He is important, not as the first known experimental dissector, but as a general theorist. At the very start of his text to Brontinus and others, he drew a programmatic distinction between divine and human knowledge. Neither its wording nor translation is entirely certain, but he states, in my tentative rendering and punctuation, that 'concerning unseen things, [which] concern divine things, clarity is what the gods have, whereas for men there is inference, albeit of one thing after another'.[52] On any view of its meaning, this opening statement relates directly to the bold statements about knowledge and belief which were being made by other Greek thinkers in his lifetime and which qualify them as the world's first philosophers.[53] In keeping with Alcmaeon's remark about knowledge, nothing he is reported to have said about the human body refers to the intervention of a god or goddess. So far as we know, he reasoned more generally, in terms of abstract items. On his own view of knowledge, they were inferences, made by a human observer. They are very different from Homer's or Hesiod's assumptions about an unseen world of divinely sent diseases.

Aristotle credits Alcmaeon with stating that 'most human things are pairs', meaning, it seems, that paired opposites underlie the make-up of humans and, presumably, their world. Pythagoreans, too, emphasized the role of opposites, but Alcmaeon differed in believing that there was not a defined number of opposites to consider.[54] His opposites included very many – white and black, bitter and sweet and so forth.

Like Anaximander before him, Alcmaeon wrote about 'nature' in prose, not poetry. He was generalizing boldly, and for the first time he presents historians with a theoretical view of disease and health. Summaries of his views are given in two later collections of earlier authors' opinions, and although they are beset yet again by problems of textual accuracy, it is clear that he related health to an 'equality' between paired opposites, hot and cold, dry and wet and so forth. Sickness, by contrast, he presented as the 'monarchy' of one element over all others.[55] This theory may have been paraphrased by later authors, but its central terms, equality and monarchy, are political terms of the later sixth century BC, Alcmaeon's own lifetime, and are surely his own.[56]

By 'equality' he meant, not equal measure, but an equal distribution of power. He was projecting abstract terms of political language onto the human constitution and explaining its health and sickness as a balance and imbalance of opposites. It is the first example of a way of thinking by political analogy which was to become prominent in Plato's *Republic* about 150 years later. In Alcmaeon, for the first time, a Greek thinker can be seen setting out a comprehensive theory of disease which owed nothing to the interventions of a Homeric god or *daimon*. On such a principled view of disease, treatment could be based without invoking the random actions of a divinity. In their absence, prediction of a disease's course also became possible.

Alcmaeon is credited with ascribing the 'active cause' of disease to an excess of the hot or the cold, two paired opposites, and the 'occasion' of disease to an 'excess of food or lack of it'. He even located causes of disease, it seems, in the blood, the brain or the marrow.[57] The phrases for 'cause' and 'occasion' may be his later summarizers', not his own, but the direction of his thinking here is quite different from anything implied by Homer or Pindar. He is thinking of diseases as processes, not as unchanging Hesiodic entities. In the light of the new idea of abstract nature, he is assuming regularities and rules, notions quite absent from Homer and his heirs. He is even credited with ascribing disease to 'external causes', not perhaps his very words, but evidently implied by what he wrote. They are not gods and goddesses: he is said to have cited 'waters or places or exhaustion or necessity' as explanations of illness.[58] This statement is extremely important. Similar thinking will resurface a generation later in the medical texts which will be the subject of the second part of this book. Alcmaeon is the surviving thinker who allows us to see when it had become current.

The question remains, who may Alcmaeon have been? He announced himself in the first words of his book as Alcmaeon of Croton, son of Perithoos. Neither name is otherwise attested in the Greek west. An Alcmaeon or two and a Pirithoos are attested much later in the Aegean and western Asia and there is an Argive Alcmaeon in Greek myth related to west Greece, but these two personal names are known much more famously in one and only one place: Athens.[59] Perithoos was best known as the legendary companion of Theseus, whose fame proliferated in Attica from the later sixth century BC onwards. Perithoos even received cult there as a hero.[60] Alcmaeon, too, was a famous figure in Athenian versions of their past. An aristocratic Athenian clan, the Alcmaeonids, teemed with Alcmaeon-names, some of which were borne by the most famous rogues and rebels of sixth-century Athenian history. Like other Greek cities in south Italy, Croton was a welcoming bolthole for émigrés and losers, cast off the treacherous ladder of Greek political life. Alcmaeon may therefore be an Athenian nobleman by birth, who found a safe home in Croton after political turmoils had driven his clan away from sixth-century Attica and the rule of its tyrants' family. Those turmoils came to a peak in the last two decades of the century, a time when the words 'equality' and 'monarchic rule' were publicly discussed ideas. Alcmaeonids professed 'equality of power', to be exercised, first, among fellow-aristocrats, but then, as their own political influence faltered, to be extended, one of them suggested, to all male citizens in reforms which were the birth of democracy.[61] Alcmaeon did not, it seems, call disease a 'tyranny'. Nonetheless, his projection of terms for political constitutions onto another constitution, the human body, makes beautiful sense if he himself was an Alcmaeonid of Athenian birth, who had lived among these political slogans in the years before he migrated to south Italy. Settling at Croton, he explained health and sickness in the political terms which had governed his own experience.

5

The Asclepiads

Onesilas on Cyprus, Democedes on his travels, that Rhodian with his cups and enemas and Alcmaeon theorizing in south Italy are as far as we can go in tracing individuals' engagement with medicine in the later sixth and very beginning of the fifth centuries. They are a crucial prelude to the fifth-century texts which then proclaim medicine as a proper craft, not a hit-and-miss activity, and regard it as a new one with a new method. Socially, too, there was to be an important continuity: as in Homer's Odyssey or in Sombrotidas' and Democedes' careers, individual doctors of the new craft would still travel from city to city as mobile men of talent. Even in the most contrarian Greek community, Sparta, travelling doctors became stock characters in an early form of comic farce. The Spartans liked to be entertained by a simple sort of mime in which a foreign doctor appeared and was ridiculed as a pretentious fraud.[1]

How were these pretentions formed and transmitted? So far we have followed individual practitioners, poets and thinkers, but when Herodotus narrates the exploits of Democedes, he observes that the Greek city of Cyrene in Libya, together with Croton in south Italy, was the top centre of medicine at that time.[2] Like Croton, Cyrene was a Doric-speaking city-state: in later Greek comedies, wandering doctors are ridiculed for speaking Doric Greek, evidence of the persistent link between medics and this distinctive dialect. Herodotus' awareness that there were medical clusters in particular places is important, even though we cannot name a single Cyrenaean doctor in the years from Homer to the fifth century and only one surviving monument in Cyrene seems to honour one.[3] A now-worn stone relief, dating to the fifth century, perhaps as early as 470–450 BC, shows, on the left, a

Cos and Cnidos: home of the Asclepiads

N

Ephesus

SAMOS

Samos

Maeander

IKAROS

Miletus

CARIA

KALYMNOS

COS

Cnidos

Syrnos

ASTYPALAIA

Ialysos

Rhodes

RHODES

KARPATHOS

0   15 miles

half-naked male, bearded and seated on a chair, with both of his hands on the shoulder of an adult man, bearded but also half-naked. The scene has been convincingly read as a doctor treating the shoulder of a patient.

Conversely, Herodotus does not mention in the late sixth century or at any other time the two Greek medical centres which are most famous for modern historians. They too were Doric-speaking. Both were located on or near the south-west coast of Asia Minor, one being Cos, the island home of the great doctor Hippocrates, the other Cnidos, originally sited at modern Burgaz, about fifteen miles east of a tip of the Asian mainland that is separated by a mere twenty-mile sea journey from the south-east coast of Cos.[4] They were to become especially important for posterity, because many of the authors of medical texts in the fifth century, including those which proclaim a new invention of medicine, are men of Cos and Cnidos. One possibility is that Herodotus omitted these places because neither of them had become medically preeminent while Democedes was a young man. If so, their rise was indeed one of the fifth century, a new prominence which needs more explanation.

If their claims to be the top medical centres were relatively recent, they soon made up for them with a tale of their very distinguished origins and ancestry. It is first known to us from the historian Theopompus, who was writing c. 350–325 BC, but it was surely older than his mention of it.[5] He presented 'the doctors of Cos and Cnidos as Asclepiads', people who claimed to descend ultimately from Asclepius, the god of healing, and told how 'the first of them came from Syrnos, being descendants of Podaleirios', the famous healer with the Greeks at Troy. To us Podaleirios is a figure of legend, but, nonetheless, his descendants are assumed by Theopompus to have travelled south, presumably after that other legend, the Trojan War, and to have settled in Syrnos, a site so obscure that for many years it could not be relocated on the Aegean coast. Since the 1950s it has turned out to be at modern Beyir on a hilly tongue of land which juts out south-westwards from Asia and looks south to the northern tip of Rhodes, the island whose mainland dependency it became. It had a shrine of Asclepius, attested by an inscription in the second century BC but probably in existence already when Theopompus wrote.[6] His surprising claim that the famous doctors of Cos and Cnidos arrived from this little place is best explained by local rivalry. By asserting that the Asclepiads settled first at Syrnos and only then crossed to Cos and Cnidos, the temple staff at Syrnos could glorify their own shrine of Asclepius. In Theopompus' lifetime, Syrnos may already have been closely linked to the important island of Rhodes. As Rhodes too claimed to have had Asclepiads, the story of their initial settlement at Syrnos would exalt Rhodes' Asclepiads above their more famous neighbours on Cnidos and Cos. Fictitious mythical ancestries were often claimed by Greek sites, shrines and social groups for such reasons. Little Syrnos even claimed in due course that Asclepius had had a wife called Syrna, a local princess, further dignifying their temple.[7] None of these claims is likely to be true.

Both on Cos and on Cnidos, therefore, there were doctors who claimed to be members of a long-existing clan or lineage. Theopompus does not state that all doctors on Cos and Cnidos were Asclepiads, but descent from Asclepius continued to be an important claim. A clan, the Asclepiadai, is named in Coan inscriptions, and this family-based name is likely to go back to a grouping at least as old as the sixth century BC.[8] It persisted for many centuries. In the 320s BC one of the doctors of Alexander the Great was still known as an 'Asclepiad by family descent'.[9] In AD 53, when speaking before the Roman senate, the emperor Claudius digressed on Asclepius' Coan descendants, 'referring to them by each one's name', wrote the historian Tacitus, who had access to a summary of Claudius' speech, 'and by the era in which each had flourished'.[10] As Claudius was proposing honours for his personal doctor, also an Asclepiad from Cos, he had no doubt drawn on the man's local knowledge. It could have been extensive, allowing Claudius a typically long-winded digression. In an inscription of the Roman imperial period a man of Cos identifies himself as 'a descendant of Asclepius in the thirty-fifth generation'.[11]

It is tempting to imagine the Coan doctors and their near-neighbours the Cnidians as rivals in two opposing schools, but an inscription found at Delphi suggests otherwise. Datable around 360 BC, it refers to a common association of Coan and Cnidian Asclepiads and states that only those who are direct descendants in the male line can enjoy certain privileges at Delphi.[12] These Asclepiads were predominantly doctors, but inevitably they had expanded beyond the direct male line of a single clan or family. Sometimes, males would have been adopted into the family, and at other times the honoured name Asclepiad might have been assumed by doctors who had been trained by an Asclepiad but were not connected to the family by birth. As a result of this wider usage, the direct descendants on both Cos and Cnidos were concerned to maintain the identity which they believed they shared.

Their mythical ancestry was false, but the underlying reality was not, that medical skill had once been transmitted in families. As there was a good livelihood in the craft, medically able fathers wanted to pass it on to their sons. Neither Cos nor Cnidos had a formal medical school which could impose a syllabus or exams on students: the very notion of 'schools' of thought was a later one in Greek history, formulated most clearly by the third century BC onwards. Study was with individual doctors, but in due course they would start to teach pupils from outside their own family, partly because they might be more talented than blood relations, but also because they could be charged fees. Fee-paying pupils presumably served as the doctor's assistants, resembling apprentices more than classroom students. As they had to pay and as a proper training was not a matter of a few lectures and a few months, none of them could have come from a lowly background. They did not have to be of noble birth, but they were not sons of everyday labourers.

Older scholarly assumptions that Coans and Cnidians held very different views of their art have also been modified.[13] Both types of doctor wrote texts, and although a Cnidian text is only once mentioned by name in the classical period, it is a very telling one. It was called the *Cnidian Sentences* and is known because aspects of it were criticized by a later author, evidently a man of Cos.[14] In its oldest form, he complained, before others revised it, the *Cnidian Sentences* did not say anything worthy of notice about detailed regimes for patients. It prescribed very few remedies, mostly made of milk and whey. Galen remarks that 'Cnidians', apparently authors of the *Cnidian Sentences*, subdivided diseases much too prolifically, distinguishing 'twelve diseases for the bladder', 'four diseases of the kidneys' and so forth.[15] In modern terms they were splitters, not lumpers.

In their original form, wholly lost to us, these *Cnidian Sentences* may have been one of the earliest Greek medical texts in existence, perhaps even going back into the later sixth century when the prescription of a regimen or *diaita* was still not a widespread medical practice. The name 'sentences' suggests that they were made up of short statements intended as a guide for practitioners rather than being written as a general defence of medicine's claims to be a principled skill. Even so, they mark a new phase in the transmission of Greek medicine. Thanks to the *Sentences*, its outlines could be learned from a textbook and no longer only from oral lessons and demonstrations by a recognized teacher. As a result it could extend even more readily beyond the families who had a long history of practising it. This same progress from clans to a wider range of practitioners is traceable in the history of crafts elsewhere, painting in Japan in the mid-eighteenth century being a neat example, when written manuals began to spread the craft beyond the lineages of master artists in Edo.

Despite the uncertainty about their exact date, the *Sentences* were probably not the first prose guide on a technical subject to be written in Greek. The credit for that innovation goes to architects, two pairs of whom wrote texts in the mid- to later sixth century about famous temples, one in Ephesus, one on Samos, which they had respectively designed.[16] Each of the architect's texts discussed only one item, the particular temple which the author had erected. The *Sentences* were rather different: they appear to have addressed a range of topics in their general field of competence. A pithy style of short sentences is implied by the title: it had a parallel in the sentence-like structure of texts by followers of the master Pythagoras, also in the later sixth to early fifth century.

These medical sentences are an important item in the invention of Greek prose, or at least of prose texts written at length in a scroll. They enabled summary statements to be transmitted across time and distance: no doubt the very need to write them made their author formulate them more clearly than he had before. Other 'ancients' are said by fifth-century authors to have written medical texts too, and here, also, it would be good to know exactly when they did so.[17] Although the *Cnidian Sentences* were primarily texts for teaching and studying, there may also have been an element of rivalry with texts written by their Coan neighbours. The very word 'Cnidian' by which they are identified means that their origin was emphasized and seen as important.

The *Sentences* had been revised and updated by later Cnidians before the Coan author criticized them. Since then, in the shadow of Cos's great doctor Hippocrates, Cnidian doctors have largely been obscured from modern scholars. One of them, Euryphon, was remembered to have been active in the fifth century and was even credited by Galen with the *Cnidian Sentences*, whereas he was more probably one of its revisers. Later readers in antiquity mention some of his ideas.[18] He recommended that adult sufferers from consumption should be fed on a mother's milk, a drastic extension of the *Cnidian Sentences*' favour for milk remedies. The patients had to be suckled directly at the breast because milk, Euryphon believed, lost some of its quality if it was exposed away from its source.[19] He also claimed that diseases are related to digestion. 'When the stomach does not expel food which has been taken', a later summarizer of his views remarks, 'excess residues are created which rise up to the regions round the head and bring about diseases'. Later, a fellow Cnidian, Herodicus, took a more complex view of digestion's role: he is a reminder that Cnidians did not always agree among themselves.[20]

Another Cnidian, as we will see, composed a whole cluster of medical texts, several on women's health and reproduction and a remarkable one on glands, aware of that silent bodily wonder, lymph.[21] These texts were all composed in the later fifth century, but the Cnidian best known to modern historians is not their author: he is Ctesias, so famous that he was hired to be a doctor at the court of the Persian king.[22] In 401 BC he even accompanied King Artaxerxes II to the battlefield against his younger brother, Cyrus. His long text on *Persian Matters* is known only in small extracts and paraphrases, but they include some medical observations and once again show the diversity of views among Cnidians. They do not conform to what modern scholars consider to be distinctively Cnidian opinions.

As for Cnidian remedies, Cnidos was a member of the Greek shrine, the Hellenion, on the Nile delta in Egypt, whereas Cos was not.[23] This base could have given Cnidian doctors constant access to Egyptian drugs and to any medical techniques which Greeks contrived to learn in Egypt. However, access to Egyptian imports was not restricted to the Hellenion's members, and attempts to link Cnidian medicine very closely to Egyptian medicine have yet to succeed beyond dispute: the one plausible interrelationship is literary, the form in which the Cnidian author of part of the text called *On Diseases* listed and then discussed each particular disease.[24] In fact, Cnidian doctors are best known for a remedy which had nothing to do with Egypt at all. Greek medical texts refer in the fifth and fourth centuries to the therapeutic properties of the 'Cnidian berry'.[25] It had obviously been discovered locally by Cnidians, and although it did not remain exclusive to them, it is said to have been used by them often. It is the firm red fruit of an evergreen daphne, *Daphne gnidium*, a low shrub. Even though it is not confined to land around Cnidos, it is a fine example of Greeks' close observation of plants in wild nature. Its berries are powerfully inflammatory if taken in quantity. Realizing this fact, the ancients used them more moderately as pills or in combinations with honey and other fluids. The Cnidian doctor-author even prescribes them, peeled, in compounds for cleansing problematic wombs.[26] Recently, distilled compounds from this daphne have been tested and proposed to be effective against a different problem, breast cancers. However, these compounds are obtained not from the plant's berries, but its roots.[27]

Apart from this daphne, the best surviving memorial to Cnidian medicine may be a famous marble relief, made available through the antiquities' market in 1969.[28] It is incomplete, surviving only in eleven fragments, but the pieces show a male figure who is touching his pointed beard with his left hand and holding a long staff in his right hand. His fingers are very elegantly carved. He is seated on a collapsible chair with his feet on a low stool. He is wearing a tunic, a pleated mantle and a cap: in a later Greek comedy, such a cap is considered to be a distinguishing mark of a doctor.[29] He is approached by a young boy, naked, who has raised his right hand, perhaps to offer him something. The boy's left hand holds the sheath of a knife and what seems to be a cupping glass. Above the older seated figure are two more cupping glasses, confirming that he is indeed a doctor. The boy is perhaps his assistant, but unfortunately, the loss of part of the relief makes it impossible to know what he may have been proffering.

Close comparisons of the robes and other attributes have allowed the style of the relief to be dated *c.* 500–480 BC and traced convincingly to the south-east Aegean, where well-considered parallels have supported an origin at or near Cnidos.[30] The *Cnidian Sentences* are lost to us, but this relief, now in a Basel museum, may be a lasting tribute to a Cnidian doctor and his expertise.

6

Hippocrates, Fact and Fiction

I

Coan doctors have remained much more famous than Cnidians because of their great master, Hippocrates, but the first one known to us by name was not so admirable. In the 440s BC Apollonides the Coan was hired by the Persian king Artaxerxes I and became the trusted doctor of the family of one of the Persian princesses, a daughter of King Xerxes. In his book on *Persian Matters*, Ctesias, his Cnidian successor at court, claims that when Princess Amytis' husband died, she continued to 'have relations' with men, just as her mother had done before her. Sexually active Amytis was already about fifty years old, we can calculate, and had previously been suspected of adultery during her husband's lifetime, but when she fell ill with a slight fever, her doctor Apollonides, according to Ctesias, told her that she was suffering an illness of the womb and would be cured of it if she had sex with a man.[1] This drastic remedy is indeed sometimes commended in Greek medical texts, but Apollonides' proposal was out of proportion to the princess's fever and was only made, says Ctesias, because he had fallen in love with her.[2] His plan worked and Amytis was up for it, but when she continued to fade away (from her illness, presumably), he stopped. Near to death, she went to her mother and told her that Apollonides had been forcing her. Her mother told the king. Apollonides was cruelly punished for two months and then buried alive; Amytis herself died naturally.

There may be some malice in Ctesias's story, a Cnidian's about a Coan, but no such behaviour was ever ascribed to Apollonides' fellow-Coan Hippocrates, still admired as the 'father of medicine'.

On modern Cos he is commemorated by an elderly plane tree under which he is supposed to have taught, by a statue supposedly of himself and by images associating him with the big shrine of Asclepius near the modern city. The plane tree does not go back to anywhere near Hippocrates' lifetime. The statue is not based on his portrait. The temple and the city did not exist when Hippocrates was alive.

Though a figure of the fifth century, he remains remarkably elusive. The earliest mention of him to survive is by Plato, writing no earlier than the 390s BC, who mentions him in two of his dialogues, one of which is set in the mid-430s, perhaps in 433. In it Socrates cites Hippocrates as one of the Asclepiads from Cos and suggests that someone could go to him and pay him a fee and learn medicine from him.[3] In the other, the *Phaedrus*, Plato's Socrates ascribes to Hippocrates the Asclepiad the view that the study of the body requires 'study of the whole'.[4] The scope of this assertion has been much disputed, and will need to be considered later, but it too shows that, for Plato, Hippocrates was a significant teacher of medicine, the one most apt for citation by Socrates and his dialogue partners in the 430s.

Hippocrates, Plato knew, was a man of Cos and an Asclepiad. This much, and his being alive *c.* 433 BC, are certain, but they are a very meagre harvest about a man whose fame became so great. Eventually, other biographic details about him circulated.[5] Two pseudo-biographical texts purport to give his 'family and life'. Two rhetorical speeches of a late date add more details, one claiming to be by his son Thessalos, another by Hippocrates himself. Twenty-four letters are ascribed to him too, and yet more information survives in short entries in much later Greek lexicons or dictionaries. All of this material was composed long after his lifetime. One response has been to dismiss it as fiction and to conclude that 'nothing is known about Hippocrates except that he lived'.[6] Another has been to regard this dismissal as 'undue scepticism' and to try to salvage a date and the outlines of a career for him from what these late sources contain. Before we address the many texts which eventually became ascribed to him, this issue needs to be resolved as it bears directly on the dating of his life.

Fictitious 'lives' of famous Greek poets and thinkers are a well-known genre, all written much later than the person they purport to present.

The fullest text on the 'family and life of Hippocrates' was not written earlier than the first century AD: it cites an author who wrote then.[7] The second text on his 'origin, life and teaching' survives only in bad Latin. It is shorter and contains manifest fictions, including the claim that Hippocrates went to the court of the Persian king Artaxerxes. It gives a list of Hippocrates' pupils, but that too may be an invention.[8]

Fictitious speeches and letters were frequently composed as literary exercises and ascribed to famous persons in antiquity. The two speeches relevant to Hippocrates are examples. Scholars in Alexandria seem to have been unaware of them: they are works, therefore, of the late Hellenistic period, *c.* 150 BC at the earliest.[9] The long one claims to be a speech by Hippocrates' son Thessalos as an ambassador to the Athenians, in which he pleads for the island of Cos, which, he claims, they are treating like a slave and are regarding as theirs, fit to be looted. It is full of unhistorical claims and, like other such fictions, it tries to rewrite or expand what was already known in the histories of Herodotus and Thucydides.[10] At best, some of its material may go back to rhetorical fiction in the third century BC. The other speech, supposedly delivered by Hippocrates himself in Thessaly, is unhistorical from start to finish.[11] The letters ascribed to Hippocrates are no better, none being by himself nor deriving from a collection in his lifetime: their date of composition remains uncertain. Most of them concern two famous people whom Hippocrates probably never met, the Persian king Artaxerxes and the famous Greek philosopher Democritus. One purports to give Hippocrates' unlikely advice to his son: learn geometry and number. Like the others, it is late and fictitious.[12]

In the fictional life, Hippocrates is given a precise birth date, the twenty-seventh day of the eighth month of the Coan calendar in the year 460 BC under the monarchy of Abriadas. Soranus of Cos is alleged to have found this date in the city of Cos's archives, but the claim is thoroughly implausible.[13] Greeks never had birth certificates, and although a clan or a local subdivision might perhaps hold evidence of the age of some of its members, a Greek city would not. The city of Cos was re-founded on a new site in 366 BC and is most unlikely to have preserved historical records from outlying local sources before that date. The office of monarch, mentioned in the supposed birth date, is also problematic. It may have existed locally on Cos in the fifth century but it was probably not used for dating each year until the new city had been founded in 366. Soranus is problematic too. The name was borne by a well-known doctor in the second century AD, but he was a man of Ephesus, not Cos. Soranus of Cos is a bogus authority, invented to make a fictional text seem credible.[14]

In several late sources, including the one on his 'family and life', Hippocrates is given a long genealogy. It stretches back through as many as eighteen named ancestors to Asclepius himself at the time of the Trojan war.[15] Such lengthy genealogies are occasionally cited for and by a member of a Greek family, but even the 'family and life' can only credit this one to much later authors. One of them, Pherecydes, has been mistaken for an Athenian author of the same name who wrote a text on genealogies in the mid-fifth century BC: if so, he would be important evidence that Hippocrates' long genealogy was known in Hippocrates' own lifetime. However, this Pherecydes is nothing of the sort: he is a Hellenistic author, Pherecydes from Leros, and, like the other scholars named with him, he was writing in the late third or second century BC.[16]

By then, the desire to know had fabricated its own tradition. The early names in Hippocrates' supposed ancestral line are manifestly fictitious: some of them relate to legends about the history of Delphi which had only proliferated in the fourth century BC, long after Hippocrates' lifetime.[17] Even the most recent ones are historically problematic. The name of his mother or grandmother is sometimes given as Phaenarete, but as this was the name of Socrates' mother, and, as Plato presented her as a midwife, it was borrowed by genealogists in the absence of factual evidence. It linked Hippocrates to a philosophic pedigree and validated texts about obstetrics and gynaecology which had become ascribed to him.[18] His father's name is said to have been Heracleides, indeed an attested Coan name, but for that reason an obvious guess when no other evidence was available. One of his uncles is said to have been Aineios. As Aineios was a famous doctor c. 500 BC, commemorated on that marble disc found in Athens, his name was another neat borrowing despite being a little too early to be Hippocrates' real uncle, on the usual view of his birth date. Disagreement also surrounds Hippocrates' age at his death. Even the text on his 'family and life' is uncertain: it gives 85, 90, 104 and 109 as possibilities. As the truth had been lost, guesses were introduced to allow Hippocrates to survive long enough to write the various later texts which had become ascribed to his authorship.

Even the place of his death and burial is uncertain. The 'family and life' claims that his tomb lies in northern Thessaly by the road from Larissa to Gyrton. Later sources repeat the claim and in the mid-1830s it led to a supposed discovery of the very inscription which, allegedly, commemorated Hippocrates. This claim and its eventual refutation are a fascinating episode in their own right, involving Greek doctors in Larissa in the 1830s, keen to locate the father of medicine, and prudent inquirers, unsure whether or not to believe them. The local will to believe remained strong, so much so that a marble monument to Hippocrates was put up on the alleged site of his burial as recently as December 1978. The remarkable story of the inscription's fabrication and its eventual discrediting was finally pieced together by Bruno Helly in 1993.[19] It is still uncertain if Hippocrates died near Larissa, let alone where his burial was sited.

Hippocrates named one of his sons Thessalos, but as Cos was credited with a close mythical kinship with Thessaly and Thessalos was a name linked with Cos in Homer's Iliad, the naming does not entail that Hippocrates himself ever went there.[20] Nonetheless, colourful stories were invented to explain why he left his home island. According to one Andreas, probably a court physician in Alexandria c. 220–175 BC, Hippocrates left Cos because he had burned the cache of medical writings which doctors had amassed on rival Cnidos.[21] Others claimed he had burned down the temple of Asclepius on Cos itself. These slanderous stories were later countered by a tale in the 'family and life' that Hippocrates left Cos and settled in Thessaly because of advice from a figure, surely Asclepius, whom he saw in a dream.[22] All of these tales were fictions.

The 'family and life of Hippocrates' credits him with yet another destination, a visit to the court of the Macedonian king Perdiccas II. Perdiccas' father, it says, had just died there, and Perdiccas was wasting away, apparently with consumption. Hippocrates is said to have diagnosed that Perdiccas was suffering from love for one of his

father's slave-concubines.[23] The story presupposes a visit by Hippocrates to Macedon soon after the death of Perdiccas' father, in 454 BC
therefore, as other historical sources allow us to establish, but the
birth date in the fictitious 'family and life' means Hippocrates would
have been only six years old at that date. Nonetheless, the tale had a
long literary afterlife. It even shaped a late Latin poem, composed c.
AD 500, in which Perdiccas was said to have fallen in love with his
own mother before Hippocrates came to attend him.[24] In ignorance
she held a beauty parade of the girls at court in order to find out
which was the object of her son's desire, not realizing that it was she
herself.

This visit to Perdiccas' court in Macedon is an invention. It is
matched by another fiction, possibly its model, which also involved a
doctor and a love-sick Macedonian prince, this time in the early third
century BC, who had fallen in love with his stepmother.[25] The story
about Hippocrates and King Perdiccas is certainly not evidence that
Hippocrates went to Macedon, let alone as late as 413 BC. It is no
more true than the story of his visit with a slave girl to the philosopher
Democritus. In the ancient story Democritus showed remarkable
powers of insight about the slave girl's sexual status, whereas in the
version current on Cos around 1900 Hippocrates had become the
insightful observer.[26.] Once, the legend now said, Hippocrates greeted
a shepherdess as 'maiden' when she walked past, but as 'woman' when
she walked by later, causing her to blush deeply. It emerged that she
had lost her virginity in the interval to a young shepherd who had
been waiting for her in an ambush. Hippocrates is said to have
explained that he knew by the way she walked, because the walk of a
virgin differs from the walk of a girl who has had sex. This long-lived
belief among boys is most unlikely to have been a belief of the father
of medicine himself.

II

These negative results discredit all the details which fictions ascribe
to Hippocrates and the date which they give for his birth. All that
remains as external near-contemporary evidence for him is the

testimony of Plato, who assumes he was alive and famous in the 430s
BC. Nonetheless, the hunt for the true Hippocrates has attracted
modern scholars for more than a century. Later in this book I will
return to it, with a new suggestion based on a markedly new date for
a classic medical text, one of the many ascribed to him personally in
antiquity by later readers and copyists, usually for very tenuous
reasons. It will conflict with his supposed birth date in 460 BC, but
that date, I have shown, is fiction.

What, though, about the most famous item associated with his
name, the Hippocratic Oath?[27] It still serves as a charter for many
medical institutions. It lives on in many doctors' minds as a shining
statement of the principles of doctoring, particularly lustrous as it is
the first such statement known in the world. It has the further attrac-
tion of originating within the doctors' own profession: it was written
by a doctor for fellow-doctors. It has even been esteemed as the essential ethic of Western medicine. It is a remarkable text, but, although
it is an oath, its claim to be linked to Hippocrates is eminently
contestable.

The first clear reference to it dates to the AD 40s, in appreciative
words of the doctor Scribonius Largus, who was writing in the reign
of the emperor Claudius. Largus refers to Hippocrates transmitting
the 'principles of the discipline' in an oath which, he thinks, was so
far from permitting hostile treatment that it forbade the giving of a
medicament to a pregnant woman to 'shake out' what she had conceived.[28] The exact wording of the Oath itself is still debated. The
text which is usually printed is based on Greek manuscripts copied
from the tenth century AD onwards, but a bit of it is also known from
a fragmentary text on papyrus, written c. AD 300, whose wording
varies in a few places from some or all of the later manuscripts: it is
not always preferable.[29] Even earlier, some of the words in the Oath
were listed in the AD 60s by the grammarian Erotian in his glossary
of words in Hippocratic works.[30] Later, Galen wrote a commentary
on the Oath, probably in the early third century AD near the end of
his life, and as usual quoted bits of the original. His Greek text is lost
but it was translated first into Syriac and then into Arabic in the ninth
century AD. Later Arabic authors quote extracts from it and recently,
and most remarkably, an early, but fragmentary, Arabic text of what

is probably the second part of Galen's commentary has been re-discovered.[31] Consequently, a full critical awareness of the Oath's original wording requires a foray into Arabic as well as Greek.

There is also the problem of translation, as the Oath's Greek is not always straightforward. My translation, which follows, at least shows the blend of ethics and worldly undertakings in the Oath's conception of a doctor's life. My separation of the clauses is modern and not attested in any original.

I swear by Apollo the Physician, by Asclepius and by Health and Panacea and by all the gods and also goddesses, making them witnesses, that I will carry out according to my ability and judgement this oath and this written contract:

To regard my teacher of this craft as equal to my parents and to share in partnership my livelihood with him and to give him a share if he is in need of necessities and to adjudge his male offspring equal to my male siblings and to teach them this craft, if they want to learn it, without fee and written covenant, and to impart precepts and lectures and all the rest of the teaching to my sons and the sons of my teacher and to pupils who have made a written agreement and also sworn an oath in a customary medical manner [or: 'according to medical law'], but to no one else.[32]

I will apply all regimens for the benefit of the sick according to my ability and judgement, but if they are for their harm or injustice, I will obstruct them according to the best of my judgement.

I will not give a drug that is deadly to anyone at all [even] if asked nor will I suggest such counsel, and likewise I will not give women an abortive pessary.

And in a pure and holy way I will guard my life and craft.

I will not cut at all those suffering from the stone, but I will withdraw in favour of men who are practitioners of this [sort of] practice.

Into as many houses as I may enter, I will enter for the benefit of the sick, keeping far from all voluntary wrongdoing and other corrupting behaviour, especially from sexual acts, both on women's bodies and men's, free and slave. And whatever I may see or hear in the course of treatment, or even outside treatment, concerning the life of people, which is not right ever to be gossiped about outside, I will remain silent, considering such things not to be divulged.

If, then, I render this oath fulfilled and do not confuse and confound it, may it be that I enjoy the benefits both of a livelihood [or: 'life'] and the craft and be in good repute among all people for all time, but if I transgress it and perjure myself, may I receive the opposite.

Even though some details of this translation can be disputed, the Oath has inarguable qualities. It is manifestly an oath which a doctor would swear personally, presumably at the end of his training rather than the beginning. It was accompanied by a written contract ('this contract') which he probably had to sign. Several of its main provisions are still highly pertinent after more than 2000 years: the aim to benefit patients, to avoid harm when prescribing regimens, to remain discreet about personal information and not to harass male or female patients sexually. The Oath even declares those ubiquitous recipients of sexual abuse, slaves, to be out of bounds.

Its taker swears repeatedly to behaviour which requires him to exercise his own discrimination: he is to regard others as equal to his own brother, to prescribe 'to the best of my ability and judgement' and to undertake to live 'in a pure and holy way'. He is a doctor, but is he distinct from a surgeon? If he foreswears all cutting, on one translation of the Oath, he is, but it is likelier that he merely forswore difficult and painful cutting for kidney stones, leaving it to others whose 'doings' he regarded as rather dodgy.[33] The scope of the words about not giving 'deadly drugs' nor 'abortive pessaries' touches on sensitive modern topics and remains controversial too. By forswearing deadly drugs, is the doctor swearing never to help requests to assist suicide, or only requests to bring about euthanasia or, most likely, never to kill or assist in killing somebody with, for instance, poison? When he forswears deadly pessaries, is he forswearing all means of abortion or only the dangerous use of pessaries, tampons which were indeed soaked and applied by doctors for that purpose and were known to be very dangerous to the mother? When he swears to guard his life and craft in a 'pure and holy way', is he swearing to guard his entire life at that demanding level or only the professional part of it as a doctor?[34] Other undertakings, including the one about whatever houses he enters and whatever he hears outside treatment, support the view that the Oath indeed covered the doctor's entire way of life.

In a brilliant study published in 1943, one of the great medical historians, Ludwig Edelstein, argued that the Oath bore the stamp of Pythagorean philosophy. Although it was not composed by a philosopher, he considered that 'it is quite possible that a physician, strongly impressed by what he had learned from the Pythagoreans either through personal contact or through books, conceived this medical code in conformity with Pythagorean ideals.'[35] As Hippocrates was not a Pythagorizing physician, Edelstein inclined to date the Oath as late as the fourth century BC. Pythagoreans were always a small and abstruse minority in the Greek world, and so his study marginalized what had previously been regarded as an Oath for all good Greek doctors. To support his theory, Edelstein understood the oath not to give a 'deadly drug' as an oath not to assist suicide 'when asked' by a patient: opposition to suicide was a distinctive precept of the Pythagoreans. However, his translation here is an unlikely limitation of the meaning of the Greek words.[36] An oath by all the gods is also most unlikely for someone 'strongly impressed' by Pythagoreans, as they were said to swear by quite other divinities.[37] The Oath is indeed solemnly phrased and ethically minded, but it is not obviously influenced by Pythagorean ideas at all.

Manifestly, it originated at a time when the transmission of medical training to family members was still strong. The doctor swears to teach the offspring of his own teacher, if they so wish, and to regard them as equal to his own male siblings. He will also regard his teacher as equal to his own parents. He is not being adopted into his teacher's family; nor will he adopt his teacher's children into his own. He is an outsider, not a family member, and, as the Oath prescribes, he is therefore someone who has been taught according to a written covenant and an oath. These undertakings had plainly obliged him to pay fees: only the family members, who do not swear to the Oath, are explicitly exempted from paying. The Oath, therefore, arose when doctors were teaching pupils outside their own family.

To maintain the teacher and his family, all outside pupils were being obliged to swear to support him henceforward with a share of their livelihood and, if the teacher could not do so, to teach his children and transmit the craft to them for no fee at all. The Oath, therefore, is a text imposing a restrictive practice, utterly alien to modern medical training. The obligations to support the teacher and his sons are onerous commitments but the swearer, himself an outside pupil, had good reasons for assuming them. One was that his training was so valuable. Another was that he himself was to become a teacher and so he could expect his own pupils to swear to honour and maintain him likewise.

In the Commentary on the Oath ascribed to the great Galen, the Oath is explained as arising when there were not enough family members of the Asclepiads on Cos who wished to carry on the medical tradition. In response, Galen states, Hippocrates set out terms to be sworn by fee-paying outsiders who would promise to teach their skill to the children of an Asclepiad teacher.[38] The Oath, therefore, would maintain a strong Asclepiad family continuity, something which we know from that Delphic inscription to have remained important to them. However, the Oath does not mention Asclepiads at all, and although Plato assumes that in c. 430 BC Hippocrates was teaching outside pupils on Cos for fees, it is wholly uncertain when this practice had begun.[39]

Since Galen, dates for the Oath have varied from c. 500 BC to the Hellenistic age c. 300–100 BC, with much discussion as to when an oath to 'guard one's life in a pure and holy way' could possibly have been formulated and when it would fit what we otherwise know of Greek ethical thinking and language. Texts which once seemed decisive for one date or another have been called in question and the solution is still elusive.[40] As for the Oath's language, it is Ionic Greek, but as Ionic is the frequently used dialect for medical prose, it does not show that the Oath originated in the eastern Greek world, in Ionia. Several of its words are not attested anywhere else in the Hippocratic Corpus: can others be dated by their usage in medical texts and other oaths? The sample is too small, and so all Greek texts must eventually be searched for parallels.[41] The restrictive clauses about teaching one's teacher's children for free and teaching only pupils who swear the Oath suggest to some scholars an early date for the Oath, up in the fifth century BC. However, its language and its clauses about a 'holy life' suggest to other scholars a date in the Hellenistic era, maybe c. 300 BC. I tend to favour the later date, but not with conviction.

How widely was the Oath known and observed? If the oath not to practise cutting refers to cutting of any kind, it conflicts with the advice of several medical authors who wrote from *c.* 400 BC onwards. If it refers only to cutting out stones, it is less problematic, but the advice of such medical authors is still at odds with the oath 'not to give a pessary to a woman'.[42] Even if the Oath was composed by one of Hippocrates' followers, few of the other authors who were later grouped by ancient scholars as his 'Hippocratic' heirs obeyed or, presumably, swore it.

Positive evidence is later in date. A verse epitaph, inscribed on Corfu in the second to third centuries AD, speaks for a doctor, Thrason, who declares that he honours his teacher of medicine 'equal to his parents'.[43] This esteem is compatible with the Oath's prescriptions and implies that Thrason knew its clauses: perhaps he had actually sworn to them. Three papyrus texts, all found in the modest Egyptian town of Oxyrhynchus, attest to the Oath's continuing currency.[44] In the late third century AD one of them contained a text of it written in a hand which papyrologists consider not to be the work of a literary scribe. It suggests practical use of the text, and perhaps, as its first editors suggested, it was a copy made for the purpose of an oath-taking ceremony.[45] In 2009, another Oxyrhynchus papyrus, also of third-century date, was published, whose author, otherwise unknown, expresses his belief that 'for those young men who are being introduced to medicine in a systematic way . . . it is proper, at least as I see it, to make the beginning of learning from the Hippocratic Oath, since it was established as a most just law and one extremely useful for life.'[46] The author does not copy out the Oath, but he is fully aware of it. As in the later schools in Alexandria, the Oath is recommended here for teaching, but the words 'at least as I see it' make plain that others still disagreed with its use in that way. The writer states that the Oath's text must be studied. He does not insist that it must be sworn.

Those who profess reverence for it need to consider it in its entirety. Unlike Apollonides at the Persian court, doctors still generally endorse the ban on sexual harassment. They try, at least, to avoid gossiping about private details of their patients, but not all of them are now religious and very few indeed, outside a religious order, are committed to living their entire life in a 'holy way'. Even fewer of them endorse the giving of a share of their livelihood to their medical teacher for the rest of his life. In the Oath there is one further element at odds with modern practice: it assumes that all doctors are male.

III

Famous though the Oath now is, an even shorter programmatic text, generally known as the 'Law', survives in many more manuscripts which date to the Middle Ages: they attest its great popularity in that period, even greater than the Oath's. Once assumed to be a late-Hellenistic work, it has recently attracted renewed scholarly interest and its date has become an open question.[47] Its title 'Law', in Greek *nomos*, was already known to Erotian in the mid-first century AD, as he included it in his list of works by Hippocrates, listing it directly after the Oath. Might it be a fifth-century work too, from a similar milieu?

It begins by stating that 'medicine is the most brilliant of all the crafts, but that because of the ignorance of those who apply it and of those who judge such people at random, it falls far short of all other arts at the moment'. The author thinks that the reason for these pseudo-doctors' errors is that, alone of all the crafts, city-states have no defined punishments for medical bad practice. In a striking analogy, he compares these ignorant so-called doctors to extras in the theatre who adopt the same poses, dress and masks as the actors, but are not actors themselves.[48] This parallel between doctoring and acting is unique, but, given the history of Greek drama, it could have been written any time from the early fifth century BC onwards.

The author then emphasizes the preconditions for being a proper doctor: a natural aptitude, learning in one's early years, teaching, a place to learn, a love of hard work, and time. Those who then know the craft truly will, he says, be considered doctors in deed, not only in name, when they travel between city-states.[49] The author assumes, therefore, that good doctors are likely to be travelling doctors, the class whom we have encountered from the Odyssey to naughty Apollonides. The text then concludes very strikingly: 'matters which are holy are shown to men who are holy, but it is not right to show them to those who are profane before they have been initiated into the holy rites of knowledge'. Medical instruction is here equated with initiation into a mystery-cult whose secrets are revealed only to devotees.[50]

Does anything in the Oath help to date this Law? The Oath refers near its beginning to those who have made a written contract and sworn an oath in 'a doctorly *nomos*': is this *nomos* an allusion to the actual *Nomos* or 'Law'? Some think it is, and that the Law is therefore closely interlinked with the Oath, but I take the Oath's words to mean no more than 'in a customary medical manner'.[51] Some of the Law's choice of imagery and phrasing and its interlinking of 'craft' and 'experience' have also been considered to point to an origin in the later fifth century BC, but here too a later author could echo at a much later date concepts and language known in fifth-century texts.[52] The crucial point is the Law's concluding presentation of medical knowledge as an initiation into holy mysteries. Such imagery is not readily paralleled in the fifth century and, in my view, points to a late origin for the entire text. Like the Oath, it is more likely to belong in the fourth century or later.

Its context is also elusive. It belongs in connection with the teaching of medical pupils, genuine doctors in the making. It has been considered to be a graduation address to them at the end of their studies, but no such graduation ceremonies are known in medical schools, late arrivals in antiquity, or in any ancient centre of higher education.[53] The statements about the necessary conditions for genuine pupils, from youth onwards, would sound odd in a graduation context. Perhaps the Law is a statement made by the leader of a school about his vision of true medical teaching, as opposed to the ignorance of other so-called doctors outside it. Certainly the author writes in the first person – 'it seems especially to me' – and contrasts knowledge with ignorant opinion.

In a fine flourish he compares the learning of medicine to the cultivation of 'what grows in the ground'. Our nature is like the earth, he says, and the firm opinions of teachers are like the seeds: love of effort is like the working of the soil and, in both spheres, time is needed. A comparison between sowing seeds and educating the young is first attested in the late fifth century in Athens, but the Law's author develops it in much more detail.[54] Whatever his date, he has a special distinction: he is the first person known to have compared medical teaching to farming, or even gardening.

7

The Hippocratic Corpus

Although details of Hippocrates' life remain a blank to us, texts ascribed to him have helped to maintain his fame. There are many more than the Oath and the Law. Galen (*c.* AD 200) remarks that the Cnidian doctor Ctesias (*c.* 400 BC) disagreed with Hippocrates' method for setting a dislocated hip.[1] Ctesias is most likely to have known this method from a text, and if he really did refer to something by Hippocrates himself, it might be our text called *On Joints*. However, Galen may have identified it as a text by Hippocrates, whereas Ctesias had not been so specific. In the later fourth century, Aristotle, the son of a doctor, discussed the view that diseases are caused by 'breaths', a view which he assumed to be Hippocrates' own and which he too must have known in a text ascribed to him, possibly the one which survives as *On Breaths*.[2] However, as the author did not identify himself in either of these texts, one or both of them may have been ascribed to Hippocrates wrongly by copyists who then misled Ctesias and Aristotle.

The great man surely wrote medical works for his pupils' use, some of which may even be preserved, but by *c.* 300 BC many more texts were being ascribed to him than had any solid claim to be his. Our first evidence that texts under his name were being studied as a cluster is that some of their words were 'unfolded' by a grammarian from Cos who was present in Egypt's Alexandria *c.* 300 BC.[3] Another scholar in Alexandria, Baccheios, then worked on at least eighteen such texts in the mid- to later third century BC. He too assumed they were by Hippocrates and he too was more interested in their vocabulary than in editing the full contents of the texts themselves.[4] There is no good reason to suppose that these eighteen had been acquired for

Alexandria's Library as an existing collection, let alone as a single batch from a medical library on Hippocrates' home island, Cos. They had been assembled piecemeal. Meanwhile there was scope for yet more texts, preserved anonymously, to become ascribed over-optimistically to Hippocrates himself. By the mid-first century AD texts under his name had grown to more than fifty, the number which was assessed then for authenticity and searched for rare words by Erotian.[5] They were 'Hippocratic' in the sense that readers now believed, mostly for tenuous reasons, that they were indeed Hippocrates' own work.

In the Greek world, on into the fifteenth century, scribes copied texts which they ascribed to Hippocrates and thereby preserved them in manuscripts. At least three such manuscripts were then assembled in a single volume as *All the Works of Hippocrates*, printed in Venice in 1526: it included fifty-nine texts, the Oath and the Law among them.[6] It had been preceded in Rome a year earlier by another 'All the Works of Hippocrates', a collection of those which had been translated into Latin from Greek. Neither volume closed the matter. Other texts ascribed to Hippocrates continued to surface, especially in Greek manuscripts which the 1526 edition had not used. From 1839 onwards superb work was done by the great French scholar Émile Littré who edited, collected and translated many of them.[7] The name 'Corpus' then became standard for them from the early twentieth century onwards, when they began to be edited collectively under that name. Currently up to seventy-two texts are included in the Corpus, though other scholars would limit the number to fifty-one or so, partly by dividing or uniting the texts in different ways.

The preceding collections and the Corpus are culminations of a centuries-old inclination to identify texts as Hippocrates' own: most of the texts they included were originally composed no later than *c.* 300 BC, at least so far as the evidence of language, thought and content implies.[8] The consensus, hitherto, is that the earlier ones were written sometime between 450 and 400 BC, with the most famous ones falling in the second half of that broad span. The 'Hippocratic Corpus' remains a fascinating labyrinth, far the most valuable evidence for Greek medicine, and has not been exhausted by scholars even after 150 years of study.

A basic distinction needs to be kept in mind here. Hippocrates' name appears nowhere in any of the texts in the Hippocratic Corpus. Modern scholars call its texts and their authors 'Hippocratic', as I will too, but only because they are present in the Corpus, not because they necessarily had a link with Hippocrates himself and his island of Cos. Language, style and content show that the Corpus's texts are very varied, but among those most readily datable to the fifth and early fourth centuries are the ones whose authors explicitly regard themselves as exponents of a new method and craft, making them truly inventors of medicine. They are crucial to the scope of this book and include the enigmatic texts on which I will focus in most detail. To put them into perspective, the texts called 'Hippocratic' need to be addressed first: their originality and the shared approach, if any, which caused them to be grouped in an ever-increasing cluster from at least *c.* 300 BC onwards.

None of these texts survived with its author's name, but apart from those ascribed to Hippocrates several were quite soon ascribed by ancient scholars, including Aristotle, to younger members of Hippocrates' family, to his son or his son-in-law. These attributions were without independent evidence and were a way of keeping texts which had a distinctive tone close to Hippocrates' circle without crediting them to Hippocrates himself.[9] Modern scholars prefer to look for texts closely related by thought and language and to attribute them on those grounds to one author but to leave him without a name. The text *On Joints*, possibly for Ctesias a text by Hippocrates himself, was by the same author as the text *On Fractures*, both probably belonging in the fifth century. Two famous texts, on the *Sacred Disease* (epilepsy) and on *Airs, Waters and Places*, are almost certainly by a single author in the later fifth century too, though an exact date for him there is impossible.[10]

The texts called 'Hippocratic' in the main modern Corpus are not even unified by one place of origin. Only one of them names Cos in its title, and although at least another six are plausibly ascribed by modern arguments to authors from Cos, many others may not originate from Hippocrates' home island at all.[11] One, *Places in Man*, has recently been proposed to be a text from Greek Sicily. Another, *On*

Sight, is in rough Greek and has even been suggested to be by a doctor for whom Greek was not a first language, perhaps someone in Libya.[12] A word in another text, *Diseases 1,* was identified in antiquity as local to Chios.[13] If we had the dates and authors of everything in the Corpus there would be some surprises.

One recent surprise, still not absorbed by non-medical historians, is that critically controlled study has isolated at least six texts in the Corpus as essentially the work of one writer, a doctor from Cnidos.[14] He compiled one text on the *Diseases of Women* and one on the *Nature of Women.* He composed all or most of at least four more texts, including a vivid one on the *Diseases of Girls* and that masterly one on lymph (*On Glands*).[15] Women's bodies and their problems were particular concerns of this seldom-recognized author, one whose name is unknown.

Another clue to a Cnidian origin has been somewhat neglected and misjudged. Four texts in the 'Hippocratic' Corpus give a prescribed dose in terms of an Aeginetan measure.[16] Weights and measures from the Greek island of Aegina were never in use on Cos, but were used, strikingly, in Cnidos. Their use there might perhaps go back into the later sixth century BC, but it certainly prevailed during much of the fifth century. Three of the four texts which specify it also specify a dose in an Attic measure.[17] These measures were based on the weights of coins in local circulation, against which doctors could measure the right quantity of ingredients on a pair of scales. At some point in the later fifth century the Athenians imposed use of their own Attic coins, weights and measures on their subject-allies, of whom Cnidos, using Aeginetan measures, was one. The dating of the decree which imposed them remains very controversial, some scholars favouring a date as late as 414 BC, others preferring a date in the 440s or in the 420s. As a doctor on Cnidos might have continued to refer to the old Aeginetan measures even after the Athenians had ruled to discontinue them, the decree's existence is not a decisive dating point for their use. However, mentions of Aeginetan measures certainly belong in Cnidos, not Cos, and the naming of them specifically as Aeginetan sits most easily there between *c.* 430 and 405 BC.[18]

The texts which refer to Aeginetan measures are all gynaecological, including a text on pregnancy and childbirth, now known from its first chapter as *Superfoetation*, the formation of two embryos in succession in the womb.[19] Any of these texts might have been worked over by later editors and revisers, but no reviser would have inserted a reference to Aeginetan measures after 400 BC when Cnidos had certainly abandoned them. Cnidians, therefore, were not only authors of central parts of our texts about women's bodies. They wrote in the later fifth century, and so medical texts about women were written no later than texts about men. Though classed later as 'Hippocratic' they were not by Hippocrates or his fellow Coans at all.

Another puzzle about the texts now in the Corpus confronts anyone who sets out to read it through: there are very close links between some of them. Sometimes, one text has used another directly, though it is not clear which of the two came first. Sometimes two texts are so similar that they appear to have drawn independently on one and the same original, though it is lost to us.[20] An important part, very influential for posterity, is made up of texts which give short statements one after another as summary bits of wisdom, or aphorisms, or as recipes for remedies and treatments. The aphorisms tend to draw on longer texts known elsewhere in the Corpus and to abbreviate them, presumably so as to make them more accessible for teaching and memorizing. They were also liable to additions and accretions as generations of users put more such aphorisms into their working copies.[21]

Among this diversity and complexity what impelled authors of the early texts in this assemblage to write so much more than previous doctors?

One possible impulse would be a breakthrough in treatment. Six of the texts on surgery seem to be a coherent group and may go back ultimately to a linked collection.[22] They are impressive for their observations of the body, but probably do not contain many new solutions to old problems. One text addressed *War Wounds and Missiles*, but is only known to us fragmentarily from later quotations. It discussed tools and skills for extracting weapons from a wound, but most of its advice was probably being practised already in Homer's lifetime, even if his poems simplified the procedures. One advance may have been the skill of sewing up the wall of the abdomen if a weapon penetrated it.[23] Otherwise the techniques were familiar: cutting out weapons and applying kindly drugs to the affected outer area, the skills of Homer's doctor Machaon.

Wounds in the Head discusses trepanning as a main treatment, one which was certainly no innovation. *Fractures* and *Joints*, by contrast, describe several techniques which are probably innovative, including a method for addressing dislocations of the shoulder, still known nowadays as the Hippocratic manoeuvre.[24] Their author is aware of other so-called experts, whom he criticizes as impostors, implying a new sense of his craft, but his differences from them are probably mainly ones of detail. His most famous piece of equipment is what has become known to moderns as the Hippocratic bench, on which patients with bone problems could be stretched. It too may have had a pre-Hippocratic history, despite the author's refinements: he warns that it has power if used for harmful ends, a hint of its possible use in torture, hardly a new application.[25] If only Ctesias' words about the best way to fix a dislocation could be pressed, the author of these two texts would indeed be Hippocrates himself, speaking out against impostors and implying his new and superior craft.

Discussions of surgery for women, usually for their reproductive organs, are probably no more innovative. They refer to fairly obvious techniques of widening the cervix or extracting a dead foetus with probes or excising it altogether.[26] Mothers might also be shaken to reposition their unborn baby, but that was surely not a new technique, either.[27] One extreme treatment was the tying of a woman upside down by her legs to a ladder if she was suffering from a prolapse and then shaking the ladder and replacing the slipped womb by hand. This rough remedy was ascribed to the Cnidian doctor Euryphon (c. 450 BC), but he may not have been its inventor. The fifth-century text *On Joints* describes it as an old treatment already used by many doctors for dislocated limbs.[28]

For other traumatic wounds and swellings, the surgical texts refer to treatments which we have already traced among post-Homeric doctors: bandaging and cupping, cutting, trepanning, cauterizing (a last resort) and enemas, though they describe an enema tube only once.[29] Texts in the Corpus were no more original when they recommended routines of exercise, food and drink: regulated routines, as we have seen, had been recommended long before 450 BC, by those important experts, athletes' trainers. Hippocratic authors merely refined them, adding novel details. Bathing, exercise, diet and even vomiting were specified and the word *diaita* was applied to them, the origin of our word 'dietetics'. Some of them stress nature's power to heal, if restored to internal balance, and here at least their words were to have an influential afterlife.[30] The authors were certainly not committed to homoeopathy, but in the 1830s their emphasis was understood in a way crucial for the founding of modern homoeopathic medicine.

Were the remedies proposed and the thinking behind their choices new, impelling these authors to write these dietetic texts? Some of them, especially those about women's health, give long sequences of recipes which combine animal, vegetable and mineral ingredients: there are about 1500 such recipes in the Corpus, a major body of information.[31] Among them are ingredients, gruesome to us, which had surely been used by pre-Hippocratic healers, even the dung of cows and dogs. There are also mixed recipes, combining ingredients, but they were not a fifth-century innovation, either; nor were recipes based only on plants. One text, *On Affections*, is addressed to general readers (probably *c.* 380 BC) and refers to a separate 'remedy text' or *pharmakitis*, apparently a written list of recipes which existed independently.[32] Such lists had probably circulated anonymously for many years beforehand. If so, they are an early use of literacy in the interests of healing. 'They cannot be called Hippocratic', their latest scholar, Laurence Totelin, well concludes, 'they became Hippocratic when the treatises in which they were included were ascribed to the physician of Cos.'[33]

In the remedies which Hippocratic texts propose, plants preponderate, at least 380 different species being mentioned.[34] Some of them are given a geographic origin, 'Cilician hyssop' or 'Egyptian crocus', and two of the most frequently attested ingredients were not local to Greeks' territory, nor cheap to acquire: frankincense, used specially in cults of the gods, and myrrh.[35] In Totelin's view, therefore, the plant remedies attest a sort of 'haute médecine', fit for richer clients only: she means by this concept, modelled on our haute cuisine, a medicine 'characterized by variety, by the mingling of expensive "fashionable" remedies with reinterpreted folk remedies of all kinds'.[36] However, study of all the plant-based remedies in the Hippocratic texts has now shown that forty-four particular species account for nearly half of the references to plants, forty of which were quite readily available in the Greek world.[37] Of these forty, nearly half were familiar from cooking: Hippocratic doctors were not always proposing remedies with exotic components, only available to the rich. Some of them might travel with a dried stock of rarer ingredients and then sell them, no doubt profitably, to clients, but the majority of items could be grown in gardens easily enough. When doctor Onesilas was given that garden at Idalion, plants like garlic, rue, coriander and even cumin (originally at home in India) could be grown in it to support his medicines very cheaply and easily.

Parallel evidence implies that the 'Hippocratic' doctors' refinement was one of selection more than exotic inclusion. In the later botanical texts by Aristotle's pupil Theophrastus (compiled c. 315 BC onwards), local Greek 'root cutters' are also credited with particular uses of plants for healing. Those which happen to be named do not always overlap with plants in the Hippocratic recipes, implying there was not a total takeover of pre-Hippocratic plant recipes by the new doctors.[38] Some of the doctors' faraway plants were also quite recent arrivals in Greeks' awareness, plants like the enigmatic silphium from Libya, only known since c. 620 BC, or the 'Indian spice which Persians call "peperi"', a name not known, therefore, before c. 530 BC.[39] By c. 430, however, most of them may no longer have been highly expensive. There is not a word about cannabis, though Herodotus, c. 430 BC, knew of its use, albeit for pleasure, among Scythians on the steppes beyond the Black Sea.[40]

What was the thinking behind the variety of proposed remedies? Even if some of them were never actually applied, they imply a range of underlying ideas. Some indeed trade on the exotic or rare, items like the powdered horn of a stag or the corpse of a dog, eviscerated, boiled, bottled and then burned so that the smoke could fumigate a woman by passing up through her.[41] Like ingredients in modern beauty products, they impressed innocent clients by their rare contents. Others were symbolic, the fruits of a pomegranate being full of neatly arranged seeds and therefore considered useful in recipes to promote pregnancy.[42] Others, however, had active ingredients, even if their activity was not properly understood: the Cnidian daphne's berry is one example, pennyroyal another. There was also a principled difference. Whereas non-medical spells had often worked on a principle of 'like being attracted to like', Hippocratic doctors tend to deploy the contrary principle, no doubt in opposition to such traditions: the text *On Breaths* (*c.* 420–400 BC) states summarily that 'opposites are cures of opposites. Medicine is taking away or adding, taking away of what is in excess, addition of what is deficient.'[43]

At a simple level, this principle of adding or taking away could cause a traditional remedy to be used more aptly. Hellebore is a good example, the third most-mentioned ingredient in Hippocratic plant recipes. It is prescribed in sixty-three of them, but Ctesias, himself a doctor, remarks that doses of hellebore had been inexact and frequently fatal in the time of his grandfather and great-grandfather, back, therefore, in the early fifth century BC. In his lifetime, he says, the quantities had been refined.[44] Hellebore was extracted from a hellebore's roots, not seeds. It did not come from modern gardeners' beloved Lenten roses, but from *Helleborus cyclophyllus* on Greek mountainsides where herdsmen were said to have observed it as a plant which their flocks avoided. It helped to purge sick patients, but its extract is very inflammatory and needs more careful calibration than even the Hippocratic recipes tended to give. It is not surprising that in later case histories doses of it still caused choking and, sometimes, death.

Though their surgery and medicaments were not startlingly new, were the fifth-century doctor-authors impelled to write by a new understanding of hidden inner organs and events? Texts about women are particularly telling evidence here. None of the Hippocratic authors was a woman, but ten titles in the Corpus address women's health and the problems of childbirth: one of them, *The Diseases of Women*, amounts to three books. All of them contain major misconceptions. The authors believe that, as women's bodies are spongy and much more absorbent than men's, they store up more blood during a month and hence need to expel it by menstruation. Most of them believe that the womb wanders around inside a woman's body; they all accept that a single superhighway runs up inside her from her vagina to her head, and as a result suppressed menstrual blood may travel upwards, even to the breasts, where it becomes milk.[45] Hippocratic authors also believed in a sort of thin veil which they imagined to be covering a girl's womb, but there is no evidence they recognized the existence of the hymen.[46] However, a woman's role and pleasure in sex were defined in rather more detail. According to the male author of *On Generation*, that Cnidian who probably wrote in the later fifth century, when a woman's private parts are rubbed during sex and her womb moves, she has a sort of tickling sensation in these parts 'and it gives pleasure and heat in the rest of the body'.[47] She ejaculates, either into her womb or, if her womb is very wide open, externally. If she is eager for intercourse, she ejaculates before the man does, but then for the rest of the act she no longer has pleasure 'similarly'. The effect resembles the pouring of wine onto a flame which at first flares up, then dies away. If she is not eager for sex, her pleasure is cut short. It ends with the man's when his seed is released into her womb: 'it is as if someone pours cold water onto boiling water' and her 'heat and pleasure spurt, and then cease'. Even so, the author states, women feel less pleasure during sex than men do (not all Greek men agreed), but they maintain it for longer.[48] These observations and prejudices were surely not new or unformulated by men until *c.* 450 BC.

Like other Hippocratics, the author of *On Generation* becomes prescriptive on the topic: he believes that sex makes women healthier. It helps them with their periods, he states, and stops their wombs from becoming dry.[49] Other Hippocratic authors recommend that sex should be most frequent in winter, moderate in spring but very limited in summer, for reasons of the body's composition.[50] It is hard to believe that anyone took their advice seriously. There was even a view that very fat women were unlikely to become pregnant: they should therefore be given slimming prescriptions to help them conceive.[51] The author of *On Girls* states graphically why women who are old enough to marry but do not do so risk all sorts of problems, including delusions and a wish to kill themselves. They need sex to cure them, because without it their menstrual blood becomes dangerously blocked inside them, and anyway women are 'more prone to despair'.[52]

Childbirth and pregnancy were, as ever, major topics of discussion. There is no sign that authors in the Corpus ever realized that the womb contracts when labour is in progress: labour pains were explained as the kicking and movement of the baby. Twins were correctly realized not to arise from two separate sexual acts, but notions about their positioning in separate parts of the womb were fanciful.[53] Boys, being superior, were believed to be imparted by the right testicle and to mature on the right side of the womb. They supposedly cause a mother's right breast to become bigger and fuller in anticipation and her right nipple to become hard. Girls, by contrast, arise from, and affect, the left side, the inferior one.[54] Both the man and the woman were generally assumed to contribute seed, but the outcome varied according to the preponderance of a male or female force in it. According to the author of *On Regimen* (*c.* 400 BC), if both parties' seed is predominantly male, the child will be male and manly. If the woman secretes male seed and the man female seed, and the male element prevails, the child will be a 'man-woman', or *androgyne*, 'correctly so called', the author adds.[55] This text has been taken to be the one reference to homosexuality in the Hippocratic authors and to imply it relates to nature and inheritance. However, the author does not say anything about the sexual preferences of these men-women, any more than of the manly women he goes on to discuss, and this modern interpretation is unfounded.

In the entire process there was no knowledge of the ovaries or the existence of a female's eggs. One empirically minded author, perhaps writing in the fourth century, recommends that in order to bring about pregnancy in a woman who is having difficulty, she must first be classed as 'bilious or phlegmatic'.[56] To do so, her periods must be examined. Fine, dry sand should be spread out in the sun and then the menstrual blood must be poured onto it. If she is bilious, it will turn green, but if she is 'phlegmatic, it will look like mucus'. She can then be cleaned out inside according to the category to which she belongs.

It is not too surprising, then, that doctors, even Hippocratic doctors, continued to be only one resort for the childless. The author of a text on 'sterile women' even discussed the problem of pus in the womb and prescribed that a woman should treat it by inserting a dose of mare's milk into her vagina.[57] As mares' milk was drunk by faraway Scythians, but not by Greeks, this prescription was not exactly easy for a Greek woman to carry out. Greek personal names testify meanwhile to other options. People were named 'the gift' of this or that god (Apollo-dotus and so forth) and were often so named for the god who was considered to have had a role in initiating them, usually in answer to prayers or vows: among the last pagan philosophers in the fifth century AD there was to be a cluster of people called Asclepio-dotus, gifts of Asclepius, the god of healing, whom their parents still worshipped in an increasingly Christian age.[58] People who wanted children resorted to prayers and vows to the gods, not just to doctors who claimed to have a new medical craft.

Had dissection given the new doctors new knowledge? Among the discussions of pregnancy in the Hippocratic Corpus, one has become especially famous. The author of *On the Nature of the Child* describes how a kinswoman of his owned a slave girl who was very valuable because she was a musician but who 'resorted to men' and became pregnant, which she should not have done.[59] She told her slave-owning mistress and the tale came to the notice of the doctor, a family relation. He told the slave girl to jump up, hitting her buttocks with her heels, a movement, we know, which was practised by girls in a special dance in Sparta.[60] On the seventh jump, the seed ran out of her and 'when she saw it she gazed at it and was amazed'. The doctor also examined it and found evidence of a pregnancy after only six days.

Significantly, this autopsy was an accident, not an experiment, and arose by chance from other advice, not from dissection or anatomy. It is indicative of a wider truth: there was not a paradigm shift in active treatment in the fifth-century 'Hippocratic' texts, neither for women nor for men, which would cause their doctors to write a flurry of texts. The breakthrough, rather, was conceptual.

8

The Invention of Medicine

I

When Hippocratic authors looked back on 'ancients', now lost to us, and criticized their choice of words or, in the case of the *Cnidian Sentences*, the inadequacies of their treatments and understanding, they had a real sense of progress. Their skill, several of them state, is a new craft with a distinctive, effective method. No healer in Homer's or Sappho's lifetime could have formulated such a statement of principle. According to *On Ancient Medicine*, certainly a mid- to late fifth-century text, 'for medicine everything exists, a starting point and a way which have been found, in accordance with which the things that have been discovered have been many and very fine over much time, and the remainder will be discovered if someone capable and with knowledge of the discoveries conducts his inquiry by starting from them.'[1] The author was arguing a conservative case, stressing that the past had value too, but he was admirably confident of progress if the correct principles and method continued to be applied. The author of *On Places in Man*, of roughly similar date, was even more assured: 'medicine seems to me already to have been discovered totally'.[2] Fortunately, he was over-optimistic.

One of the Hippocratic authors, datable in the late fifth century, asserts that this medicine is used by all but a few of the Greeks but not by 'barbarians'.[3] On present evidence from Egypt or Babylon he was correct. Underlying this self-consciously Greek invention was the belief, already apparent in Alcmaeon and other early thinkers, that man is part of a natural world which is explicable in terms of under-lying elements and forces and their interrelationships and regularities.

Doctors datable to the fifth and fourth centuries frequently stress the importance of an abstract 'nature' in a way that no healer in the age of Homer or Sappho ever could.[4] It is not always Nature in general, what we would spell with a capital 'N', but often it is the 'nature of man'. The three books of On Regimen (composed c. 400 BC) are a good example. They discuss drink, food and exercise, but before their author gives any practical advice on these practical subjects, he begins by asserting, 'I maintain that anyone who sets out to write correctly about human regimen must first know and discern the nature of man in its totality: he must know from what it has been constituted initially and must then discern by what parts it has become dominated.'[5] Only then will he be able to diagnose and prescribe. Nonetheless, the primary elements of man's nature, the author insists, are fire and water. Conversely, the author of On Ancient Medicine states that 'in order to have any clear knowledge of [man's] nature, medicine is the only source'. He has no time for random postulates like heat and wet and he prefers to work from previous medicine's findings. However, he too insists that a general nature of man is relevant.[6]

This stress on the nature of man was a consequence of that Greek philosophical revolution of the mid-sixth century BC but it was, as ever, a broad category: if anyone thought it through, diseases too were a part of nature, and as each one had its own nature, how did they fit into the nature of man in general? Skating over this complexity, Hippocratic doctors stressed the natural health of the body and saw diseases as disturbing it. This new thinking had a very important consequence: it caused most of the fifth-century Hippocratic authors to exclude the gods from their discussions of health and treatment. The assumptions of Homer or Hesiod, Solon or the many unhealthy worshippers of Asclepius in his shrines were decisively rejected.[7]

This striking shift needs to be pinned down. At least in Herodotus' tale, Democedes is never said to have prayed to the gods or encouraged his royal patients to do so. However, Herodotus is telling the story up to eighty years later, and may have simply cut out, or no longer known, things that Democedes did. Alcmaeon had certainly discussed diseases and their origins without simply assigning them to the agency of a god, but he was only writing a theoretical text, On

Nature. Unlike him, some of the Hippocratic authors recorded actual cases and practice, and there their lack of reference to vows, prayers or appeals to the gods is especially striking. It prevails even when they discuss uses of traditional plants in their remedies. Local root cutters combined prayers, rituals and so forth when picking or preparing their plant-based remedies, but no Hippocratic text recommends such accompaniments.[8] These silences do not mean that the doctors were atheists or that 'science' or a new method could only emerge when religion had been boldly consigned to the dustbin. Some of the greatest scientists are, and were, believers in God or gods. For the early Hippocratics 'belief' did not really come into it. When some of them discuss the general role of the gods, they simply assume that the world is presided over by them. The extent of divine micromanagement was quite another matter.[9] Most of the doctor-authors probably considered their craft to be god-given, but that it had its own momentum according to regularities which could be worked out and then exploited. It was not dependent on the old Homeric world view, one of gods' random acts of assistance. In Homer's lifetime I suggested that everyday sicknesses were not actively and primarily regarded by their sufferers as visitations by gods, ubiquitous though those gods were in the ups and downs of human life, and that they pointed in a direction into which reasoned medicine might expand. In the fifth century it expanded there, but only in texts by the proponents of the new principled medicine. Without the new idea of nature and the consequent changes in views of the body and its processes, that expansion would never have occurred.

Medical proponents of the new craft were aware of this divide: they distinguished themselves from incompetent practitioners whose pretensions and religious hocus-pocus they denounced. There was another related consequence. As their medicine projected a new professionalism, it began to use technical terms, a category whose early presence in the Greek language has sometimes been contested.[10] Words like *apostaseis* ('abscessions', in modern translations), *pepasmos* ('concoction') and others are clearly medical jargon. Doctors' language marked them off as professionals.

This jargon was a symptom, not a cause, of their new writings. Much more important for their formulation was a new context for their authors, the lecture. In the fifth century, travelling thinkers, often of east Greek origin, were to be found speaking before audiences in a lecture context.[11] Some of them talked, and then wrote, about the elements underlying Nature: they even related them boldly to general health and medicine. At least nine of the Hippocratic texts contain internal evidence that they were delivered as speeches: some of them are independent lectures, whereas others answer from a doctor's point of view generalizations which had been stated by non-medical lecturers on other occasions. One of them, *On the Art*, even begins by attacking those who 'have made a skill of vilifying skills' and who belittle all experts, presumably then too in public speeches.[12] *On the Nature of Man* starts by refuting those who claim in speeches that an unlikely element like fire or earth is a basic constituent of humans but cannot even agree which element is the one. They cause their hearers to approve now one, now another, at different times.[13] The author addresses his polemical reply as if to an audience and then champions an equally mistaken answer, that the body has four humours.

In these texts, the authors use the first person 'I' and vigorously advance their own views, as no medical text in Babylon or Egypt ever had. Even so, use of the first person in a prose work was not these doctors' invention: it had precedents in the late sixth century BC in the philosophical writings of Heracleitus (active *c.* 510–490 BC) and in the broadly geographic writings of Hecataeus (similarly dated).[14] However, some of the medical lecture texts are particularly forceful in their use of it: is their assertive way of speaking related not just to lecturing but to a new political context, democracy?

Since its first adoption in Athens in 508 BC, democracy had become ever more widespread in Greek city-states and was giving a new role to personal argument by speakers in public debates. However, doctors' use of the first person was not a consequence of this change. It is not even clear that any of the doctors' first-person arguments were delivered in a democracy. If most of the mid- to late fifth-century medical authors came from Cos or Cnidos, neither place is likely to have been democratic when they matured there. By the late fifth century, doctors might indeed be invited to make oral presentations to an assembly meeting when a city was wanting to appoint a public doctor, but appointments of public doctors preceded the era of democracy, as Democedes' career in Athens shows.[15] None of the polemical lecture pieces attaches to such a context.

The very fact that the doctors wrote and preserved their texts distinguishes them from democratic political culture. In democratic cities, political pamphlets were circulated only by anti-democratic critics. In the fifth century, no speeches to a democratic assembly were ever published in writing.[16] The Hippocratic texts are no exception. Those which are most obviously addressed to an audience fall into two groups.[17] One group, the smaller, begin with a rhetorical introduction and, in two instances, need about half an hour to be delivered, too long for an assembly speech. These talks may have been addressed to a small group of listeners, like the lectures which famous visiting thinkers delivered in grand Athenian private houses. Such lectures could perfectly well have developed in a pre-democratic society.

They were delivered, nonetheless, in city-states: is the city-state, then, the crucial factor which accounts for them?[18] However, city-states had existed for centuries by *c.* 450 BC: they do not explain why argumentative texts on medicine began only then. The new lecture culture was more relevant, but the medical texts which are related to lectures fall into two broad categories. Some are argumentative general lectures, but others begin and end abruptly and use language which suggests that they were designed for a basic practical purpose: teaching an audience of aspiring doctors. Texts of this type in the Corpus take longer to be recited, running to an hour or an hour and a quarter at a reasonable rate of delivery.[19] They were then written down, presumably for the same purpose, so that they could be circulated, repeated and preserved by students.

At least twenty texts now in the Hippocratic Corpus refer to themselves as written.[20] Like most of the lecture texts, many of them are best explained by the same practical need: teaching, within the framework of the newly styled craft. They, too, fall into two broad categories. Some were written explicitly to instruct the general reader, the world's first surviving examples of this long-lived genre. The author of *On Regimen* (*c.* 400 BC) states that he is giving directions for the majority, for people who cannot devote time and detailed care to their regime. Only then does he go on to advise those who can. He himself may not have been a full-time doctor at all: his text has recently been interpreted as a sort of self-help manual which an interested layman could apply to himself.[21] The author of *On Affections* begins by stating explicitly that as any man of intelligence should have an understanding of medicine, he will write in a way which intelligent laymen can understand.[22] Like many authors aspiring to write general books, he does not always succeed in doing so.

Others are much more specific and technical. Their exact dates have not yet been fixed, 'mid- to late fifth century BC' being the modern consensus for most of them, but even then they would be among the first teaching texts on any subject which survive in Greek. There had already been those two specialized texts on the building of an individual temple and by the 460s or 440s another, the *Canon* by Polycleitus, on sculpting the ideal form of a male naked body. However, there were no texts as yet on warfare or on public speaking. There were no texts, either, on the crafts of the other skilled migrants whom Homer's *Odyssey* had picked out, the carpenters, prophets or poets, though each of their skills depended on teachers.

II

Is there, then, a distinctive common core to the medical texts which relate to lectures and to teaching, one which caused ancient scholars from Baccheios (*c.* 280 BC) to Galen (*c.* AD 170–200) to assemble ever more of them into a cluster and connect them all to Hippocrates? The problem here is that the range of 'Hippocratic' texts is very diverse, making it hard to define a single answer. The expert historian of ancient medicine, Vivian Nutton, has even proposed that this diversity is itself the distinctive quality, one which is 'arguably greater', he suggests, 'than that in any other comparable block of classical Greek literature'.[23] However, prominent themes and standpoints emerge from many of the early ones, helping us to understand why they, at least, were later considered to be a group.

The exclusion of any reference or appeal to the gods is one such theme, widely but not universally observed. The principle that opposites cure opposites is another, but also not universally present. Another emphasis is a principled attention to relations between climate, health and particular illnesses.[24] It had already been visible in Pindar, writing about summer and winter's effects on health, but in some of the new doctors' texts it had a much greater scope: it too related to the idea of a natural world, of which mankind is a part. In some of the Hippocratic texts, a disease which affects a community is related to bad air carrying 'emanations' or 'taints', whether from stagnant water or even from the stench of dead bodies, that un-Homeric reality.[25] One of their prescriptions, therefore, is that, in an epidemic, people at risk should reduce their diet, become thin and breathe as little air as possible.[26] Contagion was still a concept none of these new doctors formulated. This absence, to us remarkable, made them no more comprehending in a genuine epidemic than their predecessors or the purveyors of mumbo-jumbo, seers and priests who remained a major resort for anxious contemporaries. Germs were completely unknown and so, of course, were viruses. Social distancing, let alone the lockdown of an entire city, never occurred to them as a counter-measure; nor did face masks or hand-washing. Though more right than a Homer or a Hesiod, the new doctors were still seriously mistaken.

They made inferences, nonetheless, about invisible elements in the body, whose balance or disorder they considered to account for health and illness. Alcmaeon of Croton had already stated similar views, but the Hippocratic texts discuss them more precisely than his notion of an unlimited number of opposites at work inside each of us. In antiquity, the doctrine that there are four underlying humours in the body came to be considered the very hallmark of Hippocratic medicine, a view of it which persisted influentially into the modern world.[27] However, the doctrine that there are precisely four humours is emphasized in only one text, *On the Nature of Man*, in the entire Corpus. It is itself a polemic on behalf of that view, composed *c.* 450–420 BC, and the polemic shows that its view was not then widely accepted.[28] The more widely expressed notion was that juices and fluids were influencing the body's health.

Did this sort of thinking improve treatment or rates of survival, thereby helping to establish a special 'Hippocratic' core of writing? It is most unlikely. Some of the most admired fifth-century Hippocratic texts are not even concerned with diagnosis or treatment. Rather, they emphasize the value of prognosis, or prediction, about the likely course of an illness. Correct prognosis confirmed a doctor's claim to possess a skill with a method. If he could predict a dire outcome correctly, he was unlikely to be blamed for it personally when it happened. Prognosis also helped him to foresee which cases were better not taken on or attacked with potentially damaging treatments.[29] In Greek culture, the context of prognosis was different from ours, as we will see, but in pursuing it doctors had not invented an entirely new concept. However, some of them pursued it in a distinctive way: they based it, they believed, on rational human calculation and assessment. Like their notions of purging and pollution, they detached prognosis from a religious context and from links to divination and inspired prophecy.[30] For such people the healing prophets of early Greek myths were dead ends.

Even more striking are their statements about ethics. In several Hippocratic texts, not just in the Oath, ethical principles are explicitly prescribed, including treatment for no pay. Maybe the Hippocratic text known as the *Precepts* was stating a principle at odds with most other doctors' practice, but its very statement remains impressive:

> I urge you not to introduce too much inhumanity [into your dealings], but to take regard of both the abundance of material existence and the mere existence [of any patients]. Sometimes give your services for free, recalling a previous good turn or [your?] present good repute [the Greek text is uncertain and unclear here]. But whenever there is an occasion for ministering to someone who is a stranger and without means, particularly assist such people. For whenever love of man is present there is also love of the craft.[31]

In the early fifth century, Onesilas on Cyprus had shown such humanity to patients without charging them, but they were war-wounded in a siege. The *Precepts* urge it to be shown especially to strangers who are destitute: it is to be shown in all circumstances. In other cases it is combined with a typical Greek respect for reciprocity (that previous good turn . . .), but it goes beyond reciprocity, magnificently.

Medical authors did not invent ethics, but previous doctors, in the early sixth century BC, are most unlikely to have been exposed to such a barrage of ethical advice. Unlike the ethics in the Oath, it was not related to the gods or to a holy way of life. A text now known as *The Doctor* even gives advice on personal presentation and professional manners for the doctor himself.[32] A doctor's leading role, it says, means that he should be 'of a good colour with a nice fleshiness'. He should be respectably dressed, pleasantly scented, clean and not too prone to laughter and making jokes. He should be serious and exercise strict self-control with patients as they put themselves in his hands, women, young girls and 'possessions of great value', evidently an owner's precious slaves.[33] The text is probably no earlier than the fourth century BC., but one sound Hippocratic principle is that without an ethic there is no medicine, no genuine doctor.

III

Whether or not he was fleshy and nicely coloured, a pattern emerges from these texts about the practice of a fifth-century 'Hippocratic' doctor. He was more likely than a doctor in the time of Solon or Sappho to try to predict a sickness's course, to prescribe a regulated routine of diet and exercise, to relate a patient's condition to nature both invisible and visible and then to observe how the condition progressed. The adoption of a healthier lifestyle would help, but from the patient's perspective, 'Hippocratic' treatment of an illness was not obviously an advance on pre-Hippocratic hit-and-miss. Its main techniques, cupping and so forth, were nothing new; nor was its use of plant-based remedies, however differently the ingredients were combined. Its sense of method and reasoned observation rightly impress historians of thought, but its practical prescriptions were still littered with ignorance and error. In a viral epidemic it would have noted the coughing, the headaches and the difficulties in breathing. By observing individual cases it would have come up with quite a sound grasp of the days on which the condition was likely to intensify and its overall duration, but it had no idea of the causes or the mode of transmission.

The texts on observable defects and external disorders are better founded. The surgical text *On Fractures* and its companion *On Joints* contain some excellent observations, including a fine description of dislocations of the elbow and of club foot.[34] *On Regimen* (*c.* 400 BC) is admirably concerned with diet, drink and exercise, including the effects of over-exercise: its author even has an explicit concern with what he calls 'pro-diagnosis', his own new word and idea, as he himself emphasizes. It makes him the first person known in history to be concerned with preventative medicine.[35] However, his own theorizing then leads him astray. He correctly considers that a balance is necessary between food, drink and exercise, but he regards it as enhancing the balance between those constituents of all humans: fire and water. As a result his recommendations are sometimes rather bizarre.[36]

As for Hippocratic texts about women, their male misapprehensions are readily ridiculed. They have little to say about midwives or female attendants at a birth, but sometimes they cite 'women with experience' for details of childbirth and pregnancy. However, one author, at least, was more inquisitive.[37] His *On Fleshes* is variously dated between *c.* 400 and *c.* 280 BC by modern scholars, the earlier date being the more likely, and expounds his unconvincing theory that the life of man is linked to units of seven.[38] In support, however, he cites new knowledge, what he has learned from prostitutes and from midwives. 'Public prostitutes', he has found, 'know when they have taken a man's seed into their womb' and then destroy it: 'the conception falls out as flesh'.[39] 'If you place it in water', as the author evidently had, 'and inspect it there, you will discover that it has all the limbs, places for eyes and ears and its members. The fingers of its hands are visible, its legs, its feet and toes and its sexual parts and all the rest of its body.' All these features are visible, he writes, within seven days of seed entering the womb, a starting point which experienced women know because 'they feel a shudder and a warmth when it is taken in: they grit their teeth and feel a spasm between their legs and a heaviness comes over their body and womb'. Some people, he well adds, may be surprised how he knows this, but he had been chatting 'often' to prostitutes and inspecting their abortions.[40] He also knows that a child born at seven months is viable, 'having seen it many times'. Doubters, he says, can go to midwives and verify it.[41] His personal research and verification befit the new medical craft, but his sources of information are unparalleled in the entire Hippocratic Corpus.

None of the prostitutes is named. When most of the Hippocratic texts discuss examples, they discuss particular conditions, not identifiable cases. A good example is a long text, *Coan Prenotions*, which was probably not compiled until the later fourth century BC. It includes summary advice on no fewer than 644 specific problems, evidently observed by the authors and their sources. 'Little boys who are seven years old: weakness and loss of colour, breath rapid on the streets, they crave to eat earth; a sign of the spoiling and dissolution of their blood.'[42] In only one case is the name of an individual patient given: 'Spasms in hysterical disorders [related to the womb], without fever, are easy to manage, as in the case of Dorcas', a woman otherwise unknown to us.[43]

The exceptions to this anonymity are all the more remarkable, seven books in the Corpus which go by the title *Epidemics*. They give the names of individual patients and describe their conditions and sufferings across a length of time. Two of the seven belong to a genius at work, whose thinking I wish to relate to other great names in the fifth century BC, placing him where he has never been placed before. His texts, I believe, stand nearer to the beginnings of what other authors call the new medical method. They belong just after Onesilas, Democedes and their like. As I will show, they are the earliest Hippocratic writings we can date, a discovery which fills a dark hole in history and may put Hippocrates, too, in a new light.

PART TWO

The Doctor's Island

On Thasos, during autumn around the equinox and at the time of the Pleiades, many rains, continuous, gently, among southerly winds. Winter southerly; slight north winds; dry spells, on the whole a winter such as spring becomes. Spring southerly, rather cold, slight rains. Summer for the most part cloudy, periods without rain at all; etesian winds scarce, slight; they blew sporadically.

The beginning of *Epidemics* 1.1

After I had not found in Books what might satisfie a mind desirous of Truth, I resolved with my self, to search into the living and breathing examples: and therefore sitting oftentimes by the Sick, I was wont carefully to search out their cases, to weigh all the symptoms, and to put them, with exact Diaries of the Diseases, into writing: then diligently to meditate on them, and to compare some with others; and thus began to adopt general Notions from particular events.

Thomas Willis (1621–75), aware of the Epidemic books
and referring to the case-books he himself kept,
beginning with patients in villages around Oxford
where he had been at Christ Church and trained as a doctor,
later becoming Professor of Natural Philosophy: from
Thomas Willis, *The Remaining Works of
That Famous and Renowned Physician*, translated
by S. Pordage (London, 1681), 55

9

The Epidemic Books

I

The Hippocratic books now known as the *Epidemics* are entitled in
Greek *epidemiai*. This title does not refer to epidemics as we now
painfully recognize them, individual diseases which are spread widely
through a population, whether by touch, inhaling, contact with
wildlife, eating, drinking, kissing (which the elder Pliny, *c.* AD 70,
recognized to be a means of transmitting diseases) or sex, while
remaining one and the same disease.[1] In the mid-fifth century BC, the
amiable Ion, a poet and author from the island of Chios, composed
Epidemiai which referred to his visits to the *demos*, or people, of individual city-states around the Greek world.[2] His title has sometimes
misled readers of the medical *Epidemiai* into thinking that their title,
too, refers to travelling doctors' visits to particular places. They refer
to such visits, but their use of the verb *epidemein* shows that for them
the word *epidemia* referred to the presence of a disease in a community.[3] It was not necessarily a rampant disease in our sense of the word
'epidemic', and it was not contrasted with diseases which were endemic,
a category the authors did not distinguish, but it was certainly a disease at large. This meaning was still correctly understood in later
ancient commentaries on the Epidemic books.

By the mid-first century AD, seven books were grouped under this
title: the grammarian Erotian referred then to 'seven books of *Epidemics*' in the list of works which he considered, over-optimistically,
to be by Hippocrates himself. The title went back to earlier editors,
probably at least as early as the third century BC, but it may not have
been used by any of the books' original authors. All seven books

share a distinctive feature. Whereas the other texts in the Hippocratic Corpus refer to patients in general, and only once, in passing, name an individual, the Epidemic books are quite different.[4] They contain individual case histories, most of which specify the very place where the named patient lived, even the house or location. They are the very first observations and descriptions of real-life individuals during a number of days which survive anywhere in the world. In Babylonia written case histories of named individuals are unknown. In China none survives until *c*. 170 BC, and even then they were presented for a different purpose, to defend their doctor-author's reputation.[5] In ancient Egypt, cases were discussed individually in the now-famous medical papyri whose contents date back into the second millennium BC, but they never name patients or describe observations of them day after day, let alone locate them at an exact address. The great Edwin Smith papyrus dates to *c*. 1600 BC, though the contents may be even older, and sets out advice on forty-three surgical cases. A typical one begins as follows:

> Practices for a sprain in a vertebra of his neck. If you treat a man for a sprain in a vertebra of his neck and you say to him, 'Look at your shoulders and your middle' and when he does so it is hard for him to look because of it, then you say about him: 'One who has a sprain in a vertebra in his neck: an ailment I will handle'.
>
> You have to bandage him with fresh meat the first day. Afterwards you should treat him with alum and honey every day until he gets better.[6]

The case histories in the Epidemic books are personalized, detailed and completely different: they never prescribe an application of useless raw meat. They refer individually to all manner of patients, to adults and children, women of varied status, and even to slaves, both male and female. In her personal poetry Sappho, *c*. 600 BC, had referred to the rapid sequence of physical effects which were induced in her by the sight of a girl with a man. The Epidemic case histories record carefully observed effects too, but always in other people. They draw on implicit medical thinking and cover a much wider range than Sappho's self-presentation. Of the 120 case histories, here are seven examples, necessarily some of the shorter ones: the interpretation of a few of the Greek words is still uncertain.

In the seventh book we learn about Parmeniscus, on whom

> even before, bouts of despondency used to fall and a desire to be rid of life, but at times, a good mood again. In Olynthus [a city in the Chalcidice in northern Greece] once during autumn he was lying in bed, voiceless, keeping quiet, trying just a little to begin to speak: he did just say something and again [became] voiceless. Bouts of sleep occurred but at times, insomnia. Tossing around in silence and distress and a hand on the abdomen as if he was in pain, but at other times, turning away, he lay in silence; without fever [the original Greek text is disputed here, crucially] on through to the end; breathing good; he said later he was recognizing those who came in. As for drink, sometimes he did not want it during the entire day and night when offered it; at other times he would suddenly seize the jar and drink all the water from it. Urine thick like a cart-animal's. Towards the fourteenth day, that ceased.[7]

Parmeniscus has been diagnosed retrospectively as a clear case of a melancholic attack, attested here before c. 350 BC.[8] However, that diagnosis requires the text in the manuscripts to be emended at a crucial point: the manuscripts read 'the fever on through to the end', but his modern diagnosers change one letter in order to read 'without fever to the end'.[9] If they are wrong to do so, Parmeniscus becomes a fever patient, not a plain case of depression: the question remains open.

In the third book we learn:

> In Cyzicus [a Greek city-state on the coast of what is now north-west Turkey], for a woman who had given birth to twin daughters and had had a difficult labour and had not been fully cleansed [presumably, a reference to her afterbirth], on the first day, fever with shivering, acute; heaviness of head and neck, with pain. Sleepless from the beginning but silent and sulky and not obedient. Urine thin and colourless. Thirsty, generally nauseous, bowel disturbed and all over the place but then drawn together again.
>
> Sixth day. At night she kept on saying much which rambled; she did not sleep. On the eleventh day she went out of her mind, but then

began to make sense again. Urine black, thin and after brief intervals, oily again like olive oil. Bowel passed much which was thin and disorderly.

On the fourteenth day, many convulsions; extremities cold; she was no longer making sense; urine stopped.

On the sixteenth day she lost her voice. On the seventeenth, she died . . .'[10]

In the *Epidemics'* sixth book, *c.* 360 BC, we meet two remarkable women, one in Abdera, the other on Thasos just across the sea. Some of the details pose problems of interpretation, but I take them to run as follows:

> In Abdera, Phaethousa, wife of Pytheas, keeper of the household, having borne several children previously, but her husband having gone into exile, her periods were stopped for a long time; she then had pains and redness at her joints; when this happened, her body became masculinized and was hairy all over and she grew a beard, and her voice became rough and hard and although we busily did everything to try to draw down her period, it did not come, and she died, having lived for not very long afterwards.[11]

Phaethousa had been looking after the household and had borne several children, an indication in the author's view that she was a good woman. Though her body was masculinized, she is not implied to have changed sex. The author continues to refer to her in the feminine gender and appears to consider her condition to be an illness.[12] Was she menopausal, a state the *Epidemics* never mention? To modern readers it may seem that she was transitioning, but the doctor appears to connect her inability to have a period to her husband's absence and implicitly, therefore, to the interruption of her sex life. In other Hippocratic books, sex is assumed to be beneficial for various female conditions: when the doctors did 'everything' to encourage a period, how far did they go?

On Thasos:

> the same thing befell Nanno wife of Gorgippus; it seemed to all the doctors, among whom I too was present, that the one hope of making her fully female was if the 'things according to nature' [her

period] were to come, but in her case too it was not possible for them to happen, although [we doctors] did everything, and she died, not slowly.[13]

Again the author continues to refer to Nanno in the feminine gender throughout, and again the doctors did 'everything'. He does not conclude that Nanno had changed sex.

Phaethousa and Nanno's cases have been keenly studied and variously interpreted by Western doctors since the fifteenth century.[14] They were, however, exceptional. On Thasos, we meet another woman, unnamed, in the *Epidemics'* third book whose troubles also related to menstruation, but were traced to a psychological origin:

> A woman, who reacted badly to misfortunes, as a result of a grief which had had a cause, was remaining upright; not sleeping; off her food: she was also thirsty and nauseous. She lived near Pylades' property . . .
>
> On the first day, as night was beginning, fears, talking a lot, discouraged, a slight little fever; early in the morning, many spasms; when most of the spasms left off, she began to talk nonsense and use disgraceful language; many pains, big ones, continuous.

Then on the third day: 'the convulsions stopped but sleepiness and dejection set in between waking again; she kept on leaping up, could not contain herself, talked nonsense often'. Then she sweated copiously all over and the fever came to a crisis. Afterwards, she 'completely regained her reason . . . around the third day, urine black, thin, with a deposit in it which was often cloudy; it did not settle down; around the time of crisis, she had a heavy period'. To the author's way of thinking this period was a good sign, as it expelled excess blood. In his view, it outweighed the bad sign, the thin and black state of her urine, and so it helped her to recover.[15]

Young men suffer memorably too. The *Epidemics'* fifth and seventh books, *c.* 360 BC, present young Nicanor and his

> affliction, whenever he set off for a drinking party: fear of the girl playing the pipe [the *aulos*]. Whenever the sound of the pipe was beginning and he heard it playing in the symposium, from the fears

[which beset him] he was physically troubled. He said he could hardly abide it at night-time, but if he heard it by day, he was not troubled. This went on for a considerable time.[16]

With Nicanor came Democles, who suffered from blurred vision and was terrified of 'crossing a bridge or the slightest depth of a ditch': he was Nicanor's companion, coming to the doctor at the same time.[17] Their cases, too, have fascinated modern readers. They have been diagnosed as neurotics suffering from phobia, that late-nineteenth-century invention: were they intensifying each other's anxieties?[18] What scared Nicanor, the rasping sound of the pipe, not a flute but an instrument close to the type of triple pipe still played in Sardinia? If so, why was he only afflicted by it at night? Was the girl the problem too? A symposium was otherwise an all-male drinking party and it was only in that setting that fears beset him. Or was it the combination of the sound and the darkness?

In the third book of *Epidemics*, another young sufferer is more familiar. He is left unnamed, but lived in the city-state of Meliboea:

A young man: after many bouts of drinking and sex for a long time, heated up, went to bed; was shivering and nauseous and sleepless and without thirst.

On the first day, much excrement passed from his stomach with much flux round it, and also on the following days; much that was watery in colour [or 'watery green'] was passing through; urine thin, sparse, colourless; breathing spaced out, heavy, at intervals; rather flabby tension of the abdomen, extending at length on both sides; palpitation of the heart throughout, continuous; passed urine like olive oil.

On the tenth day, his mind strayed, without trembling; he was orderly and silent; skin dry and stretched all round; excretions either copious and thin or bilious and fatty.

On the fourteenth day, everything was exacerbated; he was talking much nonsense.

On the fifteenth, he went completely mad; much tossing about; no urine; small drinks only were kept down.

On the twenty-fourth, he died.[19]

This case is not only a grim sequel to youthful sex and drink: it is the first surviving glimpse of an individual in Meliboea at all. The place was a small city on the east coast of Thessaly between mounts Pelion and Ossa, probably to be located at modern Hagiokampo with the best harbour on the entire stretch of coast. Its particular asset relates exactly to the young man's behaviour.[20] In the mid-fourth century some of its bronze and silver coins show Dionysus the god of revelry, while others show a nymph, surely Meliboea herself, sometimes with a bunch of grapes in her hair. The other side of these coins shows one or two bunches of luscious grapes. Stamped handles from Meliboea's wine jars have been found at various sites abroad, including Thasos.[21] The young man's repeated bouts of sex were with males or females or both, but he did not have to go far to acquire drink too. Meliboean wine was excellent and readily available.

These patients and their distress still touch our shared humanity. With one exception, the *Epidemics'* case histories concern people who are not from Athens or Sparta, the city-states which have dominated modern historians' attention. They are from what has been aptly called 'the other Greece', whether city-states in the north as far as Perinthus on the northern coast of the Sea of Marmara or towns in Thessaly, Meliboea among them, or the north Aegean island of Thasos, the major contributor to at least four of the Epidemic books. They are outstandingly rich witnesses to travelling doctors in action, heirs to the long tradition from Homer's Odyssey to Sombrotidas in Sicily. Unlike Homer's doctors, they were not travelling with armies. The authors never say explicitly they were appointed as the public doctor of any city they visited. They travelled and stopped in particular cities, not because patients were needing care, for such patients existed everywhere, but because of decisions which escape us. Perhaps social factors were the prime determinant. They themselves were not low-class artisans, humbly travelling in the hope of earning a pittance from fellow-members of their class. They were educated people who represented the 'craft' which they had learned after long years of training and expense. They fitted easily into distinguished company and after arriving in a new venue made contact with important citizens, as local coins and inscriptions will allow us to see.

Invitations to come and help 'top people' may have taken them to many of the sites on their route.

Their day-by-day notes, their idea of a critical point in a disease, their comments on urine, excreta or blood and their noting of patients' mental states: none of these observations can be documented in the surviving hints of Greek medicine from Homer to Pindar. They are results of the revolution in medical thinking in the fifth century BC. The Epidemic books set it in action before us.

I I

Ancient scholars, like moderns, agree on one point: these doctor-authors all came from the island of Cos.[22] None of them says so in his text, but the ancients' opinion carries weight, as they could read and compare other treatises which we no longer have. The Epidemic authors show none of the characteristics which were criticized in the older Cnidian books or are present in the cluster now traced by scholars to Cnidian authors. The noting of critical days during an illness may not have been distinctive to doctors from Cos, but a Coan work, the *Coan Prenotions*, is the one in the Hippocratic Corpus which most often alludes to them.[23] It is no objection that people on Cos spoke the Doric dialect of Greek, whereas the Epidemic authors wrote in Ionic Greek. As Ionic was the accepted dialect for prose-writing, these Doric-speaking, Ionic-writing doctors were like Scottish authors writing (sometimes) King's English.[24]

Though grouped together and placed under one title, the seven Epidemic books were not the work of one and the same author. Readers in antiquity were well aware of their variety, as is evident when Galen discussed them in the mid-second century AD. Some readers, he tells us, had considered all seven books to be by one and the same Hippocrates, but he and others did not share this view.[25] In one of his fullest discussions, he remarks that it is unanimously believed that *Epidemics* book seven is not by the great Hippocrates at all but is a more recent work.[26] Book five, he believes, is by 'Hippocrates the younger, son of Dracon', a much later member of the family whom we know to have been active *c.* 320 BC. Books two, four and six are 'in the view of some' by Hippocrates, but others, he tells us, think that his son Thessalos worked them up from notes which his father had left at his death in a state unfit for wider circulation. Galen accepts the Thessalos theory and emphasizes that, in his opinion, books two, four and six belong closely together because they share similar ideas. Thessalos, he thinks, found the main core of them on tablets or rolls of skin written by his father Hippocrates, but added some remarks himself.[27] Others later added more, accounting for the obscurities in parts of the text.

The division of the *Epidemics* into these three groups is still accepted by modern scholars, although the views on their authorship are not. One modern refinement has been to propose an eighth book, a short text in the Hippocratic Corpus which English editions misleadingly entitle *On Humours*. It alludes to the 'cough of Perinthus', an epidemic described in book 6 of the *Epidemics*, and includes material which matches another little bit of book 6, but it gives no day-by-day case histories of its own.[28] If it was written by book 6's author, he would surely have used more of his previous work: *On Humour*'s form is very different. It lists topics of importance, very briefly, and is best understood as a compilation of notes for lectures, possibly by an associate of the author of books 2, 4 and 6. The ancient commentators were justified in not classing it with the Epidemic seven.

Their judgements about dating and authorship were made on grounds of style and content. Galen was a careful reader with the highest regard for what he considered to be by the real Hippocrates, but he never dwelt on historical hints in the *Epidemics* about their date. They are very important, and in the light of them the ancients' notions of their authorship can be adjusted. Dating all seven of them is important for dating the particular two which will be my focus: one argument for dating these two is their relation, or lack of it, to the other five. So they need to be pinned down.

Books 2, 4 and 6 contain two firm clues. In book 2, we learn of 'the man who came from Alcibiades', someone who suffered fevers and then 'his left testicle swelled' but his condition passed a critical point on the twentieth day and he recovered.[29] This Alcibiades is given no further explanation, but is manifestly the great Athenian aristocrat, pupil and lover of Socrates, populist politician, general and treacherous golden boy of the late fifth century BC. Case histories in books 2 and 6 include a long one, that 'cough' at Perinthus, up on the Sea of Marmara on the coast of modern Turkey, where the author worked at least from one early summer until mid-November.[30] Alcibiades was in this very region in the last decade of the fifth century BC. In summer 410 BC, after a fine naval victory, he sailed into Perinthus, no less, and brought it briefly into alliance with the Athenians, but the 'man who came from Alcibiades' might be an envoy sent from Alcibiades when he was absent from the city. In 408 he entered nearby Selymbria and took it, but another context for an envoy in his absence is between 406 and early 404, when he had withdrawn in voluntary exile to his castles on the Thracian coastline, one of which, at Bisanthe, was just to the west of Perinthus.[31]

In book 4 we meet someone equally revealing, a man followed by a long name beginning with 'M', whose lettering perplexed Erotian and Galen but which was correctly transmitted as 'Medosades' by four of the main mansucripts.[32] The name belongs to a Thracian ruler who was met by none other than Xenophon and the Ten Thousand Greeks on their return from the march into the Persian Empire. In early 399 BC they were waiting near Selymbria on the Sea of Marmara and preparing to cross back again into Asia, modern Turkey, on the Sea's southern shoreline. Medosades was sent to them again to try to persuade them to transfer themselves to King Seuthes, but his speech to that effect failed.[33]

In the manuscript texts, the M-name is attached to a place and, again with good reason, the place name is restored by modern scholars as the 'village of Medosades'. As Xenophon shows, villages in Medosades' ownership lay inland behind Perinthus and Selymbria, the coastal city about twenty-two miles east of Perinthus. Medosades' villages, therefore, lay in an area which the doctor-author of book 4 certainly visited. Medosades already owned these villages before Xenophon arrived in 399 BC and so the place name fits very well with the 'man from Alcibiades', active in that very same region at a point between 410 and 404, the year of Alcibiades' death, quite possibly between 406 and 405. The two clues converge persuasively in the last decade of the fifth century BC, maybe between c. 406 and 400. At one point the author also mentions a 'star', which modern scholars have over-interpreted as a comet and tried to use as a fixed point for dating: it is no such thing. At another he refers to the opinions of Herodicos of Selymbria, but as Herodicos was active in the later fifth century he does not fix the author's date, either. Nothing in *Epidemics* 2, 4 and 6 requires a fourth-century date.[34]

Thanks to the place names which the author of 2, 4 and 6 occasionally specifies, we can follow him around sites in the northern Greek world. In Crannon in central Thessaly he attended, among others, a married woman, a schoolteacher and a woman with 'a naturally large spleen'.[35] He went over to Ainos on the Thracian coast, a city which we know to have been set behind a sheltered harbour and beside a big river. There, he found that those who ate beans continually became unable to control their legs, women and men alike, apparently in a time of famine.[36] He went further up the Hellespont to Perinthus, another town set on a narrow neck of land and with a good natural harbour: its houses, we learn elsewhere, rose up and ran across its residential hill like a continuous single wall.[37] There, with a team of fellow-doctors ('we' in the text), he worked and observed during the great sickness which lasted through the summer and autumn. It was accompanied by coughing and by other consequent illnesses and may well have been diphtheria.[38] Interestingly, he avoided the kingdom of Macedon, although he had been very near to it in north Thessaly, and at the time the king was Archelaus, famously hospitable to visiting talent. Instead he went to Abdera on the Thracian coast and across the sea to nearby Thasos, the island base, as we shall see, of Epidemic doctors who worked there before and after him.[39]

This remarkable doctor and thinker was active in these places during the turbulent final decade of the Peloponnesian war. It was a time when Thasos, especially, was a centre of civil strife between oligarchic supporters of the Spartans, in power from late summer 411 BC to 407 BC, and long-standing democratic supporters of the Athenians, ousted in 411, restored in 407 but overthrown again in c. 404.[40] These cities were not free and easy places where strangers like a doctor could come and go unremarked. They had defensive walls and city gates and guards for whom the safety of the gate key was a major responsibility. At night, inhabitants were locked inside the city and a watchman with a bell would patrol along its perimeter.[41] Among his patients, our doctor attended citizens whose names match the names of men known to have been at the very top of their local society.[42] Such people could admit him, lodge him and pay for him, although his text never mentions fees.

III

Books 5 and 7 pose even more delicate questions of authorship. Galen believed that book 5 was written by a later Hippocrates, the one who served between *c.* 327 and 310 BC as the personal doctor of Alexander the Great's Bactrian wife, Roxane. He remained with her until her death, when he too was murdered on Cassander's orders.[43] This Hippocrates, Roxane's physician, is a doctor whom it would indeed be fascinating to interview, but book 5 shows no signs of being written so late in the fourth century. There is an obvious oddity about its surviving text: its last fifty-six case histories, with one exception, reappear among the 124 cases which are the contents of book 7. In the second century, Galen read a text of book 5 which stopped before these repetitions and ended with case number fifty. One possibility is that this first part, cases 1–50, was originally composed by a separate author and that the rest of book 5 and the sixty-eight other cases of book 7 were by someone else.[44] They then became combined by later scribes into the two books we now have. Alternatively, there was only one author, as the hints of travel in book 5 and book 7 suggest. They are compatible with one man's coherent itinerary. If so, his work later became divided and one part of it was copied twice over.

The author(s) give fascinating clues to their cases' whereabouts. Cases 1–50 in book 5 include one in Athens and one on nearby Salamis.[45] One concerns the 'wife of the gardener' at Elis, where the Olympic games were a more obvious hunting ground for patients than a garden or orchard. Her stomach below the navel 'was massaged vigorously with hands covered in olive oil' and then blood flowed from 'below' and she recovered.[46] The author's travels also extended into the north-west of Greece, to Oeniadae on the coast, a most unexpected venue, but rather than cross from there to nearby

Italy, the author turned east and treated a cluster of cases in Thessaly, especially in Larissa.[47] He also treated three cases in 'Homilos', a place name on which all manuscripts of the text agree. It is probably the settlement known in other sources as Homolion or Homolos, located by the River Peneios at modern Lapsochori, on the very edge of Macedon.[48]

The author of the new case histories in book 7 picks up this itinerary, implying to me that he was the same person. Unlike the author of books 2, 4 and 6, he went into the kingdom of Macedon, an easy next stop if it was indeed he who worked in little Homolos near the border. At Pella, he attended the severe fever of 'Python's slave boy' (or 'son').[49] He also attended a man from Balla, a Macedonian town of uncertain location, misunderstood by Galen but famous to modern scholars because of an influential proposal, made in the 1870s, that it was the ancient name of the site of the great palace at modern Vergina. That palace is now known to be King Philip's, father of Alexander the Great, and the site to be ancient Aigai.[50] He also treated patients on the north Aegean island of Thasos, a hotspot for Epidemic doctors. He even attended warriors from Thasos, wounded by Macedonians in a battle on the mainland opposite their island.[51] If his Macedonian visit occurred soon afterwards, the king at the time was indeed king Philip, conqueror of cities in the lowlands which adjoined his kingdom. In 354 BC Philip was attended by a doctor from Cos, Critobulus, who expertly cut out the arrow which cost him his right eye.[52] Book 7's doctor may have gone into Macedon while this fellow-Coan was at court, a reason why he made the journey, whereas his predecessor, the author of 2, 4 and 6, did not.

He continued to move in the shadow of Philip and his conquests. He treated patients in cities in the Chalcidice, the lowlands which Philip annexed to Macedon, including a patient from Aineia, a city which claimed to have been visited by Aeneas and his father during their flight west from Troy: the text is uncertain, but the doctor may have treated this patient at Dion, just inside Macedon.[53] He also visited Cardia, up near the Hellespont, soon to be disputed between Philip and the Athenians. Above all, he visited the north Aegean island of Thasos and, once again, Abdera, the neighbouring city just across on the mainland, which was soon to be captured and rebuilt by Philip.[54] A cluster of his patients have names well attested in inscriptions on Thasos: almost certainly they belonged to prominent families on the island.[55]

Books 5 and 7 have been abbreviated by their copyists. In neither of them are we reading the complete original, but, even so, there are two clues to a date. The author of book 7 observed patients in Olynthus, before 348 BC therefore, the year in which King Philip destroyed the city.[56] He also described the case of Tychon, who was struck by an arrow from a catapult 'during the siege of Daton'. Daton was founded in 360/59 BC by the Thasians on the mainland opposite their island: it lay in a valley which looks towards Mount Pangaion, rich in minerals. In 356 BC it was attacked and taken by Philip, for whom the precious metals became an important resource.[57] The catapulted arrow which struck Tychon is the first contemporary reference to Philip's feared artillery, weaponry which continued to develop under his son Alexander.[58] Soon after being struck, Tychon was afflicted with noisy laughter. The author considered that the doctor who had attended him had removed the wooden shaft but that the metal head of the weapon was still lodged in Tychon's diaphragm. Three days later Tychon died. His case is the first graphic account of the 'sharp end of battle' against Philip's Macedonians, nearly twenty years earlier than the cases of Greeks who fell in battle against Philip at Chaeronea and whose skeletons have been studied recently for signs of similar atrocities.[59] In *Epidemics* book 7, another man, unnamed, is recorded to have been hit by a stone from a Macedonian. He too may be a victim of the 'face of battle' in 356 BC, as may two other cases of acute arrow wounds.[60]

Book 7 and, I think, all of book 5 belong between *c.* 360 and *c.* 350 BC. The credit of their main, or sole, author has risen dramatically after recent medically aware readings which praise his careful observations.[61] Nonetheless, one of his comments deserves first prize for Hippocratic wishful thinking. Sex was considered by his predecessor, the author of 2, 4 and 6, to 'harden' and dry up the stomach. For the author of 5 and 7, 'uninhibited sex', surely with prostitutes, is a cure for dysentery.[62] The wilder the sex, the more, he assumes, it will dry the stomach. He has yet, I am told, to be proved correct.

10

'On Thasos, during autumn . . .'

Books 2, 4 and 6 and books 5 and 7 are separated by about fifty years. Books 1 and 3 are their ultimate model, but have a different shape, a different style, a different view of much of the human body and some notably different vocabulary. In antiquity, books 1 and 3 were universally ascribed, like none of the other five, to Hippocrates himself.[1] They were already objects of learned commentary in the third to first centuries BC, long before Galen, too, wrote commentaries on them. They are most remarkable: what are their contents and, more problematically, their date?

What we now read as two separate books was split quite early in its existence between two book-rolls, from which it became transmitted as two books, not one. There was no such division initially and claims that the 'third book' differs in range or approach from the 'first' one are wholly unconvincing. The style, outlook and vocabulary of the whole are so similar that they were clearly the work of a single author. However, during transmission by scribes over the centuries, parts of the text might might be passed on in the wrong order: what has been copied as the end of our text may originally have stood a little earlier in it. There is also the problem of small, later additions. Readers might make annotations or insertions into their own copies, especially in a text which continued in use by students and teachers. At the end of five case histories in the third year, almost in sequence in the text, a single word referring to the patient's condition ends the notes; 'burning heat' and phrenitis are examples. They were already in the text known to Galen and had probably been added by a previous reader as his one-word summing-up. However, there is no evidence that later readers added longer remarks of their own. From time to time our text contains words like 'as it seemed to me', or 'which I know' or even 'I know nobody who . . .' They are not insertions by a later reader. The author himself uses the first person, fourteen times at least, and in each case uses it seamlessly in context.[2]

He gave his work a particular shape. He began with what he called a 'constitution', the weather in all or part of a year, season by season, followed by a longer account of illnesses which were prominent in that period. Books 1 and 3 contain four constitutions, all clearly preserved. The weather notes begin in each year's autumn. They mark the changes of the seasons by changes in the heavens, 'the rising of the Dog Star' (in mid-July), 'the rising of Arcturus' (early September), 'the setting of the Pleiads' (early November) and so on. The author follows each of them with a longer survey of the illnesses which were particularly prominent. The 'sufferings', or details of these conditions, are then surveyed too. The course of each illness is summarized, although the author notes that there were variations to many of the general tendencies.[3]

An extract from the first constitution makes this pattern clearer:

when it was summer, and also in autumn, there were many fevers, continuous ones but not violently so, but they kept on happening to people who were ill for a long while but in other respects were not tolerating it with difficulty. Stomachs were upset, but for most people quite bearably, and caused no further pain worth mentioning. Urine in most cases well-coloured and clear, but thin and becoming concocted [a technical term of the doctor] around the crisis. Coughing, but not excessively, nor was what was coughed up difficult to tolerate. Nor were patients off their food: it was quite possible to give them some. Overall, those who were wasting away, but not in a consumptive manner, were only slightly ill with fevers which made them shiver, perspiring only slightly with light sweats, having paroxysms in varying degrees, erratically, but overall without intermittence, the paroxysms being those of a triple nature. The crisis used to occur around the twentieth day among those for whom the affliction was the shortest; among most people it occurred around the fortieth day, but

among many others around the eightieth. With some it was not so, but
the fevers left them erratically and without a crisis occurring. In most
of these cases the fevers did not depart for long, but returned, and
after their return came to a crisis in the same periods as before . . .[4]

The range of observation on which this survey rests is most impressive.

The first three such constitutions are given one after another and
only then do individual case histories follow, without any indication
of the year in which each case belonged. By a cross-reference, however, the first one is fixed to the third year. Have many more case
histories, those after the first and second constitutions, become lost
from the text? I prefer a simpler answer. The author did not arrange
the case histories in a chronological sequence. He began by writing
up the first three years as a single block just when they had ended: all
the cases in them occurred in the same city and so he combined them
into one group before embarking on the following year, the fourth,
which was to take him away to other cities. When he returned, he
soon wrote up that year too.

The survey of illnesses in the third constitution raises another
problem: it refers to individual patients by name. At least twenty-four
of them do not coincide with any of the forty-two named in the case
histories which follow. Had full case histories of these people existed
in the original text but dropped out of it during its transmission by
copyists?[5] Alternatively, they may have existed in preliminary notes,
but the doctor never elaborated them into case histories fit for inclusion in his final text.

The general survey of the fourth year raises a problem too: it says
much about an affliction which led to a red rash on the bodies of
those affected. In the case histories for that year, none relates to this
condition. Have case histories of this rash fallen out of the text? It is
more likely that the author did not choose to set out individual cases
of it after all that he had written about it in his survey. Cases of it
were mostly very prolonged and were better, therefore, described in
the survey than in lengthy individual histories day by day.

Did the four years he describes run in sequence, one after another?
Their most recent editor considers that they did not, arguing that
they present four different combinations of four qualities, hot and

dry, cold and wet, and were chosen for that reason.[6] If the author selected these four years out of many more, a close chronology of his years of work becomes impossible to reconstruct. He would have spent other years, unrecorded, until he hit on one with the quality he wanted to exemplify. However, this theory is not convincing. The third and fourth years, especially, do not conform closely to the type they are supposed to be representing.[7] The theory also requires a mass of work to have been left out of account. If a year began with a pattern the doctor was hoping for, he would have begun to observe it closely and he and his helpers would have started to compile histories of individual patients, the material which he needed for his eventual survey. If after several months the year deviated from the pattern he wanted, he had to abandon it, even though he had recorded many cases during it. It is unlikely that he had many such histories which he never used, not even, like some of those in the third year, as a comparison in passing: even without the right weather pattern, they were crucial evidence. The four years in his text are best understood as four consecutive years.

Invaluably for historians, their location at least is not in doubt. Each of the first three constitutions is located explicitly on the island of Thasos in the north Aegean. As a result the constitutions give a unique record of the weather and public health in one and the same Greek community in consecutive years. Nothing comparable survives elsewhere from antiquity. Twenty-six individual case histories then follow, also located on Thasos. Most of them begin with the patient's name and the very place where he or she resided. The fourth constitution follows, in a similar form to the previous three but not explicitly located on Thasos. It is followed by sixteen more case histories. However, only five of these cases are stated to be on Thasos, whereas six are located in Abdera, easily visible across the sea from the Thasians' main harbour and acropolis: it was the site, incidentally, of that trepanned skull, c. 600 BC, so there had been an able surgeon at work there long before our doctor was even born. The doctor-author spent at least four months there. Two case histories are located in Larissa, the horse-loving city in north Thessaly where Justice, Aristotle once tartly remarked, was the name of a mare who produced foals which always resembled their fathers.[8] Another case history occurred in Meliboea, that wine-producing city, also in north Thessaly.[9]

The case histories of this fourth year, as transmitted, jump around from place to place, from Thasos and then back again. They have not been sorted into a smoothly running sequence, but nonetheless most of the author's places of work fall into a neat pattern. Larissa in Thessaly was sited inland and from it a direct route ran east for about fifteen miles to reach Meliboea's good harbour. From there, Thasos's harbour, also a good one, was an easy journey and, from Thasos, Abdera's harbour on the mainland was only two hours away by boat. However, another of the doctor's destinations is an outlier: Cyzicus, about a day and a half's sea journey away on the south shore of the Sea of Marmara.[10] His journeys may not, then, have been compressed into one neat series, each stop leading easily to the next. Perhaps he travelled to and fro in this year, using Thasos as a recurrent base. As his weather notes for it cannot apply to all these differently sited places at once, Thasos was probably the basic reference point for this year's climatic details too.

In the early Greek poets, from Hesiod to Pindar, bodily conditions were related to seasonal weather, the Dog Star and so forth, in a very general and simple way. Like other 'Hippocratic' doctor-authors, this doctor is being much more systematic, following a year's weather in far more detail and relating it to diseases at large in his city.

II

The text contains another curiosity, one which greatly puzzled its readers for more than 2000 years. In what is now separated as our *Epidemics* book 3, the first case history concludes with a cluster of capital letters, at least six, without any indication of what they represent. Similar clusters of letters are repeated at the end of each subsequent case history, also without explanation. Ancient readers were already perplexed by them: they took them to be a sort of abbreviated code. The copyists of our later manuscripts continued to be baffled, and so the letters which they copied and transmitted vary in number and nature.

Here is an example. 'On Thasos for Pythion, who was residing by the temple of Earth, trembling began in the hands . . .' So begins the fifteenth case history, the first in what became divided off as book 3. On the eleventh day Pythion reached 'a crisis, with sweats, somewhat concocted matter was spat up, and rather thin urine'. Forty days later he had 'putrefaction around his seat [his backside] and an abscession [the doctor's technical term for a deposit expelled from the body] involving strangury [difficulty in urinating]'.[11] Poor Pythion, but he seems to have survived. By Galen's time, a copy of the text included at the end of this case history the Greek capital letter 'pi', written, however, with a third downstroke added to its usual two and then followed by five more Greek capital letters: they translate in our English script as 'P OU M U.' Varying letters occur at the end of each of the next eleven case histories. They also conclude each of the first eleven case histories which follow the next constitution, the fourth. After an interval the lettering occurs again at the end of one more case history of the remaining five.

Galen's commentary reviews the controversy about this lettering and gives an unwitting clue to its origin. A learned doctor called Mnemon, he explains, sailed to Egypt from his home city, Side, on the south coast of ancient Pamphylia, what is now southern Turkey. He reached Alexandria's harbour with a scroll of *Epidemics* book 3, in his possession.[12] This journey fits well with the status of Pamphylia at the time, because it was controlled by the Ptolemies, kings in Egypt, resident in Alexandria. Mnemon arrived some time between 246 and 221 BC when the ruling Ptolemy, Ptolemy III, had ordered that all book-scrolls discovered to be entering at the harbour's customs posts should be seized and taken away for copying.[13] His aim was to enhance Alexandria's famous Library. Each original scroll was kept for filing in the Library and then the copy, but not the original, was returned to its owner. Scrolls acquired in this way were marked 'from the boats': Mnemon's scroll was said to have been marked accordingly. Some said his name was added as the scroll's former owner, others that it was added with the word 'corrector' after it, confirming that he had worked on the text personally.[14]

Mnemon's scroll contained lettering at the end of its case histories. Very soon, bitter controversy arose between doctors of rival medical schools in Alexandria about the origin of these mysterious inserts. Before 200 BC, one doctor wrote a clumsy text declaring them to be the work of the great Hippocrates himself, only to be rubbished, correctly, by a rival who ascribed them to Mnemon. Thereupon the argument raged to and fro and, as the historian of the Ptolemaic city, P. M. Fraser, remarks, became 'the most flagrant battle of books in the annals of Alexandria'.[15] Some assumed, surely rightly, that Mnemon had arrived in Alexandria with a scroll which already contained these letters added by himself. Others argued that after his scroll's confiscation he had borrowed it from the Library and then inserted this lettering in black ink. This claim alleged that he had defaced a library book, the first such accusation in history.[16] It was made in order to discredit him.

Galen's knowledge of Mnemon and his doings derived from Zenon, a keen participant in the argument: Zenon wrote a long text entitled *The History of the Characters*.[17] There was also dispute as to what the lettering signified. By all parties it was assumed to refer to Greek words and Greek numbers. By some it was considered to be a sort of shorthand. However, no other user of the scroll could have understood it and the devising of a system of shorthand which could not be understood by anyone else seems most implausible. Attempts to crack it rely anyway on the contents of the accompanying case histories, making it hard to see why an impenetrable shorthand summary was needed. Others, including Galen, assumed that the letters stood for the first letters of Greek words or numerals (Greek numerals were written with alphabetic lettering). In his view, the 'P OU M U' after Pythion's history abbreviated words and numbers which meant 'Clear. Mass of Urine. 40 [the fortieth day of the case]. Health.' However, Galen is not consistent. His explanations ascribe different meanings to the same letters when they occur in other examples. He also makes assumptions about how best to group or separate the lettering. They are not convincing.

In 1983, Johannes Nollé, a scholar of the inscriptions of Side and its neighbours, proposed a compelling solution.[18] Side, Mnemon's home city, used a language and script which were not Greek. When

Alexander the Great passed through the city in 333 BC, envoys told him that its original settlers, from Greek Cyme near Troy, had forgotten their Greek speech after living among non-Greeks on the site. As Alexander was rewarding cities in the area which advanced a claim to be of ancient Greek origin, this submission was well-judged.[19] Whatever its historical status, it shows that in 333 BC Side used a non-Greek language: it is attested on the city's coins of the classical period and by at least six local inscriptions extending down to the 150s BC.[20] Unlike Greek, its script was written from right to left: it expressed a west Anatolian language which is no longer intelligible. Mnemon, Nollé realized, had used his home city's Sidetan script and language for the notes which he inserted into his scroll of *Epidemics* book 3. Some of the Sidetan letters are indeed close to the Greek alphabet, but the quarrelling doctors and scholars in Alexandria wrongly assumed that all the letters were pure Greek. The Sidetan lettering then became muddled when it was recopied by Greek scribes over the centuries. We cannot recover Mnemon's Sidetan notes in full, but they were present in his text when he arrived with it in Egypt. Our manuscripts of *Epidemics* book 3 go back ultimately to Mnemon's scroll, deposited in Alexandria's great Library in the late third century BC. In it, the text had already been split wrongly into two, our books 1 and 3. Perhaps the sequence of some of its case histories had been muddled too.

III

At the start of the first constitution, the author helps us to picture him at work on Thasos. He refers to the 'medical room', what we would call a surgery.[21] In the Hippocratic Corpus, a later text, *On the Doctor*, sets out due comportment in one: it prescribes the doctor's necessary attendants and tools, the need for properly positioned lighting, where and how he should himself sit or stand, presumably while acting on the patient's body, and even the correct length for his fingernails (no longer than his fingertips).[22] As our doctor was probably not a citizen of Thasos, he was unable to buy a property in the city himself, land ownership in Greek cities being confined to citizens. His medical room had to be rented, but it is likely to have been near the busy centre of Thasos's city as it was to be his habitual place of work.

After giving details of a widespread disease in the city in the first year, the doctor-author remarks that 'as for the other [complaints], all those in the medical room, [sufferers] came through without being diseased [with it]'.[23] This terse observation means that patients who were unwell but not bed-ridden were seen often in the medical room itself. The case histories do not record these people: they record only patients who were elsewhere in the city and were already too ill to be moved.

When the doctor travelled out from his crowded room to find them, he had plenty to do. This widespread disease in the first year is one which, for once, we can diagnose with certainty.

> Swellings beside the ears on one side or on both. Most of them, with-out a fever, still standing; some of them became slightly hot too; in all of them, swelling subsided without damage; there was none of the suppuration which attends swellings which arise from other manifes-tations. The swellings' type was as follows; soft, big, spread out, without inflammation, painless. In all patients they disappeared with-out a sign first. These swellings happened to adolescent boys, young men, men in their prime, and of them, the majority were those who frequented the wrestling school and gymnasiums; few women had them. In many patients there were dry coughs and they brought up nothing. Voices became hoarse not long afterwards; but in some of them after a time there was inflammation in one or other testicle, or for some, in both. Fevers for some, for others, not. These effects were very painful for most of them.[24]

Diseases are famously flexible across long stretches of history. They mutate, they die out, but this one has not: manifestly the men on Thasos had caught mumps. The doctor recorded it very carefully: 'in one or other testicle, or for some, in both'. He made an inadvertent contribution to social history. Boys and men in the wrestling schools and gyms were especially affected: they exercised naked, and their bodies came into close physical contact. Girls and women were excluded, and, as they seldom caught the disease, kissing and close physical contact between young men and girls were also rare in Thasian life. The boys and men had each other, as explicit rock-cut inscriptions about named males' sexual charms still testify: some of the most explicit were carved on a rock face overlooking a beach on the south of the island.[25] When the doctor describes case histories of young men who had over-indulged in sex, their partners are very likely to have been males.

The doctor-author cannot have been working alone. Twice, he writes '*we* were getting to know thoroughly' and then describes the sources of this knowledge, including treatments applied and who was applying them, evidently someone who was not the doctor himself.[26] His year-surveys exemplify the wide and diligent range of information at his disposal. When he describes the third year's fevers during spring and early summer, the consequences of losses of blood, of shivering at the moment of crisis, of upset stomachs and so forth are all carefully stated. They are not generalized merely from the few patients whom he names. Observations about them are presented as applying to a much wider sample, 'most people', or 'the young and those in the prime of life' or those of one or other sex: 'many women fell ill, but fewer than men and they died less often: most of them had difficulty giving birth and after the birth fell very ill . . . most of them began a period during their fevers . . .' These widely based generalizations rest on many more cases than a single doctor could personally have observed so closely.[27]

Perhaps he used local doctors, already on Thasos, as aides and taught them what to look for before sending them out to attend patients in the city. Perhaps he brought his own pupils and assistants with him or maybe he used both types of help. Their social origins may have been varied, but there is no sign that any of the doctors was female. It is notable that on Thasos the author attended illnesses and complications which arose soon after a woman had given birth, but never describes himself as attending a woman while she was in labour. Other sources confirm that in the Greek world the attendants during labour were usually women, including midwives and female slaves.[28]

Whereas other famous texts in the Hippocratic Corpus state a bold thesis and then argue for it, the single author of *Epidemics* books 1 and 3 worked in exactly the opposite way. His concern was to record accurately and then to try to generalize from observed cases about the predominant diseases during each year. In no way was his prose text written for a lecture, like some of the other fifth-century texts now grouped in the Hippocratic Corpus. Nor did it arise from public debate in the competitive arena of a Greek city-state. The 'invention of prose' among the Greeks has been linked rather grandly to a 'struggle over authority' in the fifth-century Greek city and has even been presented as 'constructing an authoritative voice for the writer in a contest over who has the right to speak for and to the citizens', especially citizens in a democracy.[29] The doctor's prose text has no such aim or impact. Its author already had authority in the medical group for whom he wrote. He had no further need to construct it. He wrote not for a citizen audience but for fellow-practitioners and pupils, many of whom would be travelling doctors like himself, moving from city to city. His aim was to teach and to bequeath a prose text which he and they would consult and use in their own careers. Its natural life was in medical groups who shared many of its presuppositions.

His enterprise required notes and reports to be collected and kept in quantity, probably on papyrus or wax tablets. What we now read as the individual case histories are not simply notes written by the patient's bedside.[30] Like the generalizing surveys of each year, they were worked into a coherent sequence of days, conditions and outcomes later. The data which went into them had to be kept meanwhile, and presumably it was kept in the medical room, a crucial ally for the task. Even so, some of the case histories do not resemble the conditions which the doctor began by picking out in his general survey of their year. As so many notes and so much data had to be collected and arranged, these inconsistencies are understandable.

Maybe Democedes or Aineios, 'best of doctors', in Athens had sometimes kept notes on their patients' progress, but they are not known to have used them as a basis for a text of generalizations, designed to help future students. The author of *Epidemics* books 1 and 3 was yet another wandering doctor, but his writings could not have been written on the move, as if from notes in a backpack. His remarkable text relied on the existence of a rented medical room, the basecamp for him and his database.

At odds with a modern view of the 'invention of prose', his work now needs to be dated and located in time and space. How early is it among surviving Greek prose works? Was it collected and composed in a democracy? Was a 'crucial frame' for it really 'the audience of citizens that democracy creates'? There is, meanwhile, a fundamental point: like the other Epidemic authors, the author of books 1 and 3 names patients and their exact whereabouts and gives details of their past behaviour. He freely records those who had over-indulged in sex and drink. In the 'Hippocratic' Oath, doctors had to swear never to reveal anything about the life of people 'which it is not right ever to be gossiped about'. The Epidemic doctors not only reveal such things: they perpetuate them in writing. The most admired Hippocratics in the fifth to mid-fourth centuries pay no attention to the 'Hippocratic' Oath.

11

The Thasian Context

I

In general terms this doctor-author is easy to place because he names the site of most of his patients: the island-city of Thasos. It was to occupy him for three full years and part of a fourth. After him, the author of 2, 4 and 6 was active there too, as was the author of 5 and 7, many of whose patients can be located on Thasos, not least because of their distinctive personal names. This location is an unusual gift for historians trying to set the doctor in a general context. Thasos is not at the centre of modern histories of the Greek world, which mostly focus on Sparta and Athens. It is known, nonetheless, through archaeology, the work partly of the Greek archaeological service but especially of the French school, based in Athens, who have been active there with local assistance for more than a hundred years. When the author of *Epidemics* 1 and 3 went there, was he visiting a rather backward island which promised to be a fascinating museum of sickness and disease? Or was it a flourishing place which could offer him a comfortable social existence despite what he had to do? Certainly, it was not a newcomer. Its history and assets bear directly on his date and context.

Thasos is the most northerly of the Greek islands in the Aegean, lying off the coast of what is now north-eastern Greece, to the northeast of Mount Athos and the Chalcidic peninsula. Its northern shore looks across to modern Kavala on the mainland, once the site of Neapolis, 'New City', founded by the Thasians themselves. Even without it, Thasos is the twelfth-largest Greek island, with a surface area of about 150 square miles. It is also the greenest, still unsurpassed for woods and trees, despite recent forest fires, especially in the steep centre of the island. In the early seventh century BC, a querulous arrival, the poet Archilochus, described it, with memorable distaste, as 'sticking up like a donkey's backbone, crowned with wild woods', and complained: 'There is no good land here, nor lovely nor desirable.'[1] He was writing for effect, expressing how he missed his home island, Paros. In 1864, the visiting archaeologist Emmanuel Miller was more just in letters home to his wife: 'It would be impossible to tell you the effect the island produced on us; there was an extraordinary freshness of vegetation; the plain was enamelled with flowers of all kinds and colours; the nightingale was beginning to sing . . .'[2]

The first Greeks to settle on Thasos were not its first settlers. Among others it already had Thracian inhabitants and some settlers from the Levant.[3] Among the reasons which had drawn them were the island's precious metals, including gold and silver which they began to acquire from two mines on the island's eastern side. Greeks then arrived by sea from the island of Paros about 250 miles to the south, at a date which is still uncertain but which is best placed, in my view, before c. 680 BC.[4] As the island's southern coast is the nearest to Paros, the Greek arrivals may have put in there first, perhaps at the point later named Demetrion, in honour of the goddess whose rites they brought with them. They then found that the northeast coast proved more hospitable. At modern Limenas, it recedes into a bay which is well protected by a natural promontory. Greeks were quick to settle beside it, although the land beyond its shore was not unduly favourable for farming, and on some of it the water table was very near the surface. About five miles away to the southeast, the well-watered plain of Panagia is nowadays graced with olives and plane trees and would have been a better agricultural choice. However, the site at Limenas led up to a hill which was a potential acropolis, always attractive for Greek settlers, and on one side of its upper slope there was a mine yielding gold and iron, which was probably already being worked.[5] The mine was one good reason for settling nearby and the hill was another. It was defensible and a vantage point, the nearest point on the island to the mainland opposite. As in the Bay of Naples or off the coastline of Libya, Greeks preferred to settle on an island just off a mainland before

establishing themselves on the mainland too. The first Greek settlers may have chosen their site with a similar end in view. They
called it Thasos, a name which was then extended to the whole
island. Thasos has been aptly described as part of a northern Eldorado whose mines and metals extended over to the mainland.

Thasos's city, therefore, had a long Greek history when the doctor-
author arrived. The settlers had had to fight off rivals, including other
Greeks, in order to establish it, but it was not in the least a disorderly
sort of frontier settlement. Like other Greek colonists abroad, the
Thasian settlers had replicated items from their mother island. They
copied the main civic magistracies they had known on Paros. They
adopted Paros's calendar and, initially, its style of Greek lettering
when they inscribed their public decrees. They introduced some of
the cults they knew on Paros.[6] The two islands may even have maintained common citizenship between their communities, and, if so, it
was a further reason for these similarities. In the sixth century BC, a
senior magistrate of Thasos proudly recorded that he had held the
senior magistracy on Paros too and that he was the first person ever
to hold both. His claim has been much discussed, but on balance I
take it to mean that the odds had been against any one person holding the highest offices in both places, rather than that previously it
had been impossible to do so.[7]

The Thasians' history was not confined to their island. By the
mid-seventh century BC, they had begun to lead settlers across to the
mainland, eventually founding up to seven settlements, mostly along
its coastline, which became a long network of trading posts, or *emporia*, for their mother city. By *c.* 500 BC they were already using one
and the same weight standard for their coins.[8] These settlements
rank as the first empire in archaic Greek history, long before the
Athenians' empire in the fifth century BC. One of them, Eion,
included people from Paros, as did a later one inland, Berge, which
was founded with a sharp eye on local sources of metals and their
control.[9] Links between Thasos and Paros had indeed remained
close.

On the mainland opposite their city, the Thasians' foundation of
'New City', beside modern Kavala, was already a gateway to gold
mines inland, most readily to one which became known as the

'Dug-Out Forest': recent surveys have confirmed that Dug-Out Forest was located just to the north of Neapolis on the southern slopes of Mount Lekani, a mere three miles from the sea with an easy view across to Thasos itself.[10] It was to be a long while, however, before Thasians founded a settlement further inland beyond the barrier of Mount Symbolon and began to exploit the mines on the mountain slopes of Pangaion and its opposite range. Probably the local Thracians were too hostile for them to settle there. Meanwhile, Dug-Out Forest became highly productive, yielding more revenue than all the mines on Thasos itself. When Herodotus visited the island, probably in the early to mid-450s BC, he learned that Dug-Out Forest produced gold to the value of eighty talents a year, an amount whose modern weight would be about 2080 kg.[11] A vast quantity of spoil, thousands of cubic metres, had to be dug out yearly to produce this amount of ore. In Attica, the city state owned the subsoil of a mining area but leased the working of it for a fee to individuals. If Herodotus's figure was for the sums paid yearly to Thasos for concessions to work the mines, the value of gold extracted annually would be far higher. As the mine's name suggests, trees had to be felled too, not least as fuel for the refiners' furnaces.

Westwards along the mainland, the little city-states which the Thasians founded were located on or overlooking the sea. The nearest two to Neapolis lay beside small rivers, but when these rivers dry up in summer their narrow beds lead only up onto the mountain directly behind. Perhaps this siting made them less vulnerable to hostile Thracians, but it makes it harder to see how they functioned as trading posts for people and goods in their hinterland.[12] The most westerly of the settlements were rather different. Eion, settled jointly with Parians, lay beside the big River Strymon.[13] The Strymon was certainly navigable up its course, and by the mid-sixth century BC another such settlement, Berge, lay inland about thirty miles up it, near local metals and resources which could be readily transported down the waterway.[14] There was more good land beyond the river, but the Thasians' westward network stopped on its eastern bank. There was a good reason. The land beyond belonged to settlements which had been spearheaded by Greeks from the island of Euboea, travelling heroes of an earlier age. There was a clear demarcation between the two spheres, the Euboean first-comers and the later arrivals, Thasians with Parians.

In the other direction, to the east of Neapolis, the coastline was not at all under the Thasians' control. Just across from their harbour city lay the big city site of Abdera, also to be visited by our doctor. Soon after its settlement in the mid-seventh century BC, it had a big stone perimeter wall on its northern side, surely built as a defence against Thracians.[15] It lay in the flat Thracian plain, less than a mile from the sea and just to the east of the big River Nestos, which was navigable only for a short way up its course and, even so, only for small boats. The Thasians avoided Abdera and also Maroneia, located on Cape Molyvoti near the valuable resource of a salt lake, a major change to our understanding of the coast's political geography as it was much nearer Thasos than previously thought.[16] Both of these city states had been settled by Ionian Greeks from the east, so the Thasians could only found Stryme to the east of Maroneia, a third-best after the other two big coastal sites.

From Eion in the west across to Stryme in the east the Thasians' foundations were a very valuable asset. Herodotus implies that they yielded about 60–160 talents of annual revenue on top of the 80 talents acquired yearly from Dug-Out Forest. In some years the Thasians' total revenue, he states, was 200 talents, but in good ones as much as 300.[17] This income was enormous, equal in a good year to two thirds of the total tribute payments which many Greek city-states were paying in the 470s BC to their big Hellenic alliance.[18] The Thasians' public finances were so strong that they did not need to tax their island's crops, a fact Herodotus notes because it was exceptional for a Greek city-state. The mainland trading posts also gave access to a basic resource, slaves. The working of the mines on Thasos and on the mainland required a big slave workforce, as did extraction of the island's valued marble. The trading posts facilitated their import: according to Herodotus, Thracians were willing sellers of their children for export.[19]

In the years before the mid-fifth century, therefore, Thasos's main city was not at all an impoverished backwater. By *c.* 470 BC how big was its population? As usual for a Greek city, exact figures are unavailable and there are only two hints on which informed guesses can be based. The area enclosed in the city's newly built wall was about seventy hectares. The slopes of the acropolis may have been less densely occupied and there was a marshy area, liable to flooding, between the city's main meeting place in the later fifth century and the sea.[20] Perhaps half of the enclosed area was occupied by houses, and in it a density of 150–200 people per hectare is a relatively conservative estimate. If so, the population inside the wall might have been between 5000 and 7000.[21] There is also a hint about the island's naval manpower. In a crucial sea battle in 463 BC against a besieging Athenian fleet, the Thasians lost thirty-three warships, enough to cause them to surrender.[22] If their total fleet had been about sixty warships and if each was a trireme using 200 men, a figure attested for triremes later in the century, their fighting forces amounted to at least 12,000 men. That figure might seem to clash with the estimate for the city's residents, but in such a crisis fit male inhabitants from the rest of the island surely joined the fleet and rowers were probably also hired from the mainland settlements. This fleet, then, did not require a bigger population for Thasos's city than 5000–7000.

Where, relatively, would this estimate rank it? In 480 BC, the island of Corcyra, Corfu, was also able to put sixty ships to sea: it was a rich and important island.[23] The walled area of Thasos city was smaller than the walled area of the city of Cos, home of the Hippocratic doctors, but it still ranks in the top four of walled city areas known to archaeologists on other Aegean islands.[24] The doctor is not a 'micropolitan', a small-town participant in a provincial culture. He does not happen to refer to visits to cities as big as Miletus or Ephesus, but on Cos he was part of a cluster of leading doctors and there was no high culture in another city which outclassed his. On Thasos he was working in a city of substance, with plenty of people for him and his team to attend.

Its resources are more solidly attested. As an island it had access to a wide range of fish, while on land its nuts were singled out in the Hippocratic Corpus and its red wine was widely appreciated. By the fifth century BC it was being exported far and wide, up the Dardanelles and on to Greek city-states on the northern coastline of the Black Sea. In due course, farming sites excavated on the south-west side of the island give good evidence of the local cultivation of vines.

The wine's quality was a public concern of the island's ruling oligarchy, and the best of it was admired for its bouquet.[25]

The island also had excellent marble. At Saliari there were beds of a very fine-grained variety, and at Aliki, on the island's south-east coast, there were beds of a blue-grey one. The finer marble, Saliari's, was already being exported in the sixth and fifth centuries BC. Its neighbouring beds had provided the stone blocks for the city of Thasos's wall, one which the doctor noted in a case history. Saliariz's has recently been recognized as the marble for three famous, but enigmatic, Greek sculptures of the early and mid-fifth century, two of which were transported centuries later to Rome and one, at an uncertain date, to faraway Persepolis, centre of the Persian kings.[26]

In the Aegean, very few islands were self-contained, and Thasos was no exception. The artistic styles of objects, painted pottery and jewellery found in its city attest to its wide contacts.[27] So does the foreign circulation of its silver coins.[28] The yearly income from the mainland settlements implies that their revenues were high too, if they were paying 60–160 talents annually to the Thasians, much of which may have been taxes on trade and transport. The island was not only linked to the networks which passed it from east to west and back again. As the visiting doctor would have found, the main city presented an unusually exuberant image to the world. The gates in its walls were sculpted with impressive renderings of individual gods, including Dionysus brandishing the branch of a vine and being followed by his maenad girls.[29] From *c.* 510 BC onwards, the city began to strike silver coinage, whose bigger pieces showed the god's attendant Silenus, one of the highly sexed satyrs, kneeling to grab a maenad who was being held across his lap.[30] Another of the city gates was sculpted, *c.* 500 BC, with a big image, eight feet high, of this self-same Silenus, whom Attic playwrights presented as Dionysus' tutor: he is holding a wine jug high in his right hand and advancing with the most massive erect penis to survive in Greek sculpture. Like the silver coins' imagery, this remarkable sculpture is best interpreted as a tribute to Dionysiac revelry, 'singularly in harmony with the most primitive promptings of humanity'.[31] It was another tribute to the city's excellent wine, whose bouquet was 'like the scent of apples', a comic poet made the god Dionysus say.[32] Dionysus' prominence in the city's imagery is understandable.

Even so, he was only one of many gods honoured publicly in Thasos's city. During the doctor's visit there was evidence of active worship all around him. The citizenry was divided into 'brotherhoods', or *patrai*, some of which were under the patronage of a god, others of heroes known in tales of the Trojan war ('the Priamidai' were one).[33] As the city's calendar shows, there were major festivals of a deity in each month of the year.[34] The sanctuaries teemed with items vowed by individual worshippers to one or other god in thanks for, or hope of, divine favour.[35] Although most of the major cults went back to Paros, the first settlers' home island, the city's important cult of Heracles was an exception: it was probably adapted by the first Greek settlers from a cult they already found on the island. The obvious candidate is the Phoenician god Melqart, introduced to Thasos by even earlier settlers, Phoenicians from the Levant: Greeks often identified Melqart as Heracles.[36]

This culture of worship, drink and sex was highly socially stratified. It was ruled by a steady male oligarchy, whose governing body initially numbered only 360, one for each day of the year, later a group of only 300.[37] By *c.* 500 BC, the island was integrated under their rule. Every autumn, a festival of Apollo 'of the villages' was celebrated in the city.[38] There is no evidence of local magistrates or self-government anywhere else on the island. A road ran round the island and interlinked it, because the centre, being mountainous and forested, was less easily crossed. Such settlement as can be detected there is consistent with political dependency on the city's oligarchy. In the south especially, some of the most conspicuous features are man-made towers, vantage points which look over exploitable land, sometimes with rooms for storage beside them: 'it was a place of hard labour', on one modern view, 'and servile labourers, the very reverse of the urbane city.'[39] The visiting doctors sometimes attended slaves, though mostly in the households of propertied clients. When they went on visits, they never record treating a vine-dresser or a slave working out in the mines or the beds of marble. They preferred to work with urbanity.

II

During the fifth century BC, Thasos underwent important changes, the results of contact with two bigger empires, the expanding empire of the Persian kings and then the emerging empire of the Athenians. They are important for dating the doctor's text and setting it in a context. Fortunately, they are known through the two great historians, Herodotus and Thucydides, both of whom visited the island. They illuminate Thasian priorities and political tensions.

Each of them, at different dates, mentions the city's wall, and twice the doctor-author refers to one too. By 500 BC Thasos's city had a wall, as we know from the failed attempt, c. 494, of a rebellious Greek, once high in Persian service, to capture it. Thereafter, the Thasians 'made their wall stronger', Herodotus states, and 'built long ships' for future war.[40] In 492, a Persian advance force then marched west along the Thracian coastline towards Greece and the Thasians submitted without a fight. A year later, neighbouring cities on the mainland, perhaps maliciously, informed the Persian king, Darius, that the Thasians were planning to rebel. Darius told them to destroy their city wall and bring their warships across to Abdera. They obeyed at once.[41]

The Thasians' obedience is explicable by their wish to save their trading posts along the coast, source of their exceptional riches. In 481 BC a Persian army returned and the Thasians had to host King Xerxes for an extremely expensive visit: about twenty-five years later Herodotus learned from informed sources on the island that the cost of the occasion had reached an exorbitant 400 talents.[42] Dinner for a Persian king regularly required food and drink for a huge number of hangers-on: a dry wit in Abdera was remembered for urging his fellow citizens to go and thank the gods for keeping half of the catastrophe from them, because it was not Xerxes' custom to take two meals a day, one in the morning as well as one at dinner, a demand that would have required them to run away before he arrived or to remain and endure the most miserable fate of all.[43]

This obsequious hospitality worked to Thasos's best interests: the Persians made no attempt to deprive it of its mainland dependencies.

Xerxes' army was then defeated at Salamis in Greece. He retreated with part of it, going north again and stopping at Abdera, just opposite Thasos, the first place, its citizens later told a disbelieving Herodotus, where he stopped in his hurry and dropped his robes to relieve himself.[44] In July 479 BC, a Greek fleet then defeated the remains of Xerxes' fleet off the coast of Asia Minor. Neighbouring Greek islands and mainland city-states were able to join the victorious Greek alliance, but the Spartans and other Peloponnesians in the fleet made a further proposal. They suggested that the mainland Ionian Greeks should all be moved out of Asia and settled in the trading posts, or *emporia*, of cities which had taken the Persian side in the recent invasion.[45] This proposal has puzzled modern historians, who have tended to regard it as an unrealistic fancy, floated by Spartans who had no interest in continuing the war overseas. In fact, it makes excellent sense as a proposal to settle the Ionians in the trading posts of the Thasians, who had indeed sided with the Persians. The Thracian coastline and its resources had long interested Greeks from Asia Minor, including the Milesians and the major island of Chios, whose settlement at Maronea was on the mainland within viewing distance of Thasos's own acropolis. Such east Greeks knew very well how valuable Thasos's trading posts were: they would fit very nicely with their own. Since the mid-sixth century BC, Spartans had had a history of regarding Asia as the Persian king's sphere but Hellas as a separate entity, one whose Greeks were to be free of his interference.[46] If the Ionians could be evacuated to Thasos's trading posts, Europe and Asia, geographically distinct zones, could indeed remain separate.

The proposal was rejected and the Athenians assumed sole naval leadership of the Greek alliance instead. The Thasians joined it, but fourteen years later the issue of their mainland possessions resurfaced. According to Thucydides, who was well connected locally, the Athenians contested with the Thasians 'the trading posts on the Thracian mainland and the mine which they were working'.[47] This mine must be the Dug-Out Forest, three miles inland from Neapolis/Kavala. The causes of this dispute are unclear, but one possibility is that Thasos had been claiming that the mine was not in their island's territory and so they were not to be assessed for yearly tribute on it by the Athenians and their allies. The many Ionians among those allies were not averse to a campaign to capture the mine and the Thasians' mainland possessions: in 479 BC the Spartans had already proposed much the same to them. For three years, from 465 to 463, an Athenian fleet with allied help besieged Thasos's main city, but was unable to take it quickly despite being led by Cimon, the Athenians' most capable besieger. Since the Persian defeat in Greece, the Thasians had evidently rebuilt a solid circuit of walling around their city, using their local marble as blocks.

In 463 BC, Thasos fell and lost both its mine on the mainland and its financial and political control over some of its mainland settlements. The Thasians had to pay a lump sum and then pay yearly tribute to the Athenians. They also had to destroy their city wall.[48] From 463 onwards, the Athenians installed a suitably obedient government, probably a democracy like their own. No wall protected the city any more.

Defeat had severe consequences. Up on their acropolis, by 470 BC the Thasians had begun to build a big temple in shining white marble for the goddess Athena, one of the city's guardians. After 463 the project ceased, leaving only an impressive terrace wall and the temple's foundations. Some of the former oligarchs and property-owners were exiled, and in due course rich Athenians acquired estates on the island, using their political dominance to over-ride the usual limiting of land-ownership to local citizens of a Greek city-state. One such upper-class Athenian opportunist, Adeimantos, a friend of the flamboyant Alcibiades, acquired two estates on the island, known to us because they were later sold up in Athens when he became involved in political scandal in 415 BC. However, in May 411 the political wheel turned again.

A visiting Athenian commander worked with oligarchs on Thasos to overturn the island's democracy, encouraged by a disastrous move towards oligarchy by a few Athenians in Athens itself.[49] The Thasian oligarchs set about building a wall round their city and then, 'in the second month', broke with the Athenians altogether. For nearly four years, until spring 407 BC, they held out against Athenian attempts to recapture the island. The Athenian fleet seems to have briefly recaptured part of Thasos, perhaps the southern coast, late in 410 BC, leading to yet more political turmoil.[50] However, Thasos city did

not fall, evidence that its newly built wall was effective. After two
more years the revolt ended in spring 407 BC.[51] Despite its rebel oligarchs' best efforts, Thasos had failed to recover all its former
settlements on the mainland during nearly four years of fighting. The
most important of them, Neapolis, had remained fervently pro-
Athenian in order to escape being under Thasian control again.[52]

These events in the fifth century BC relate directly to the doctor's
time on the island. In his third year there, he attended one Philiscos,
who was residing 'beside the wall'. He was lying in bed, suffering
from 'acute fever', and on the first day he 'sweated; at night, much
discomfort' and so on until the sixth day, when he died.[53] In one of
the doctor's first three years, one Hermocrates was also seized with a
fiery fever. He 'began to have pain in the head, loins' but on the eleventh day 'everything seemed to be lightened'. On the seventeenth he
relapsed and on the twenty-seventh he too died. He was lodging by
the 'new wall'.[54]

What was this new wall? Since at least 1934, it has been related by
historian-scholars to the wall begun in summer 411 BC when the
oligarchs on Thasos rebelled and 'in the second month' broke free of
Athenian rule and 'set about building a wall', as Thucydides remarks.[55]
They presumably finished their wall by January 410 at the latest. If this
new wall is the one the doctor mentions, he was active on Thasos in
a year which belongs somewhere between early 410 and spring 407,
before the island was retaken by the Athenians and the new wall was
demolished. However, this wall is not the only candidate. Back in 491
BC, the Thasians had demolished their city wall on King Darius'
orders.[56] After the Persians' defeat in 480/79 BC they had evidently
built a new one, as their city was then able to resist an Athenian siege
for nearly three years from 465 onwards. The doctor's 'new wall'
could just as well be this earlier wall, built between 478 and 465,
before it was destroyed in 463 on Athenian orders.

Orders to destroy a wall did not require the annihilation of every
single trace of it. So long as no long defensible stretch remained, a
destruction was considered satisfactory. Archaeological research suggests that at Thasos very short stretches of walling beside some of the
city's gateways were left unscathed, as were the gateways themselves.[57] If the doctor's words are to be pressed, when he describes

Philiscos as living beside 'the wall', not the new wall, one possibility is that he might be referring to a very short section of the older wall, left standing uselessly since 491 or 463 BC. Alternatively, in my view preferably, Hermocrates was ill 'by the new wall' in the first or second year of the doctor's presence, when the wall was brand new, and Philiscos, in the doctor's third year, was ill by 'the wall', the same one but no longer brand new.

Here, one false turning for historians needs to be eliminated. The phrase 'new wall' in the doctor's text is indeed chronologically significant. It does not refer to a wall which was new when first constructed but no longer new when the doctor remarked on it. Examples of long-lived 'new' entities have been cited as if to blur this fact, including New College, my academic home in Oxford, founded in 1379 but still known by this name 640 years later. New College is not a relevant example: it was founded as a college of the Blessed Virgin Mary, whose statue, as yet unremoved, stands above its main gate. The name 'new' was enshrined in its title by statute to distinguish it from an existing Oxford college of the Virgin Mary, Oriel College.[58] The name persists, because the name 'New' is entrenched in its statutes and because its predecessor, Oriel, persists, so far, too. The 'new wall' on Thasos was quite different. It was not a once-new wall, still lingering on: the word 'new' was not formally entrenched in a founding statute and there was no other full length of city wall still standing from which it continued to need to be distinguished. There had been city walls before it, but there had been no effective and continuous wall round Thasos between 491 and c. 478 BC or 463 and 411 BC. It was, then, quite unlike the Via Nova in ancient Rome which continued to be called 'Nova' ('new') to distinguish it from the older, persisting Via Sacra. It is also unlike the Pont Neuf in Paris, so called to distinguish it since 1609 from older bridge points across the Seine, or, in British politics, New Labour, so called to distinguish it from Old Labour, which, its leaders believed, it had reduced to an irrelevant minority.

Commenting on the doctor's text, Galen assumed, rightly, that its new wall was indeed recently new.[59] The doctor was certainly not noting and writing at a time somewhere between 463 and 412 BC: in those years there was no new wall for him to notice. One possibility is that he noted Hermocrates by the new wall in years between 412/11 and 408/7 BC, when he coincided with a time of political upheavals on Thasos, wars and eventually famine. The other is that he visited in the 470s or early 460s BC and noted Hermocrates by the new wall then. If so, he was present in years when the island still controlled its mainland settlements and when the lucrative gold mines of the Dug-Out Forest were enhancing the yearly revenues. Thasos was a contributing member of the Hellenic Alliance, led at sea by the Athenians, and even after the expense of entertaining Xerxes in 481 BC it was still very rich. A modern scholar of Thasos's fine arts has called the decade after 479 BC the island's artistic *belle époque*.[60]

These options are not just of antiquarian interest. Whichever is correct, the doctor was certainly noting and observing his patients during a period of oligarchy. His text owes no debt to democracy for its setting, intended audience or outlook, yet it is the pearl among the Epidemic books, those masterpieces of method and observation. If he was working and noting between September 412 and September 408 BC as usually supposed, his aim, to teach pupils, was one which other fifth-century medical authors had already addressed. If he was working c. 470 BC, his text becomes much the earliest datable example of a medical text for practitioners and students, possibly preceded only by the *Cnidian Sentences* whose form, presumably sentences like bullet points, was very different. It even becomes the first Greek prose text which survives in full, a cardinal item in any grounded discussion of the invention of prose among the Greeks. Even more strikingly, if written c. 470 BC, its terse style, its implicit reasoning and its author's emphasis overturn simple linear views about the progression of Greek thinkers and authors from an 'archaic' to a 'classical' age, traditionally located from c. 450 BC onwards. There is nothing 'archaic' about this author.

For dating his text its language, thinking and local context are crucial. Thanks to masterly work by their most recent editor, Jacques Jouanna, it is clear that books 1 and 3 of the *Epidemics* use a cluster of words which are unique in the entire Hippocratic Corpus.[61] They do not reappear even in the other five Epidemic books. They include vivid words like *blēstrismos*, used for the 'tossing around' of patients, both males and females, including that young man from Meliboea, broken down by sex and drink. It was a useful word for doctors, but none of the late fifth-century BC medical texts uses it.

The author's expressed thinking about the human body is also distinct from the thinking expressed by other doctor-authors in the late fifth century BC, surviving in the Hippocratic Corpus. He never refers to the existence of four humours, the idea which the author of *The Nature of Man* postulated and made famous from the late fifth century onwards.[62] However, as that theory was the author's special one, not taken up by others, the Epidemic author's silence is not decisive evidence for his own earlier date. More suggestive of one is his vocabulary. He uses the word *chumos*, or 'juice', only once, referring to blood; the next Epidemic author, the author of books 2, 4 and 6, uses it eight times and considers these juices, often translated as 'humours', to be important signs for study or even encouragement.[63] Books 1 and 3 use the word *hygros*, meaning 'wet' or fluid', but only for a runny stomach or for a cough and the wet matter coughed up.[64] They write about bile, blood and phlegm in the body and are aware, of course, of the importance of movement and excess in such fluids, but they do not dwell on 'fluidity' as a named concept, whereas 2, 4 and 6 do.

They also differ in what they pick out in the environment. Their author stresses the importance of a city's weather and wind patterns but unlike the late-fifth-century BC author of *Airs, Waters and Places* he does not dwell on the significance of a city's water supply and its general location.[65] The crucial comparisons are with his heir, the author of books 2, 4 and 6. There are major differences between these two masters, two doctors with totally different aims and understandings of the body and what to do with it, as we will see.[66] Such differences might exist at any one time between doctors working in different settings, but the basic point is that these two were not working in an unrelated tradition or place. Like our doctor, the author of 2, 4 and 6 was a Coan and worked partly on Thasos. He used the same framework of seasonal climate and day-by-day case histories as the author of 1 and 3, but his deployment of them, as we shall see, was quite different. Most importantly, he was working by 406–405 BC (that 'man from Alcibiades' with the swollen testicle), perhaps even in 410: on the prevailing dating of books 1 and 3, their author, 2, 4 and 6's model, finished his fourth year of work at most only two or three years earlier. The author of 2, 4 and 6 was not his master's dissident young pupil. He wrote as a mature doctor of long experience, summing up his considered view of the medical craft and its application. If his predecessor was working as late as 412–408 BC, the author of 2, 4 and 6 had been contemporary with him in the very same places, and yet his views, his method and approach were radically different. Unlike his senior mentor he did not even write with the fundamental aim of aiding prognosis, or foresight, about the course of a disease.

On Thasos, the author of *Epidemics* 1 and 3 gives not a hint of any cases of wounding, any impact of a severe siege or, eventually, shortages of food. By contrast, the author of books 5 and 7, writing in the shadow of King Philip's wars, recorded battle wounds which, inevitably, he attended, and the author of 2, 4 and 6 recorded physical consequences of a food shortage at Ainos on the Thracian coast.[67] *Epidemics* 1 and 3 survey the constitutions of each year, its weather season by season, and its patterns of disease. If major political upheavals were going on simultaneously, it is hard to credit that they received no mention at all in the accompanying surveys of disease or in one or other of the case histories, relevant though their impact was to the annual public health. If their author was writing about four of the years between 412/11 and 408/7 BC, as has been hitherto believed, he lived through the political coup by oligarchs on the island, the return of exiles from abroad and the break with fifty-two years of alliance with Athens. These changes caused bloodshed, as did a returning Athenian fleet in 410 BC. So did the continuing struggles on the mainland to recover Thasos's former possessions, including, unsuccessfully, Neapolis. Thasians needed medical attention during all these battles: as the historian Xenophon later summed up, it was a time on Thasos of 'wars and revolutions'.[68] They culminated in the arrival of an Athenian general with a fleet which retook the island in spring 407 BC when it was weakened by a period of famine. If he began work in autumn 412, perhaps he slipped away from Thasos in mid-408, his fourth year, seeing the situation deteriorate. If he began work in autumn 411, his fourth year on Thasos ran from 408 onwards, a very traumatic time indeed. Whichever date is preferred, his account of his fourth year never refers to these major items. When he reviews its diseases at large, he dwells on 'burning fevers' or a red-rash disease, without any comment, even an incidental one, that the political, military and economic context was contributing to the public health.[69] In the case histories, he never mentions attendance on anyone who had gone hungry or who had been wounded, though many had been, and the skills of himself and his team would have been in urgent demand from the casualties.

If he wrote in the later 470s and perhaps on into the early 460s BC, none of these silences is problematic. Internally the island was at peace, recovering from the Persian presence. The wall was new but there was little distinctive to record about each season apart from variations in its winds, its hot spells, and its intervals of cold and snow. If he wrote then, sixty years separated him from his follower, the author of books 2, 4 and 6. It is not, then, surprising that their outlooks and vocabulary were so very different.

This early dating has never been considered for *Epidemics* 1 and 3. It solves significant problems in the text's contents and their relation to 2, 4, and 6, their followers: it is on these points that my re-dating rests. Is there also external evidence which supports or even requires it? The likeliest would be inscriptions, those ever-growing sources of our knowledge about individuals and Greek cities. To them we must now turn and to the microhistory which they make possible.

12

Building Blocks of History

I

Secure connections between inscriptions and literary texts are some of the most potent beams of light which historians can cast into dark corners of ancient history. The doctor's text is full of patients' names: can any of these people be traced convincingly in inscribed evidence and dated there? In many Greek cities one type of this evidence is coinage. Sometimes a city's coins are inscribed with the name or symbol of the magistrate in whose tenure of office they were issued. Thasos's coins have no such inscriptions, but just across the sea the coins of Abdera are very early examples of the practice. Abdera is one of the places where the doctor worked.

There he attended and noted the cases of four named patients and an unnamed fifth, a young girl who 'lived by the Sacred Way' and suffered from 'acute fieriness, burning heat'.[1] She had her first period and also some abundant nosebleeds: a double bleeding was always a good sign in the doctor's view of the body, as it expelled plenty of excess blood. After twenty-seven days, she survived. He also attended young Nicodemus, who had been seized by a fiery heat after repeated sex and drinking. He too recovered, but it took him twenty-four days to do so.[2]

For the dating of his text, the tantalizing cases in Abdera are his three others. One of them, Pericles, was suffering from an acute, continuous fever, but he too had a nosebleed, if only from his left nostril, and after repeatedly passing urine of varying texture, he survived: again, the expulsion of blood and fluid was a good sign.[3] Another patient was more time-consuming. Heropythos began by suffering from an acute headache and then suffered a fever which the author noted on and off during no fewer than 120 days.[4]

Do we know anything else about these people? In 1982, a lifelong scholar of the Epidemic books and their context, Karl Deichgraeber, emphasized what had first been noticed in 1852, that there were over-laps between the names of patients visited in Abdera by the doctor-author of *Epidemics* book 7 and the magistrates' names inscribed on the city's silver coins.[5] The inscribed names are abbreviated but it is not difficult to supplement them.

Book 7 was written in the mid-fourth century BC: what about books 1 and 3, written in the fifth century? Here Deichgraeber briefly referred to two names 'worthy of mention'. One issue of Abdera's silver coins is marked with the abbreviated name 'Hero-' and another with 'Peri-'. Deichgraeber suggested that these magistrates' full names were Heropythos and Pericles, the names of two of our doctor's patients in the city, but as he believed that these coins were dated *c.* 520–500 BC, he proposed that the names on them belonged to grandfathers of the doctor's like-named patients.[6] In Greek families, grandsons were indeed quite often named after their grandfathers.

In 1984, just after Deichgraeber wrote, another silver coin from Abdera was discovered, marked with the letters 'Apol-'.[7] Another of the doctor's patients in the city was Apollonios, sufferer from an enlarged spleen and pain in the liver. After an ill-advised consumption of large quantities of goats' and ewes' milk and a bad accompanying regimen prescribed by other doctors, Apollonios deteriorated.[8] By the time our doctor was called in, apparently after a fortnight, Apollonios was worsening. He died on the thirty-fourth day. The abbreviation 'Apol-' on the silver coin is consistent with his name, Apollonios.

Since Deichgraeber made his suggestions, the coinage of Abdera has been reclassified and decisively re-dated. The Peri-coin belongs not in 520–500 BC but between *c.* 500 and 475 BC. The Hero-coin belongs between *c.* 475 and 450 BC. The Apol-coin was found in Lycia in Asia Minor, far from Abdera in a hoard which was buried *c.* 460 BC: it is dated nearer 475 than 500.[9] Of course, the abbreviations on these coins might refer to other Greek names which began with these same letters. However, the letters do not appear at any other date on any other issues of Abdera's long-lived coinage, whether silver, gold or bronze. The three names might very well be Hero-pythos, Pericles and Apollonios, three of the doctor's patients. On Thasos, he certainly treated people in high society. In Abdera he is likely to have done the same during a stay which ran for at least four months.

These coincidences are very striking, but they are not absolute proofs of identity and date: thousands of people lived in Abdera, any of whom might have had these names and been the doctor's patients. Nonetheless, these candidates belong in the high social milieu which he attended elsewhere. For that reason they fit attractively with him being at work in years between the late 470s and the early 460s BC. They do not stand alone. Thasos's other great assets, crucial for scholars, are its many inscriptions. Among them too are texts whose names overlap with names of the doctor's patients. What could be more solid, it might seem, than names on blocks of the city's excellent marble? Repeatedly, scholars have adduced them in order to pin down the date of the doctor and his work. Their dating, relevance and context need closer study too.

I I

If an inscription is securely dated and refers to a contemporary event or person known in a literary text, it dates that text's reference. His-torians of the ancient Greek world have particular scope here, because most of the Greek cities teemed with inscribed texts, a feature which sets them apart from the sparsely textual cities of the medieval and early modern eras. Sometimes they were inscribed on bronze, some-times on stone. Bronze tended to be melted down and reused by posterity, but if a city had a local source of inscribable stone, there was a high chance that some of its texts would survive.

Thasos's various types of marble were a chiselling inscriber's best friend. As a result the island is second only to marble-rich Athens for its number of surviving inscriptions, including public decrees, dedica-tions of sculpted monuments, funerary inscriptions and rules for best practice in a temple precinct.[10] Its graffiti include male homoerotic comments, some being those examples carved with skill onto a sea-side rock face, and others inscribed on stones in or near the city's walling.[11] Other materials have preserved some rare survivors, including small plaques or tickets of bronze from the fourth century BC, inscribed, as in Athens, with the name of an individual citizen, probably for his use during jury service.[12] A clay plaque, found beside a house in the south of the city, near the gate showing Silenus rampant, has preserved a very rare survivor, a private letter, inscribed on wet clay and then hardened in an oven. In it, one Euarchos instructs another man, but unfortunately the text is fragmentary. To judge from the style of its lettering it belongs in the fifth century BC, possibly as early as 470–450, and is one of only two letters on clay to survive in Greek.[13] From another of the island's great trades, wine, hundreds of clay handles also survive, broken off the big wine jars which were used for storage and transport. From the early fourth century BC onwards, these handles were stamped with one, later two, names in the year of stamping.[14] They have been found far and wide beyond Thasos, even in Greek cities of the Black Sea region, heavy importers of the island's excellent wine.

The city of Thasos has a unique distinction, relevant to the doctor and his patients: in prominent sites on or near its main *agora*, or meeting place, it eventually displayed four inscribed lists of the holders of its two main civic magistracies.[15] These inscriptions listed most of the holders year by year, naming them and their fathers, the standard Greek practice, and marking off each year's magistrates from the next. Like a handbook of the great and good, they gave the names of hundreds of members of Thasos's ruling class in a chronological sequence which extended back to the seventh century BC. Fascinatingly, similar names recur in places across many decades, a witness to families who endured in high society.

They are extremely challenging to scholars, because none of the lists survived intact or undisturbed. They eventually fell from their original positions and in due course many of their marble blocks were reused as building materials. In 1864, some of them began to be rediscovered, a process which has continued ever since: one block was found as recently as 1983.[16] Only a few of them have been recovered unbroken, whereas others are in fragments and many others have been lost altogether. Their restoration in an orderly sequence faces formidable problems, but has continued to progress thanks to devoted work by members of the French School at Athens, active for more than a hundred years on Thasos, and also in the Louvre, where many of the blocks have been transferred. The blocks of each list have to be arranged into columns but the height and number of the columns can only be worked out by a chain of interrelated conjectures and observations. Step by step, options have been narrowed down and a workable shape for each inscribed list has been proposed.[17]

Two separate types of magistrate were inscribed: two of the lists gave the city's yearly rulers, or *archons*, and the other two gave other magistrates, its *theoroi*. For most of the periods recorded, each magistracy was held for a single year by a group of three men. Both of these magistracies are attested on Thasos's mother island, Paros, from which the Thasians, therefore, copied them.[18] Some of those who were *theoroi* also appear in the lists of *archons*, but one problem is to know in which order the two magistracies were usually held. The *archons* were evidently Thasos's top magistrates, as in many other Greek city-states: did some of them go on to become *theoroi* in older age, as the current French interpretation of the list assumes, or were *theoroi* usually younger men, some of whom went on to be *archons* later?[19] There is no decisive evidence, but in my considered view the usual pattern was the latter, that *theoroi* were younger men, some of whom eventually went on to hold the archonship. The alternative, that *archons* went on to be *theoroi*, has to accept cases of irregularity, which are problematic. In other Greek cities, magistracies, priesthoods and so forth were held in a regular, largely fixed order.[20] My view of the two magistracies' sequence, differing from the specialists', has fundamental consequences for a chronology of the names in the lists.

What did *theoroi* have to do? When Plato in old age wrote about his ideal, and repugnantly repressive, city-state, he laid down rules for special *theoroi* and their duties. They must be men over fifty, he specified, serving for up to ten years on trips abroad, whereas no private individual of his city was to travel abroad at all.[21] Plato's fantasy state is not evidence that in real life a city-state's *theoroi* had to be old men, any more than it is evidence that in real life travel was forbidden to all their fellow-citizens. In several Greek cities there are clear examples of *theoroi* who were young men.[22] On Thasos their full functions, unfortunately, are unclear. Their actions are mentioned only in a very few inscriptions other than the lists, but two certainly, and another six probably, connect them with supervising the inscription of important grants by the city, once of a lease to the 'gardens belonging to Heracles', and otherwise of grants of citizenship to named individuals. These inscriptions were to stand in major sanctuaries, the lease in Heracles's sanctuary, one case of citizenship in Athena's, the other cases in Apollo's. They were to act in these cases as the *archons* might specify, confirming that in these duties *theoroi* were subordinated to them. *Theoroi's* names in their year of office also date another inscribed enactment, which credited Apollo with confiscated property: as we will see, it has been particularly important for historians.[23]

These duties are surely not the sum of all these major magistrates did. In the big list, the earliest *theoroi* are recorded as serving 'in the time of the first *aparchē*' and then others in the 'time of the second *aparchē*'. The meaning of this word *aparchē* has been much discussed, but the likeliest answer is that it indeed had the meaning persistently attested elsewhere, not 'independent rule' but 'offering of first fruits' or simply 'offering'.[24] Old men would still travel, but if the *theoroi* had a major role in escorting an offering to a major temple overseas, perhaps to Pythian Apollo's great temple at Delphi, it was quite risky to reserve this job every year for three men all aged in their fifties or more. The latest study of the question concludes, convincingly, that the 'role [of *theoros*] on Thasos had two components: inside the polis, serving as a magistrate, and outside the polis, at least originally, representing one's city at sanctuaries abroad, the latter role a religious one, and the former a political one.'[25] As they were well-born citizens, younger holders of the office already had family connections abroad and prestige at home; enough for them to fulfil this double role.

III

Other Greek city-states kept lists of their magistrates and inscribed them in public. Thasos is unique, on present knowledge, for the number and scale of its lists, publicly displayed in central sites in the city. Their scale has become clearer through generations of work by French specialists, on whose reconstructions, still in progress, my summary depends. It gives a sense of the Thasians' sustained achievement, and its details bear on dating the doctor and his patients. The first of the two lists of Thasos's *archons* was inscribed in fine lettering on slabs of excellent white marble, a few of which survive in whole or part. Its original site is considered to have been a building, now lost, on the north-east of the city's public *agora*. The latest assessment by its French scholars credits it with at least nine, perhaps ten, columns of names, each covering 32 or 33 years. The first nine columns are considered by them to have run from *c*. 650 BC down to *c*. 326–325 BC, just before Alexander the Great's death.[26] The second list of *archons* was inscribed later on grey marble. It is known from even fewer fragments, but the current French theory is that it was probably inscribed on four big plaques, presumably attached to the façade of a building or to a wall inside one, probably also on the *agora*. It too is considered to have begun by having nine columns which gave the *archons'* names down to the 320s BC, matching those in the first list. Many more columns were inscribed later and continued the list well into the Roman imperial period.[27]

As for the two lists of *theoroi*, the earlier one was inscribed in smallish letters on blocks of a grained marble, only three of which survive intact. According to the French experts' current reasoning, its first six columns contained the names of *theoroi*, usually three for each year, thirty-six years making up the average length of each column. A seventh column was probably then added.[28] The later and bigger of the two theoric lists was much better preserved. Before its blocks collapsed, it stood on the face of a wall which overlooks what was for long described by scholars as the 'passage of the *theoroi*'. It is now a decrepit walkway beyond a rickety bridge, but archaeologists have recently inferred that the 'passage' was part of a main axial street of the city, which was laid out in the early fifth century BC. If so, the list stood beside a much-frequented street, which led up to what had been a gateway in the city's main wall.[29] It is considered to have been inscribed at first with seven columns of *theoroi*, each one being made up of eight blocks set one above the other. There were 33–4 years of *theoroi* in each column, or 99–102 names.[30] Many more columns were then added, continuing this list, too, far into the Roman period.

These lists grew to contain hundreds, even thousands, of names. They are most remarkable, but they were not inscribed when each year ended and the magistrates left office. They were inscribed much later in long sequences by a stone-cutter who worked with a pre-existing list of names arranged year by year. The *theoroi* stretched back in the French experts' view to the 570s, but the *archons* went back a little further, in the French view at least to the 650s, perhaps to the 670s.[31] Their span shows that a Greek city could indeed pre-serve detailed lists of its magistrates year by year which reached back through two or three centuries into the archaic past.[32] None of them had mythical or obviously fictitious names.[33] Why did the city's gov-erning body decide on, re-inscribe and maintain such a detailed public record? The repeated decision to inscribe it resulted in impres-sive displays, not just of past members of the city's top families, but of a long civic continuity. Thasos's history had been punctuated by periods of civil strife, siege and disorder, but these highly public lists projected continuity and a long past.

Scholars soon observed an exciting fact: some of the names in these lists overlap with the personal names of patients who are located on Thasos in *Epidemics* books 1 and 3 and in the later book, *Epidemics* 7.[34] Are they the very same people? As some of the names are wide-spread in the Greek world, it is risky to assume that they necessarily refer to the doctors' like-named patients: they might also have been borne by people outside the narrow governing class.[35] However, there are some very unusual names in the lists, known only or mostly on Thasos.[36] My guess is that they were upper-class names only. If so, a doctor's patient on Thasos who has the same rare name as a listed magistrate of the right date is even more likely to be the very same person. If the two persons are not one and the same, they are highly likely to be close relations in the same family where the rare name was current. They support what the names on Abdera's coins have implied, that the doctor-author of *Epidemics* 1 and 3 often attended members of a city's upper class. He had good reason to do so. They were the people best able to appreciate the methodical medicine which he claimed to represent. He did not confine his efforts to people who could pay him, a matter he never mentions, but these prominent citizens were those best able to do so.

IV

Crucially, one name, Antiphon son of Critobulus, occurs both in the doctor's survey of his third year on Thasos and on marble blocks from each of the two inscribed lists of *theoroi*.[37] As Antiphon is attested in all these texts with his own name and his father's name, he has been assumed to be one and the same person. He is also known in a separate text, inscribed on a wall of the sanctuary of Apollo high up on Thasos's hill, whose rediscovery is a tale with a long history.

It begins with Cyriac of Ancona, a many-sided Italian diplomat who was also a theologian, an antiquarian and a man of letters. On 11 November 1444 he set out by boat westwards along the coast of ancient Thrace, passing what he believed to be the 'shore of ancient Diomede' on which the hero, he recalled, was said to have kept man-eating horses and set them on visiting foreigners. He travelled on past the city-state of Abdera, founded, he thought, by Diomede's like-named sister, and then turned south across open sea for the brief run to Thasos.[38] Since 1434 the island had been granted by the Byzantines to the Gattilusi, a family from Genoa whose history included piracy but who had assisted the Emperors in times of faction and were to marry well as a result. Cyriac found that the young Francesco Gattilusio, the current ruler, had re-erected an ancient Greek statue near the entrance to Thasos's port, modern Limenas. Its inscribed base recorded several ancient Thasian names, and so Cyriac copied them, his usual practice. Ancient sculptures, including one of Heracles and the lion, were being displayed by Francesco to dignify his new domain. They were a fine context for Cyriac's antiquarian interests while he copied yet more inscriptions, some of them on ancient sarcophagi.[39]

On other Greek islands Cyriac was not inclined to travel very far from his place of lodging. Just over a year later, his researches on the island of Chios were undertaken 'not too adventurously or energetically', in the considered judgement of the modern epigrapher who has best followed in his footsteps.[40] On Thasos he was more energetic, thanks, it seems, to the encouragement of two Italian companions who went out with him on the steep climb to Thasos's acropolis, 155 metres above the harbour. Eventually they reached the 'very high' citadel from which they could see the 'entire circuit' of the ancient city's walls, 'not much less than 30 stades in length'. Those walls are still impressively visible, but 'what was more of a pleasure to record', Cyriac wrote in his diary, was an inscription 'at the entrance to the citadel', he believed, written in 'very ancient lettering'. He copied it out: it was dated during Antiphon son of Critobulus and his two colleagues' tenure of office as *theoroi*. The names which Cyriac copied are exactly the names which are borne by Antiphon's colleagues on inscribed blocks from the two lists of *theoroi*, found in the city below. When Cyriac made his copy these lists were unknown.[41]

After his copy, the stone which he had found remained unstudied until the early 1870s. It was then rediscovered and re-examined, but only a very few letters seemed to be legible. After the First World War, it was located again by members of the French archaeological school at work on Thasos, and in 1921 its text was published in detail.[42] Most of the letters seen by Cyriac turned out to be still visible, proof of his careful transcription. After the dating by Antiphon and his colleagues, it went on to state: 'Of these people the property is sacred to Apollo according to the resolution of the three hundred.' The names of six men follow, four of whom are evidently Thasians, whereas the other two are citizens of Neapolis, the 'new city' founded by Thasos on the mainland, beside modern Kavala. Their property had been confiscated.

The location of the inscription was relevant to its contents. Whereas Cyriac had recorded it only 'at the entrance to the citadel', it had been carved into a block of a south-west wall of what is now known to be the sanctuary of Apollo. As the text referred to the dedication of property to Apollo, it was appropriate to inscribe it on a wall of the sanctuary of the god whom it benefited. His priests and personnel could refer to it if their god's ownership of the property in question was contested.

Thanks to these pieces of evidence on stone, can Antiphon son of Critobulus date our doctor's text?

In both of the lists of Thasos's *theoroi*, the French experts place Antiphon son of Critobulus in the fifth column. They date him there to 410/9 BC, partly by the contents of the inscription dated by his name, the one found by Cyriac, and partly by linking them to three other inscriptions, one found in Athens, two others found in Thasos, where they had once been displayed in the city-centre, its meeting place or *agora* (I discuss these links and this dating in my Endnote 2, raising objections to each of them). The year 410/9 BC for Antiphon then becomes their crucial lynch-pin for dating all the names which run in columns before and after Antiphon's in the big theoric list.[43]

In *Epidemics* 3 the doctor also mentions an Antiphon son of Critobulus: he was a sufferer from a burning fever which, after forty days and some nosebleeds, he survived.[44] However, as I have argued on other grounds, the doctor's text has a very different date. On grounds of its vocabulary, its author's aims, its notions of the human body, its omission of any sign of political and military bloodshed or food crisis during its four years and its marked difference from the Epidemic books of its successor, who was writing by 406 BC, I date the earlier cases in it to the 470s. Of the various options, the one I will use, for clarity's sake, is to date the four years of cases from September 471 to September 467 (the doctor began his year in the autumn). This dating also fits the dating of those three names abbreviated on Abdera's coins, compatible with three of the doctor's patients there.

Can the names in the lists of Thasian magistrates be reconciled with this earlier dating? One valid route is to deny that the two like-named Antiphons are really the same man. In the big list of *theoroi*, similar names sometimes recur at intervals of at least forty years. Aristoteles son of Menedemos is a clear example, a name which occurs about forty years before Antiphon's and then recurs ten years later than this.[45] A citizen of Thasos could only be a *theoros* once in his lifetime, so, plainly, these two Aristoteles were not the same person. As Greeks sometimes named grandsons after their grandfathers, the explanation is obvious. The earlier Aristotle was the grandfather of the later one, his like-named grandson. Might Antiphon son of Critobulus be another example?. The one named in the doctor's text may be the grandfather of the one in the theoric list and in the text on the temple wall. If so, a dating of the contents of *Epidemics* 1 and 3 to the late 470s and early 460s BC is still valid.

The grandfather route remains open, but there is another possibility worth exploring. The columns of the big theoric inscription and the placing of its marble blocks are increasingly fixed convincingly, but the dates to be given to each column are still hypothetical. Can the date for their Antiphon's year as a *theoros* be raised and placed earlier? If so, both types of evidence, the medical text and the magistrates' list, would come neatly into line. They would name one and the same man, alive in the 460s, not the 410s BC.

The possibility has not been adopted or considered by the inscriptions' experts, but relates to a question with wider consequences. If the doctor was on Thasos in 410 BC, he was a contemporary of Socrates and the intellectuals in late fifth-century Athens. His rational method went with a way of thinking which is less surprising in the era of these famous thinkers, people whom Plato called 'sophists'. However, in my view he was no such person. He was observing and explaining in this reasoned way much earlier, as early as the 470s BC. His near-contemporaries were doctor Onesilas, who had been treating the wounded for free in the Cypriote city of Idalion, that unnamed doctor on Rhodes with his cupping glasses and other metal instruments and Pindar writing of 'natural' sores while he consoled poor Hieron in Syracuse. He was also a contemporary of the young Herodotus, who related historical events to divine anger and retribution, the mature Aeschylus, author of plays about divinely induced catastrophe, and other effusive odes in which Pindar related his patrons' sporting victories to the deeds of gods and heroes. In the 470s and 460s BC the Epidemic doctor's thinking stood far apart from these contemporaries. He becomes a significantly different figure in the history of Greek thought, at least as hitherto constructed.

13

Art, Sport and Office-Holding

I

The dating of the magistrate Antiphon, hitherto considered to be the doctor's patient, connects with two other members of the theoric list, each known from evidence outside it. Both enliven the history of Thasos, but the current dating of their magistracies conflicts with separate evidence for their lives. Art and sport will show why.

In the big list of *theoroi*, the two names to investigate are Polygnotos son of Aglaophon and Disolympios son of Theogenes. Both are very remarkable. Polygnotos is a towering figure in the history of Greek painting. Disolympios' name means 'Twice Olympian' and is unparalleled in all Greek history. His father, born on Thasos, was the greatest athlete after Milo of Croton and was active in the years from *c.* 490 to 470 BC.

The appearance of Polygnotos' name here is as exciting as if the name Michelangelo Buonarroti were to survive in a list of members of the exclusive *signoria* in Florence when only a few descriptions of his sculptures had survived and almost all of the city's other history had been lost. Disolympios' appearance is as exciting as if Daniel son of Gary Sobers were to survive in a list of Barbados's magistrates when the island's history had been destroyed except for a few cricket scores from the early 1960s, made by its greatest player at home and abroad.

One point is certain: the relative positions of their names in Thasos's list are fixed for sure. A block survives from the bigger of the two lists which spans two columns and has Antiphon's name in one column and Polygnotos' name in the column which adjoins it on the left: the length of the columns is known, and so Polygnotos son of Aglaophon was a *theoros* thirty-seven years before Antiphon son of Critobulos.[1] Disolympios son of Theogenes was a *theoros* seven years after Antiphon: the blocks inscribed with their names are interlinked in a vertical sequence.[2] The dates, therefore, of these men relate directly to the dating of Antiphon and his year of office.

Polygnotos son of Aglaophon is the greatest genius born on Thasos. He was trained there by his father, also an artist of note. He became the finest painter in the Greek world. None of his paintings survives, but they are known from texts referring to their subjects, two of which are long descriptions by an admiring visitor, Pausanias: he gave them in the Greek guide for travellers which he compiled in the AD 130s.[3] From the 470s BC onwards, Polygnotos is known to have worked in three major places outside Thasos: Delphi, Athens and Plataea. At Plataea he painted *Odysseus after Slaying the Suitors* for the temple of Warlike Athena. The temple was almost certainly vowed after the great Greek victory at Plataea against the Persians in 479 BC and paid for promptly from the spoils of that battle.[4] Perhaps it was through his work there that he entered into a close relationship with the Athenian general Cimon. In 477/6 BC Cimon led Athenian and allied troops up north to Eion, the most westerly of the mainland trading posts which Thasians and Parians had settled beside the Strymon river. It is quite likely that Cimon made contact with Polygnotos, a Thasian too, soon after he took Eion from its Persian commander.[5]

After that victory, Cimon sailed on south and conquered the rocky Aegean island of Skyros, populated, in part, by 'pirates'. Here, he found a great and timely marvel. Beneath a mound, his men dug up a sarcophagus containing very big bones, an ancient bronze spear and a sword. They identified the bones as those of the hero Theseus who was believed to have died on the island. The spear was sent for dedication at Delphi, the source, probably, of the oracle which validated the finds, but the bones travelled back to Athens on Cimon's own warship.[6] With a keen sense of publicity, he arranged for them to be processed into the city, where a shrine for them was proposed: it probably stood in what is now Plaka, to the east of what is still visible of the Athenian *agora*. It was decorated with four paintings of scenes related to Theseus' legendary career, at least one of which was painted by Polygnotos.[7]

As this shrine, the Theseion, was planned to house the hero's bones, Polygnotos' painting belonged very soon after the bones' return to Athens in 476/5 BC. He is credited with two other Athenian commissions, a painting in a newly built shrine for the heroic twins Castor and Pollux, probably at a similar time, and work in a famous colonnade, patronized by Cimon's brother-in-law.[8] It became known as the Painted Colonnade and the painting in it of *Troy Taken* was certainly by Polygnotos. This famous colonnade, or *stoa*, was open to all-comers, a place where they strolled and talked; it later gave the name 'Stoics' to the philosophers who met and discussed there. In 1981, archaeologists discovered the very building on the north side of the *agora*, where recent finds have established its impressive scale. The paintings are lost, but they had been displayed on the colonnade's inner walls on detachable panels, presumably made of wood.[9]

The date of the colonnade has been endlessly disputed but it belongs in the late 470s for four reasons. According to its recent excavators, 'pottery found against the foundations and under the floor at the West indicates a construction date around *c.* 470 BC.'[10] Inside, Pausanias explains, one painting commemorated a battle of the Amazons, another *Troy Taken* and a third the land battle of Marathon against Persian invaders, the high point of the career of Cimon's father, Miltiades. No painting commemorated Cimon's greatest victory, his win by sea and land against a Persian force in 469 BC.[11] That victory, therefore, was still in the future when Polygnotos went to work. Gossip also claimed that Polygnotos had had a long love affair with Cimon's obliging sister Elpinice and that in his *Troy Taken* he had painted her fair features as the face of a like-named captive Trojan woman. By 463 BC, Elpinice was considered to be an old woman.[12] The allegations of this love affair and Polygnotos' tribute to her presuppose that she was still highly attractive when the colonnade was being decorated: they are another reason for dating his work to the 470s. Above all, Polygnotos, an artist of noble birth, was said to have painted his masterpiece for no pay whatsoever.[13] In return for this truly aristocratic gesture, he received the immense honour of full Athenian citizenship.[14] This grant had to be made by the Athenians' democratic assembly, but it was surely brought forward at the urging of Cimon. From 469 BC he was out of the city until spring 468, and after spring 461 he was in exile.[15] A date in 470/69 or earlier fits the grant of citizenship and the finishing of Polygnotos' work in the colonnade. (I discuss further details in Endnote 1.)

The dating of the colonnade is important for the next known act in Polygnotos' career. He was invited to do two paintings for a similar columned building at Delphi. It too was intended as a space for talking and socializing. Its foundations are still visible, high up beside Delphi's theatre, where they overlook the oracular temple of Apollo and enjoy the most magnificent natural view.[16] Its patrons were the Cnidians from south-west Asia Minor, and it is fascinating to think that Asclepiads there, from medical families, may have contributed to the building's costs. Polygnotos painted two big masterpieces for it, each containing about seventy human figures. Once again, one of the subjects was *Troy Taken* and *The Departure of the Greeks*, on a bigger scale, however, than his recent painting in Athens. The other's subject was the Underworld as visited by Odysseus, an episode immortalized in Homer's *Odyssey* and in post-Homeric poetry.[17]

Polygnotos showed considerable independence from poetic sources for these subjects: he varied what the paintings' titles led viewers to expect. Since the 1750s, artists have tried to render the two of them, based on Pausanias' detailed verbal description, but no reconstruction has yet prevailed.[18] Their dating has been controversial too. Even the attribution of the colonnade to the Cnidians has been queried, as if the project was originally an Athenian one and the Cnidians put their name to it only when it was restored much later in the fourth century BC.[19] However, a prominent figure in the painting of the Underworld alluded to a local Cnidian figure of myth.[20] A Cnidian allusion was present from the start because Cnidians were paying for the project.

There is a compelling context for their patronage. In 469 BC, Cimon led a big allied Greek fleet against a gathering Persian navy on the south coast of Asia Minor and won a crushing victory over them at the Eurymedon river.[21] It was a turning point in Greek and Persian foreign relations. His fleet set out for battle from Cnidos, whose double harbour, at modern Burgaz, served as a sheltered and defensible base. Cnidians were presumably active in the subsequent battle, a very important event for them. It redeemed their previous history of 'medism', or submission to the Persians, which had begun about forty years before.[22] In the aftermath of the victory they had good reason to re-advertise themselves before a Greek public and pay for a conspicuous project in Delphi's sanctuary.

The Cnidians had an obvious personal link to Delphi and to Polygnotos the master painter. In 469 BC, curly-haired Cimon, their commander at the Eurymedon, was the hero of the moment, an admired international aristocrat, greatly esteemed at Delphi, where he had dedicated what purported to be Theseus' ancient spear. It was through him, then, that the Cnidians decided to hire Polygnotos, his friend and artist, for their Delphic monument. The commission fits neatly in 469/8, in the immediate wake of the Eurymedon victory. Archaeologists of the Painted Colonnade in Athens have recently observed that the construction and clamping of its east wall are 'very similar' to those of the Cnidian building at Delphi.[23] The similarity suggests that the buildings were indeed very close contemporaries. So does a two-line epigram on Polygnotos' painting of *Troy Taken*. It referred in verse to Polygnotos as the artist and was ascribed to the great poet Simonides, author of other commemorative poems for victors in the Persian wars. Simonides died in 468, aged eighty-eight, and although other epigrams later became connected fancifully to his famous name, he could indeed have composed this simple two-line poem in the last months of his life.[24] There was no similar epigram for the Underworld painting, presumably because, by then, Simonides had died.

In gratitude for Polygnotos' work, the Delphians granted him the honour of free meals, an honour for him personally, not as the envoy of a city-state.[25] He was already a citizen of grand Athens: what did he do next? Datable evidence for him ceases abruptly after his work for the Cnidians. Paintings by him existed in a shrine at Thespiai not far from Delphi, but he may already have painted them in the early 470s.[26] Six centuries later, Pausanias saw at least one painting by him in a 'building for paintings' on the Athenian acropolis: this building was part of the north-west wing of the great entrance, or Propylaea, built in the 430s.[27] Polygnotos' painting showed Odysseus encountering young Nausicaa and her slave-maids, a memorable scene in the

Odyssey. When Pausanias saw it, this painting was only one in a collection of works by later artists. It need not have been composed specially for this room.[28]

The plain fact is that no literary or artistic evidence survives for Polygnotos' career in the years after *c.* 465 BC. This silence is far more telling than the silence about his early local career down to *c.* 475 BC and not at all likely to be due to chance. By 465, he had become the most admired and famous artist in the entire Greek world. His paintings in Athens and at Delphi remained acknowledged wonders, visited for many centuries after his death. If any other works by him had existed after the ones at Delphi, mentions of them would have passed down to posterity. They did not, presumably because he died very soon after finishing his commission at Delphi. Consequently, he cannot have returned to Thasos after twenty years of unattested life and become a *theoros* in the year which the current interpretation of its list assigns to him: 447/6 BC.

A return by him to Thasos in older age would have been a notable fact, one which a later biographer would probably have preserved. No ancient source attests any such thing. *Theoroi*, in my view, were younger men anyway, holding the job before some of them went on later to be *archons*. Polygnotos is most unlikely to have become a *theoros* when aged sixty or more: he had been born into a rich and noble family and was an obvious candidate for office much earlier in his life. The placing of his magistracy in 447/6 BC is thus highly unlikely to be correct.[29] The reason for dating it to that year is the relation between his name in the list and the name of Antiphon son of Critobulus which is fixed securely in the adjoining column to the right. Thirty-seven years separate the two men's magistracies, and hitherto Antiphon's magistracy has been dated to 410/9 BC. That dating now struggles with what we know of Polygnotos' life.

Might Polygnotos the magistrate be a grandson, bearing the same name as the artist, his grandfather? On the hitherto-prevailing view of the artist's date and career, there is no time for such a grandson to be adult by 447/6 BC. On my view of it, which puts Polygnotos' birth about twenty-five years earlier, there is, but the notion runs into difficulties with other members of his family tree and requires him to have had a grandson as a citizen of Thasos when he already had the far grander citizenship of Athens.[30] The grandfather route is not compelling: the artist was indeed the *theoros*, but the date of his magistracy needs to change.

II

Seven years after Antiphon, Disolympios son of Theogenes is also fixed as a *theoros* in the big list. Theogenes' fame had an even bigger constituency: fans of sport throughout the Greek world. He is the supreme example of what Jacob Burckhardt named 'agonal man', the competitive type of Greek male who exemplified the sixth and early fifth centuries BC at their best and worst.[31] Theogenes was credited with the amazing total of 1300 sporting victories, including two at the Olympic games, the first one in boxing and, four years later, the second one in all-in wrestling.[32] Nobody had ever won at the Olympics before in two separate events.

Theogenes remained unbeaten at boxing for twenty-two years, a feat unmatched by any major modern boxer. In that time he averaged fifty-nine wins a year. As many of them had to be won against competitors in rounds before the final match, it has been calculated that he fought in about twelve sets of games each year, spaced out fortnightly in a sports season which lasted for six months.[33] No modern boxer has sustained such a remorseless schedule, either. He won three victories at the prestigious Pythian games at Delphi, which occurred, like the Olympics, once every four years. In the third one, nobody dared to oppose him in the final, a unique occurrence in the history of the Pythians.[34]

Theogenes' Olympic victories are securely dated by the ancient lists of Olympic victors. His boxing win occurred in 480 and his all-in wrestling win in 476 BC, the year of the great racehorse Pherenicus's third Olympic win.[35] The boxing victory was all the more impressive because it was won against another legendary fighter, Euthymus from Locri in south Italy. It interrupted Euthymus' sequence of three Olympic boxing victories and occurred in July or, more probably, August 480 BC, the very time when the vast army of Xerxes the Persian king was invading Greece and about to fight at Thermopylae against the

Spartan 300. At that pivotal moment, Greek sports fans had a very different fight in view, the boxing match of all time. Theogenes won it, but was fined heavily by the Olympic officials. He had been competing in the all-in wrestling too, but he withdrew from its final, exhausted by the boxing, and left his opponent to win in a walk-over.[36] The fine may explain why he returned in 476 at the next Olympics for the wrestling event, not the boxing, although he was at the peak of his boxing career. It may also be the reason why he took such pride in his 'twice Olympian' status. It was not just that he had won in two sets of Olympics: that feat was not unprecedented. Unlike anyone before, he had won in different events in two successive Olympic games. His win in 476 was extra-special, a victory in the very contest which he had abandoned in the previous Olympics and which had caused him to be disgraced and fined.

Theogenes achieved other remarkable records, including ten wins at nine sets of Isthmian games, one of which was truly a double victory, a win at boxing and at wrestling in one and the same year, probably also 476 BC.[37] Even so, he did not name his son 'Twice Isthmian': the Olympic victories had greater prestige. When, though, was Twice Olympian born? Unique in all Greek history, his name makes most sense if it was given in the immediate aftermath of Theogenes' second Olympic victory.[38] Either his wife was already pregnant and bore a son very soon after the games, to whom Theogenes gave this eloquent name. Or in the very heat of victory, perhaps after the sexual abstinence which competing athletes sometimes observed, Theogenes wrestled in bed with his wife and promptly fathered a son, the third jewel in his Olympic crown. If so, the boy was born in May 475 BC.

Theogenes' athletic career ended by 470 BC. Only an artful speech, delivered by the orator Dio on the island of Rhodes c. AD 90, claims that after retiring Theogenes gave himself over to community affairs and made a political enemy on Thasos.[39] This claim is absent from the other, much fuller main source for Theogenes' life, Pausanias' travel guide in Greece, and it ranks as one of Dio the orator's typical inventions, intended to make Theogenes a relevant example for his speech.[40] There is no other external evidence that Theogenes lived on after the 470s BC, let alone that he served Thasos in a political role.

For athletic reasons only, he was honoured with a statue in Thasos's city. In due course offerings were made to him there as a god, as inscriptions and even a circular stepped altar testify, found by archaeologists in the very centre of the city: the beginnings of this cult have been suggested to date to *c*. 390 BC, a time when the city received back political exiles and needed to assert a new unity.[41] It then spread most remarkably. When Pausanias, *c*. AD 130, reviewed the details of Theogenes' career, he stated that he was considered to be capable posthumously of healing the sick, not only in Thasos but in cities throughout the Greek world.[42] Worship of him as a god began some seventy years after his lifetime but it credited him with a healing power quite contrary to anything which the author of *Epidemics* 1 and 3 had been willing to countenance during his years of practice in the city.

In the theoric list Disolympios is fixed as a *theoros* on Thasos seven years after Antiphon. If Antiphon was *theoros* in 410/9 BC, as scholarly chronology has hitherto assumed, Disolympios was a *theoros* in 403/2. If he was born and named in his father's first flush of victory, in 476/5, he was aged seventy-two or seventy-three when he held his magistracy. Even if he was still alive at that advanced age, such a feat is immensely unlikely, in my view impossible. Might he have been born much later, perhaps as late as 460, when his father would have been about fifty years old? However, his unique name is less easily explained if it was given so long after the event. Some fifteen years after the Olympics, his ageing father had no need to commemorate his double victory for posterity in this unique way. It was not going to be forgotten anyway, with or without a son named after it: the Thasians had already voted Theogenes a statue in the City's main central *agora*, a unique and unmissably public honour. Theogenes is far more likely to have given the name Disolympios to his son as a celebration straight after the victory than as a commemoration many years later.

In my view, *theoroi* were appointed at a much younger age than fifty to sixty, let alone seventy-three. In the late fifth century BC, Thasians were certainly being made *theoroi* before being made *archons*, as other examples in the lists attest.[43] In the final decade of the century, the office-holding class had been thinned by years of civil strife and war. It is extremely unlikely that at this juncture the Twice Olympian, son of the richest and most famous man in Thasian history, had been made to wait for his magistracy until he was in his fifties, let alone (if alive) until his seventies. He too poses a very awkward problem for those who date Antiphon's magistracy in 410/9 BC.

The current datings of Polygnotos' and Disolympios' magistracies cause serious historical difficulty: Polygnotos, I infer, was long dead and Disolympios was far too old, if not dead too. My solution, radically, is to raise the dates of their magistracies by fifty-four years.[44] Here is why. They interrelate with the magistracy of Antiphon son of Critobulus: let us assume that this Antiphon was indeed the same man whom the doctor recorded in his text. If so, he was an adult around c. 470 BC, my date, on independent grounds, for the doctor's visit to Thasos. He also coincided with the city's new wall, the one which the doctor mentioned. Such a wall existed from the 470s until its demolition in 463 BC. Between 463 and 411 there was no new wall standing around Thasos city.

If Antiphon's magistracy is moved up to 464/3 BC, near the end of this new wall's life, Polygnotos' magistracy, fixed thirty-seven years earlier, has to move to 501/500 and Disolympios', fixed seven years later, to 457/6. These changes can be accommodated in their careers. Polygnotos would have been born c. 527, which is possible given the lack of other evidence for his early career, and would have died aged c. sixty. Disolympios, born in 476/5, was only nineteen when he held his magistracy, but, as he was the son of the most famous man on the island, he might have been given the position especially early. He was rich, very famous and an extra-special candidate to whom the usual patterns of age and career did not apply: early-career magistracies are well attested in other Greek cities' later history.

This raising of the date does not require the moving of any of the inscribed blocks or the changing of any of the columns' height and length, brilliantly reassembled by their French specialists. It merely requires a raising of the dates of each column of names in the list. At present nothing fixes them decisively to any one date rather than another. There remains the matter of the separate inscription, the one whose ruling was dated by Antiphon and his two colleagues and stood on the wall of Pythian Apollo's temple, high above the city.[45] Its ruling concerns the confiscation of named individuals' possessions and their consecration to Apollo. It has been linked, in my view wrongly, to two other Thasian inscriptions about confiscations which are usually dated to 410–408 BC, but those confiscations were not made for Apollo's benefit, and the dating, I consider, is mistaken.[46] The better answer is that the ruling inscribed on Apollo's temple was separate from them and was passed much earlier (I discuss the details in Endnote 2).[47] It was a *hados*, a word for what pleased a political, not a judicial, body, in this case an oligarchic body of 300. It belonged, I infer, in *c.* 464/3 BC, Antiphon's year of office: it recorded the oligarchy's confiscation of the properties of two men from Neapolis, among others, and, as the word 'property' readily included land, these men were house- and landowners on Thasos.[48] However, after 463/2 BC Neapolis was separated from Thasos, its mother city, after Thasos's failed revolt against the Athenians and their allies.[49] It is most unlikely that as late as *c.* 410 BC, the date conventionally given to the enactment, men of Neapolis were still allowed to own property on Thasos. The right to own a house or land in a Greek city was intimately linked with its local citizenship: to the Thasians' continuing annoyance, Neapolis, originally their colony, had been separated from their citizen body for the past fifty-three years.

'Archaism' has been recognized in the wording of this ruling, even by editors who date it after 411 BC.[50] This archaism fits neatly with a much earlier date for it, up in the 460s, but its text was inscribed on the wall in the Ionic alphabet, a script which is not known to have been used so early in Thasos's surviving public inscriptions.[51] However, another decree was inscribed in the same script beside it, suggesting that the texts were inscribed together as part of a collection of rulings relevant to Apollo.[52] Such documentary 'archives' on a temple wall are well attested in the Greek world. They were usually inscribed some while after their initial enactment: if so, their Ionic lettering is not an insuperable obstacle to dating Antiphon's year as *theoros* to the 460s. The text may have been inscribed on the wall some thirty years later than its original ratification: it was inscribed in a brief, curt summary, abbreviating the longer and more detailed original ruling.

A dating of Antiphon and the other Thasian *theoroi* about fifty-four years earlier than hitherto proposed has consequences for the dating of Thasos's *archons* too. Some of the *theoroi* reappear in the lists of *archons*, fragmentary though they are. I consider that their dates need to be raised too, about twenty years earlier than the current dating (I discuss this in Endnote 3).[53] *Theoroi*, I think, were *theoroi* when relatively young, and if they went on later to be *archons*, a gap of some 20–30 years between these people's years as *theoroi* and their years as *archons* works well enough.

These datings conflict with those currently proposed by the lists' specialists and, naturally, raise problems too, especially in later columns of the theoric list. (I discuss them in Endnote 3). However, the central difficulties remain: Polygnotos' death, Disolympios' birth, and the year of the inscription dated by Antiphon and his colleagues. My raised datings take the theoric lists back much nearer to the foundation of the city, a more natural starting point for records of its magistrates.

If my suggested re-dating of the theoric list runs into insuperable difficulties when yet more of it is fully published and analysed, my dating of the doctor still stands. I assume there were two Antiphons, the grandfather being the doctor's patient, the grandson being the magistrate. For independent reasons the completion of the doctor's text will still belong in his four years on Thasos, which had run from, say, 471/70 to 468/7.

During those years he was compiling his notes and keeping them in the medical room. I consider he worked them up into a single text while the room and its contents were still available to him, certainly before 465 BC, when the city began to be under siege. His text did not wait years to be finished. He began it with yearly constitutions for his first three years, because he wrote them up as a single block when those years had just ended. He then left in his fourth year on journeys away from Thasos. When that year finished he wrote a constitution for it too and added its case histories, surely very soon after it ended. His text shows no signs of being composed over a long period of time with a style or vocabulary which altered drastically as the years passed. We do not know what happened to the records on which the case histories were based or the practice room itself, but the doctor may have taken the records away when he left. The room, surely only rented, was probably lost during or soon after the Athenian siege in 465–463 BC.

The doctor's work on Thasos belonged to *c.* 470 BC. It overlapped, we now see, with the work of its most famous artist, busy in Athens and then at Delphi, painting at the height of his powers. It also overlapped with the island's supreme athlete, who returned home *c.* 470 BC, hung up his boxing gloves and remained at ease, I like to think, with his five-year-old son. The boy, Disolympios, was healthy enough to survive the island's diseases and grow up to hold public office. The doctor never notes that he attended him.

14

Sex and Street Life

I

Locating the doctor in time helps us to locate him more vividly in space. If he had worked on Thasos in the years from 412 to 408 or 411 to 407 BC, as generally supposed, he coincided with the building of a new wall by its oligarchic faction, but the city around him was under constant stress. Its sanctuaries and public buildings were still standing, creations of a previous generation, but there were continuing fears of a democratic counter-coup, causing people to be exiled or killed. In spring 410 BC, a visiting Athenian naval commander raised money from the island, presumably after landing on it and looting part of it. Thasians were fighting meanwhile on the mainland to try to recover their former possessions, including Neapolis, where their besieging failed. In spring 407 BC, Thasos and its city were retaken at last by an Athenian democratic force. Thasos was 'in a bad state', the historian Xenophon remarked 'because of wars and coups and famine', of which the first two, being plural, had extended over more than the most recent few months.[1] Coups and fighting cause wounds and deaths, and a food crisis compounds patterns of sickness, but the doctor's text gives no hint of such complications, relevant though they were to the public health and the cases needing his attention.

If he had arrived in 411 and left in 408/7 BC, the timing of his visit would have political implications too. His weather-surveys of each year run from one September to the next: he would have arrived in autumn 411, when Thasos's new oligarchy had broken ties with the Athenians and gone into open rebellion only two months before. He was active in oligarchic Thasos for the next three years and a bit, but he then left for Abdera, Cyzicus and coastal Thessaly when the Thasian oligarchy's circumstances worsened sharply in later 408/7. If oligarchy attracted him, its imminent fall, on this dating of his visit, did not. He was not politically inflexible: Abdera may have been democratic when he went there in 408/7 for at least four months.[2] What he disliked was political upheaval.

As I have argued, his visit belonged far earlier, in the 470s to early 460s and certainly before 463 BC, when the city's former new wall was destroyed. Its context could hardly have been more different. Since 479 BC, loss of the island's mainland territories had been averted, despite the Spartans' proposal to resettle them with Ionians from Asia. The atrocious expense of the Persian king's dinner in the city was a worry of the past. Politically, the years of the later 470s were years of oligarchic recovery and peace. Work was progessing on the big new temple for Athena on the acropolis. The doctor might even have seen some huge blocks of Thasian marble into which superb groups of sculpted figures had just been carved. One block, now in Rome and known as the Ludovisi throne, showed a goddess, probably Aphrodite, in a see-through robe being lifted by two helpers and, on a side panel, a naked flute girl with her legs crossed, sitting on a cushion: she is the first naked woman to survive in Greek sculpture on such a big scale. The other block, now in Boston and known as the Boston throne, shows a broad-winged figure, probably Eros, flanked by two older women, both clothed: he was weighing what was probably the soul of their beloved, using metal scales, now lost. After years of controversy this masterpiece is now known to be genuine, the Ludovisi throne's pair.[3] They were parts, most probably, of a big altar and are likely to have been made by Thasian sculptors, in my view so as to stand in their city. Made *c.* 470 BC, they are fine witnesses to the city's artistic *belle époque*.

His political setting also becomes consistent. When he went to Abdera in his fourth year, perhaps 468/7 BC, it may well, like Thasos, have been an oligarchy. So probably was Cyzicus and so, certainly, was Thessalian Larissa, ruled by the very few, not the many. On the coast, Thessalian Meliboea was surely oligarchic too, oligarchy being the dominant political constitution throughout Thessaly. The doctor indeed worked happily in oligarchic settings, shared at the time by his home island, Cos. His prose text is not only the first full one to survive in Greek. It has nothing to do with democracy, let alone with 'the audience of citizens that democracy creates'. It was written for future medical workers and students, whose medical knowledge it presupposed. It was not concerned to be an easy, seductive read. Its initial life was for specialized medical readers and was probably very limited.

In the years of its compilation Athens was living with the aftermath of its temples' destruction by King Xerxes and his Persian invaders. After the Persians' assault, a ground zero of desolation still disfigured its acropolis. Meanwhile, Athens' only political rival, Sparta, was, as ever, a drab cluster of villages. Not until the late 450s BC does any other text give us a glimpse of a contemporary Greek townscape. It relates to little Camarina in eastern Sicily, where Pindar effused about a 'high-branched grove of firmly standing dwellings', for a patron who had been a victor at Olympia, albeit only in the race for chariots drawn by mules. He had helped to 'glue' them together and 'bring the citizen-people from helplessness into light'.[4]

Thanks to the doctor's observations, made c. 471–467 BC, Thasos becomes the best-attested city in this hitherto-obscure period. Even its seasonal weather is known year by year, as nowhere else in ancient Greek history. It makes fascinating reading now that climate change is a major concern for historians. Here is the doctor's survey of his second year:

> early in autumn [the second week in September], storms not according to season, but suddenly wet and breaking out early among many northerly and southerly winds. These things happened until the setting of the Pleiads [c. 8 November] and also occurred during the Pleiads [when they were still visible, in September and October]. Winter northerly; many rains, violent, heavy; falls of snow; very often, intervals of clear sky. These things were all happening without the cold spells being too unseasonable. But directly after the winter solstice [24 December], when the west wind begins to blow, heavy late storms, many north winds, snow, and many rains continuously; sky tempestuous and clouded over. That persisted and did not relent until

the equinox [25 March]. Spring cold, northerly, rainy, cloudy. Summer was not too scorching; etesian winds blew continuously [winds from the north-east, still notorious as the Aegean's summer *meltemi*]. Very soon, around the time of Arcturus [at its rising, 14 September] among northerly winds many rains again.

His surveys of each year can be compared with modern records kept at the weather station on Thasos.[5] Nothing much has changed across nearly 2500 years, not even the snow in winter, but the doctor then sums up: 'the year having been wet throughout and cold and northerly, during the winter people were in most respects healthy, but in early spring, many, indeed the vast majority, were beset with illnesses.' Healthy winters, followed by sickly springs and summers, are the very opposite of what modern European cities have generally expected.

II

As the doctor walked round the city, *c.* 471–467 BC, he is likely to have benefited from the rulings of a remarkable civic decree. It had been inscribed on stone and displayed publicly, until eventually it was discarded, only to be recovered from Thasos's harbour in 1984. It is dated by its style of alphabet and lettering to the early fifth century, and it is agreed to be highly likely to belong to *c.* 470–460 BC.[6] Although its text is only partly preserved, its regulations are the earliest known example of Greek urban maintenance, a subject which was to have a long history across the next 600 years.[7] They penalize three types of abuse: encroachment, dung and sex. The doctor has never been brought into contemporary relation with their enactment, but the re-dating of his work allows it.

The decree required the *archons*, or senior magistrates, to exact specified fines from anyone who encroached on one of the city's main streets or who established a well there. The street ran in the very centre of the city, beside the 'sanctuary of the Graces', passing the very wall on which the big list of theoric magistrates was later inscribed.[8] Householders were to be responsible for seeing that the street up to their own building was clean, and officials called *epistatai*

1. Doctor Iapyx fails to heal Aeneas, whose mother, Venus, brings divine herbs instead, while his son Ascanius weeps beside him. Pompeii fresco, *c.* AD 45–60.

2. Divine Amphiaraos (*left*) heals a young patient's upper arm in his sleep, *c.* 360 BC, Oropos.

3. Papyrus-fragment of a Greek medical author praising the Hippocratic Oath and recommending it for novice students, first to second century AD.

4. Statue marking the grave of doctor Sombrotidas, *c.* 550 BC, Sicily. Right thigh inscribed 'Of Som[b]rotidas the doctor son of Mandrocles' in, probably, Samian lettering.

5. (*Left*) Silver tetradrachm, marked 'PERI-', an abbreviation of the magistrate's name, *c.* 500–475 BC, Abdera.

6. (*Right*) Silver drachm, marked 'HERO-', an abbreviation of the magistrate's name, *c.* 475–450 BC, Abdera.

7. Grave relief of a doctor, arguably a Cnidian, *c.* 480 BC.

8. Silenus striding into Thasos, sculpted in an entrance gateway, *c.* 500–490 BC.

9. (*Left*) Silver stater-coin, Silenus seizing a maenad, *c.* 513–463 BC, Thasos.

10. (*Right*) Silver tetradrachm, Heracles modelled on his archaic sculpture in a city gateway, *c.* 390–360 BC, Thasos.

11. Female nude playing the pipes, sculpted in Thasian marble, c. 470–460 BC.

12. Three Graces, sculpted to one side of a small altar, in archaic to severe style, c. 480 BC, Thasos city centre in the 'Passage of Theoroi'.

13. Banquet relief of a husband and wife, exemplifying the severe style, *c.* 470–465 BC, probably contemporary with the doctor in Thasos city.

14. Head of Philis, in fine Classical style, enlarged from her funerary *stele*, *c.* 445–440 BC. She was probably born in the year when the doctor visited Thasos.

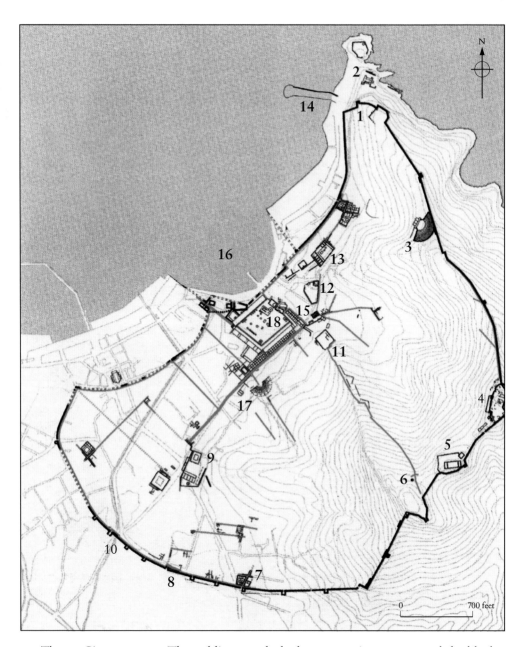

15. Thasos City *c.* 470 BC. The red lines mark the known ancient streets and the black line marks the encircling city wall, re-built in the 470s.

SANCTUARIES
1. Delian Apollo and Artemis
2. Demeter
4. Pythian Apollo
5. Athena
6. Pan
9. Heracles
11. Artemis
12. Dionysus
13. Poseidon

GATES
7. Silenus
8. Heracles
10. Zeus

FEATURES
3. Theatre
14. Harbour Mole
15. 'Passage of Theoroi'
16. Harbour
17. Main Street
18. Agora

('overseers') were to clean the street once a month, presumably with the help of slaves. They were also to clean the length of street which ran from the sanctuary of Heracles to the sea, in the south-westerly area, therefore, of the city. Here fines applied to anyone who threw out dung and caused water to flow into the street, probably by setting waterspouts into a house façade.[9]

The decree did not regulate the four-legged presence of animals, a major source of dirt and disorder in every pre-modern city. Certainly, Thasians brought them into their city, not least to kill them and offer them at one of their many altars to their gods. Instead, there were regulations against human sexual advertisement. On the length of street from the sanctuary of Heracles to the sea, 'nobody is to go up onto the roof of the public lodgings in the street so as to be seen nor is a woman to look out of the windows'. This ban has been convincingly explained as a ban on males, whether rent boys or pimps, flaunting themselves on the roofs of what were probably public brothels and on women showing themselves off provocatively at the windows to solicit sexual interest.[10] Prostitution itself was not banned here, but ostentatious flashing was: no tits at the windows, no bums on the roof.

Between several other public buildings named in the decree, 'let nobody throw out rubbish nor exhibit a prostitute [for hire].' The places named include the city's Exchange, important for its coinage, the first surviving reference to any such place in the Greek world.[11] Fines for these offences were to be paid in the city's coinage, further evidence of its currency in daily use, and are interestingly varied. Fines for encroaching on the street and digging wells were 2400 times bigger than those for each case of touting a prostitute in public near the Exchange.[12] Any woman seen at windows of the 'public buildings' and anyone making themselves seen on the roof of one would provoke a fine twelve times the fine for each day of letting water flow into the street and six times the fine for each day of not taking up fallen rubbish from it. Nonetheless, the fine for each offence of flashing was only a two-hundredth of the fine for well-digging and encroaching in the street near the centre of the city.[13]

If a woman in a house with a public brothel inside it had been seen showing herself off at a window, the male resident in the house had to pay the fine himself. The assumption here was that, as these houses were public property, there was no individual owner to be held liable for what happened inside them. The offending women were likely to be slaves, but even if not, they would have little or no personal authority over money and would be unable to pay the fine themselves. The magistrates, too, were regulated. If any overseer failed to enforce a fine, the overseers as a group had to pay a heavy penalty. Thereby, they were made collectively liable for a colleague's failing. It was a neat way of encouraging them to see that each one carried out his duty.[14]

These regulations assume that properties were often inhabited by people who were not their owners. They were rented, therefore, like the medical room which the doctor used as his base. Though he was a visitor, he too was bound by these street rules: they applied to one and all, whether Thasian citizens or not. They were not made in order to keep Thasos clean simply for the sake of appearances to visitors from abroad.[15] They were the rules of a tidy-minded oligarchy, upper-class men of taste who had a basic pride in their city and did not wish other people to turn it into a vulgar, scruffy disgrace. They are the earliest such rules known, long before those in democratic Athens, where there was eventually a proposal in the 330/20s BC that all female prostitutes who solicited in the streets would be fined and where passive homosexuals and rent boys were forbidden to enter public sanctuaries of the gods or the inner centre of the *agora*, the main gathering place.

The rules give a little more context to case histories in the doctor's text. They refer to a place beside the money exchange called the Symposion, evidently a public building in which drinking parties were held in the usual Greek style.[16] After a separate dinner elsewhere, the male guests entered to recline on couches and drink wine mixed with water, acknowledging rules set by the master of ceremonies for the evening. As the drinking progressed, they might take to courting one another or an attendant slave girl. They could reserve her for sex after she had sung or played to the guests on a lyre.[17]

In many Greek cities, honoured guests were invited to dine at public expense in a council building, the *prytaneion*, which served the council's presiding committee. A *prytaneion* is mentioned in Thasos's civic rules, but the Symposion building is separate: it is the only such building attested as yet in a Greek city. Presumably it was the building for the drinking parties which followed dinner in the *prytaneion*: at a later date, when the separate Symposion may no longer have existed, Aristotle's pupil Theophrastus, *c.* 300 BC, says that guests in the *prytaneion* were offered wine sweetened with a mixture of flour and honey, 'wonderfully pleasant', or at least so it seemed to Theophrastus.[18] In *c.* 470 BC, male diners on Thasos, like members of an orderly London club, withdrew to a separate space for their drinking after dinner, in their case a different building altogether. Their Symposion building has been interpreted as evidence 'for the degree to which commensality [sharing a table or meal] is a practice indissociable from the definition of citizenship in the early history of the [Greek] city'.[19] It is no such thing. It is immensely unlikely that most of the citizens of populous Thasos ever drank and partied inside it. It stood near the *prytaneion*, arguably on the opposite side of the main square, and was no doubt reserved for the governing body of the oligarchy, initially numbering 360, one for each day of the year, to give equal turns, a day at a time, to each member of the group. Thirty couches were the maximum for a proper symposium if it was to retain intimacy and not be turned into a mass party.[20] If the 360 were split into twelve equal groups, each to preside for one month, the members could dine thirty at a time in the *prytaneion* when on monthly duty and then go on to drink wine, perhaps already mixed with those honeyed doughballs, and recline in the Symposion as the maximum number for which a symposium could be maintained. When the governing body narrowed to 300, certainly by the early 460s BC, the dining groups became twenty-five for each month. The Symposion was by definition a building for a select and intimate group of guests. It could not possibly house all the citizens of the island, the vast majority of whom were never invited or allowed to drink inside it. In no way did it define their citizenship.

On one of the gates of their city, Thasians could see that over-sized and over-sexed figure of Silenus, Dionysus' unruly tutor, holding a wine jug up in one hand and striding forwards with a massively erect penis. On their silver coins they could see the same Silenus, ivy-wreathed, carrying off that prostrate maenad. Sex and wine were part of their city's imagery: Thasians were known for using large cups at drinking parties.[21] However, not every man in the Symposion could reckon on sex with a music girl when the party in the building finished. As guests came out, they would be susceptible to touts who were offering girls or boys for the purpose. The oligarchy's regulations tried to keep this touting out of view.

In the doctor's case histories on Thasos, one twenty-year-old, the aptly named Silenus, fell feverish after 'drinking bouts and exercises in the gym'. Another, Chaerion, was 'seized by fiery heat as a result of drinking', probably wine.[22] In the few case histories which are set elsewhere, one at nearby Abdera involved a young man who had indulged in plentiful sex and drinking, and another, at wine-rich Meliboea, involved that young man who had very frequent sex and drank very often before succumbing to a burning fever.[23] No doubt there were drunken young men in Thasos city in the 470s BC, but although Thasos, too, was rich in wine, the doctor's text records no case there which started with sex and drinking combined, let alone on the scale of those youths in Abdera and Meliboea who overdid them both and suffered the consequences.

III

The doctor's case histories take us out from his medical room and its customers into identifiable parts of the city of Thasos where he tended patients too ill to visit him: we can follow him thanks to yet more inscriptions found in the city and discoveries made by its archaeologists long after his visit. No contemporary text ever takes us on a walk round classical Athens.

Even so, we cannot always fix his whereabouts. Like historians of ancient Rome, historians of the doctor's Thasos stand 'at the entrance of a cavern of . . . unmeasured dimensions', as the great Roman historian Peter Brunt once remarked, 'much of it impenetrably dark, but here and there illuminated by a few flickering candles'.[24] The flashes of light are fascinating, but the shadows cast by the candles are part of ancient history too. The doctor mentions a gymnasium and a *palaestra*, or wrestling-space: neither has been found yet by archaeologists. Archaeology gives a sense of scale and distance to his various locations, but neither he nor it gives a full picture of the city's public buildings.

Befitting their shadowy evidence, in antiquity the streets of Thasos were not lit. Like those in other Greek cities they were not identified by signs or numbers, either.[25] The rules for street maintenance give sanctuaries as reference points, frequent orientations in Greek towns and ones which were still used until the modern era. The doctor's case histories often use them too, but they raise a delicate question: were many of his patients sleeping rough outdoors? Sometimes he specifies that patients 'were dwelling' beside a sanctuary, 'beside Heracles' or beside the 'Dionysion'. Others 'were dwelling' in the house of a named individual or even 'by' a landmark, 'the water course of Bootes' or, in young Silenus' case, either 'the plane tree' (the reading which I prefer) or the 'flat rock' (the wording of the text varied here already in the time of Galen).[26] In nearly twenty cases, however, he records that a patient 'was lying down' in or near a place: he uses a Greek verb which expresses lying on a bed or couch, a verb used in prose texts from Herodotus to the Christian gospellers.[27] Without exception, he uses a separate verb for patients who fell sick and only then 'took to their bed'.[28] Despite Thasos's rules for decency in its streets, were sick patients to be found lying all over its public spaces, suffering there for weeks on end?

If we follow him and his cases, a reasoned answer begins to become possible. A 'man from Paros', mother island of Thasos, 'was lying down' above the sanctuary of Artemis when the doctor arrived to attend him. He was seized by acute burning heat and the doctor observed him on and off during four whole months, noting his varying urine and excreta, his slight nosebleeds, his swellings and pains until, eventually, the man died.[29] The doctor did not have to walk too far to find him day after day. The sanctuary of Artemis stands beyond terraced walling on the slope above the main axial street of the city, on the doctor's right, therefore, as he walked north-east up the street with the sanctuary of the Graces on his left. The sanctuary began to be excavated in 1909 and its identity was confirmed by a find of seven female statues representing high-ranking women of Thasos from the second century BC to the second century AD: two of their statue bases bore inscribed dedications to Artemis. The goddess is called 'Artemis Hecate' in the street regulations, reflecting her relation to crossroads, Hecate's frequent sphere of activity.[30] In two inscriptions here, however, one on one of the statue-bases, she is 'Artemis Polos', Artemis the Filly. It may seem a curious title, but it refers to her as the patron of young unmarried girls. In early Greek poetry such girls were sometimes compared to young fillies, as indeed they still were by jolly English gentlemen in the twentieth century.[31] A temple of Artemis has not yet been found, and may not have existed, but since 1975 an altar has been found, as have deposits of offerings made to her and other gods. As Artemis Polos is also known on Paros, Thasos's mother island, her cult is likely to have been ancient on Thasos too, brought to it by the city's early settlers. Offerings to her show that her sanctuary was indeed in use from the seventh century BC onwards. They include notable items of female adornment, bands of gold, items carved from ivory, bracelets and so forth, fit for a rich clientele. There is also fine pottery, including painted plaques for dedication on a wall, and pieces of Parian origin, like the man from Paros whom the doctor attended.[32] He was lying 'above' the sanctuary, not 'beyond' it. He was further up the slope, therefore, on which retaining walls supported it. He was certainly not lying in the sanctuary itself.

How far towards the north of the city did the doctor sometimes go? In his survey of illnesses in his third year, he refers to Pantacles, 'who was dwelling near the sanctuary of Dionysus' and whose feverish history was known to him during a span of twenty days.[33] Dionysus was one of the city's guardian gods, and again his sanctuary has been discovered: its north-east corner has been excavated, and trial trenches have been dug elsewhere, but no temple has yet been found, and modern housing is an obstacle. An altar of archaic date was discovered with a later altar beside it, and once again sculptures confirm the sanctuary's identity: they include a fine head of Dionysus which dates to *c.* 340 BC.[34] The doctor's mention of this major sanctuary is the first textual evidence for its existence.

The sanctuary of Dionysus is a mere hundred yards or so from Artemis'. Did the doctor ever attend cases higher up on the rising ground to the east and south-east? His text does not mention anyone lying or residing high up by the temples of Pythian Apollo or Athena, both of which were on the city's acropolis. On the level ground below, however, his patients extended to the north, beyond the sanctuary of Dionysus as far as the city's new wall, as has recently become clear. He notes the case of Chaerion, the man who was gripped by fiery fever as a result of drinking.[35] His 'head became heavy and very painful; he was unable to sleep; stomach upset, with thin rather bilious matter'. On the third day, 'he was completely out of his mind' and on the sixth 'he kept on talking nonsense.' He survived a crisis on the seventh day, but relapsed on the ninth, and then on the nineteenth the fever left him 'but he had pain in the neck'. Galen's text of the *Epidemics* and one group of the text's manuscripts locate Chaerion as 'lying at Deimainetos' house', words which were therefore adopted in modern editions. However, another branch of the manuscripts located him 'near the Delian [female]'.[36] Remarkably, they have now been proved correct.

In 2002/3, a small sanctuary began to be studied just inside the most northerly stretch of Thasos's city wall by the coastal point now known as Evraiokastro. It is about a quarter of a mile north of Dionysus' sanctuary, and it opened onto an area of rock which had been artificially smoothed and levelled.[37] In October 2005, a stone block was found near its gateway and was identified as part of the temple's entrance: it was inscribed with two sets of regulations, each passed in the month of November, albeit in differing years. The first one forbade women to enter the sanctuary of 'Delian Apollo and Artemis' during the female festival, the Thesmophoria, which was held in honour of Demeter. During this celebration, widespread in the Greek world, women slept outdoors, fasted, told scurrilous stories and were free of male supervision.[38] Another much bigger sanctuary lies just outside the city wall about forty yards away to the north: it was the sanctuary of Demeter. From this nearby sanctuary, the focal point of their festival, women were not to roam and enter the smaller sanctuary inside the city wall. During their festival this smaller one, evidently the one of Delian Apollo and Artemis, was to be kept firmly shut.

In the second ruling, the penning of animals in the 'sanctuary of the Delian goddess', Artemis, was forbidden. Rules of best practice

often specified that animals were to be excluded from Greek sanctuaries, not least when there were trees and shrubs which they would
munch.[39] The ruling identifies the sanctuary by whose gate it was
displayed: it belonged to 'the Delian goddess' Artemis, and her twin
brother, Apollo, born on the island of Delos. A cult of them as the
Delians is known on Thasos's mother island, Paros. On Paros, too,
the Delian twins' sanctuary overlooked the sea and stood on the
outer edge of the settlement.[40] Once again, citizens who came from
Paros had imitated what they knew at home.

The discovery of this sanctuary shows that the correct reading of
the landmark by which Chaerion was lying sick is indeed 'the Delian'
[goddess]: he was 'lying' beside Delian Artemis' sanctuary which
overlooked the sea, beside the man-made mole which sheltered the
city's commercial harbour.[41] Galen's copy of the *Epidemics*, reading
'Deimainetos', was already in error by *c.* AD 150, an important point
for attempts elsewhere to use him to establish the true wording of the
doctor's text.

To reach Chaerion, the doctor had walked north in the city, presumably going past its gate of Hermes. It was here, perhaps, he
returned along the sea's edge, as he mentions an unnamed woman
whom he attended on 'the shore'.[42] She was three months pregnant
when a fever gripped her. 'On the seventh day, crisis', he noted, but
then after three days' respite, a relapse occurred and 'she shivered
acutely'. On the fourteenth day, after sweats and much else, she
recovered. So he had visited her during a fortnight.

Did he also attend the sick down in the south to south-west part of
the city? He locates several patients around the sanctuary of Heracles.[43] His text names three, and a fourth can be added, solving another
long-running puzzle. Heracles' image is sculpted on an inner face of
the gateway in the southern stretch of the new city wall, one which he
shared with Dionysus. It shows him with a lion-skin covering his
head, its paws hanging on his chest, himself kneeling on his right leg
and drawing a bow with his right arm. This very image was to be used
as a successor to Silenus' image on the city's silver coinage from the
early fourth century BC onwards.[44] When the doctor visited, it was
recently sculpted, a fine work of the early fifth century BC. Together
with Dionysus, Heracles was a guardian of Thasos.

Heracles had a rectangular shrine of his own overlooking this city gate, about 260 metres from it, one which was found by excavations in the early 1930s.[45] About nine metres wide and twelve metres long, it had a single chamber and was almost certainly the shrine which the doctor saw. Beside its east side ran a long colonnade, also visible when he visited, on which various works of sculpture were affixed. The temple's entrance looked out onto an altar and, beyond, to a hall building which was probably used for eating and dining after rites in honour of Heracles. They too had rules of best practice. As an inscription of the mid-fifth century BC, found by the *agora*, specifies, 'For Thasian Heracles a goat is not permitted, nor a pig [as offerings]; nor are women permitted [as participants in his rites]'.[46] This ruling probably restated existing prohibitions. The exclusion of women from rites for Heracles is not unprecedented in the Greek world, but the text's insistence that Heracles is 'Thasian Heracles' appears to mark him out, probably because his cult had ancient links with Phoenicians, though archaeologists have yet to find them.

There was nothing marginal about the site of his sanctuary, nor was it in a thinly populated sector of the city. The doctor used it as a landmark, as did the city's rules for urban propriety. He mentions Cratistonax, who was residing near to (the sanctuary of) Heracles and gives case notes on Cleanactides, who 'was lying above' the sanctuary of Heracles and was 'seized' by a fiery fever which came and went erratically during eighty-four days. He also gives notes on Pythion, who 'was lying up above the sanctuary of Heracles'.[47] Pythion had toiled hard and been exhausted and negligent in his style of life, factors which did him no favours: 'intense shivering, an acute fever seized him', persisting for ten days, whereupon he died. The doctor's noting of two patients as 'above' the sanctuary of Heracles falls into place on the site: they lived on the south-west slopes of the acropolis hill, which lies a little further back. They, too, were not 'lying' in the sanctuary itself.

The doctor also attended the wife of a man whose name, variously copied in the manuscripts, was, I will argue, Dēialkos.[48] She 'was lying' near a place which the main manuscripts transcribe as 'the flat plain' (*to leion*). It is a very vague location, and may be an error, influenced by a later case in which an unnamed woman was located on the 'flat plain, near the possessions of Pylades'. The wife of Dēialkos is not located so precisely. Her husband owned a garden or orchard, no doubt a big one, as another case reveals, but in my view the text's word *leiou* is a miscopied abbreviation of 'Herak-*leiou*', the sanctuary of Heracles. A fanciful alternative later entered the text of Galen's commentary and survived in its Arabic translation: *helieion*, or 'temple of the sun'. There is no evidence whatsoever for a cult of the sun on Thasos in the fifth century BC and there is very little likelihood that one existed then.[49]

A sanctuary of Hera evidently did, because another female patient resided beside it. An image of Hera with her husband, Zeus, was eventually sculpted on the inner wall of an important gate in the south wall of the city.[50] Her cult was very prominent, therefore, in Thasos and is attested in personal names like Her-agoras or Her-ophon, so much so that the island has been called a 'hotspot' of Hera-names.[51] Archaeology, however, has yet to locate her main cult. About a hundred yards to the north-west of Dionysus' sanctuary, another sanctuary had an altar to one side of its entrance inscribed with the rule, 'To Hera of the harbour it is not permitted to sacrifice a goat'. Hera 'of the harbour', therefore, was honoured there, but the sanctuary was primarily one of Poseidon: it was probably not the one beside which the doctor treated his female patient.[52]

Not every sanctuary named by the doctor's case histories, therefore, has been found. There are two other missing examples, reminders of the limits of archaeology in populated and silted modern sites. By a 'sanctuary of [the goddess] Earth' he noted another Pythion, the one who began to tremble in his hands and then to be beset by an acute fever.[53] On the fortieth day he had a supplementary problem: 'rotting', or suppuration, became visible on his backside. No personal name, dedication or inscription has yet given any hint of a sanctuary of mother Earth on Thasos. So far, its location is unknown. So is the whereabouts of another landmark, transmitted by copyists as the Chief Leader (*Archēgetēs*). Near him (or by his shrine) the wife of Epicrates 'was lying'.[54] She was about to give birth when 'shivering took strong hold of her: she was not warmed, so they say'. Here, the doctor relied on hearsay because he had not been present at the birth which followed the start of her shivering. He came later, as usual, and noted the woman's condition on and off until the 'fortieth day, when she vomited small amounts of bilious matter'. On the eightieth day the definitive crisis came and she recovered.

If correctly transmitted in the text, *Archēgetēs* is not a personal name on Thasos. It is a word used for the founder of a new settlement, either a guiding god or a mortal leader.[55] There are possible candidates, including one of the mortal founders whom the Thasians honoured, but in my view the Chief Leader was a god. The obvious candidate is Apollo, who was credited with encouraging the foundation of Thasos in an oracle delivered at Delphi. He was also the Delian god, and was honoured in Thasos with that title.[56] On Delos his priest was considered to have been the father of mortal Thasos, the very person, to us legendary, after whom the island was supposedly named. This priest of Apollo had his own sanctuary on Delos, called the Archēgesion, no less.[57] The 'wife of Epicrates' was probably lying sick near a monument or a sanctuary of Apollo the Chief Leader, not on the acropolis but in the lower city; if so, archaeologists have yet to find it.

The doctor gives two other locations of special interest. One is for a young man and, separately, a woman, both unnamed: it is the Market Place of Lies.[58] The young man had been 'seized' by a fiery fever after 'cramps and hard workouts and bouts of running contrary to his usual habit'. The woman had just given birth 'very laboriously' to her first child, a son, when fever seized her too. The 'Lies' here were the lies of traders and sellers and their meeting place was the commercial marketplace, where they talked up their wares. This blunt view of business and its relationship to lying was still present in Aristotle's mind many years later when he recommended that the ideal Greek city should have a free meeting place for gentlemen and a separate marketplace for vulgar traders, farmers and craftsmen. In Thessaly, he knew, free meeting places for gentlemen existed.[59] Our doctor visited Thessaly and worked in lordly Larissa, where archaeologists indeed claim to have found two civic meeting places, one for free gentlemen, the other for 'lying' traders on the far bank of the intervening River Peneus.[60]

In Thasos too the noble ruling class had no illusions about business and its ethics, banishing it to a place of lies, away from the main city's meeting place, which was intended for honest purposes only. The

location of these two areas can be made more precise. The Market of
Lies is unlikely to have included the official money exchange, a beacon
of honest standards, which stood near the Symposion and the council
building. The gentlemanly meeting place was up near the sanctuary of
the Graces, on the site later known to be the *agora*, where recent excavations have found traces of archaic buildings: the council building
was probably in the north-east corner, whereas the Symposion and the
money exchange were perhaps in the south-west. The Market of Lies
was separate, perhaps down by the harbour, served by the length of
street which ran from the shrine of Heracles to the sea, the street on
which brothels, but not flashing, were permitted: it has not yet been
discovered. There, goods for sale would include male and female
slaves, talked up like other items on offer, animals, dodgy coinage and
farmers' produce, seldom quite as ripe as sellers make it seem.

The mendacity of markets plays a notable role in Herodotus' Histories. He claims that after conquering Sardis in western Asia in 546
BC, the Persian king, Cyrus, received an envoy from Sparta who
arrived with a 'statement of the Spartans', formally voted, therefore,
by the Spartan assembly: Cyrus was to 'do no harm to any city of
Greece'.[61] So Cyrus asked who these Spartans were and went on to
say that he had 'never yet feared the sort of men who have a place
designated in the middle of their city to which they come together
and then swear oaths and cheat one another'.[62] Here, Herodotus and
his informants credited Cyrus with a view which exceeded even the
Thasian and Thessalian gentlemen's disdain for business. It was appro-
priately regal. 'If I remain healthy,' Cyrus was said to have added, 'they
will have their own sufferings to talk about, not those of the Ionian
Greeks.' Herodotus explains that Cyrus made this threat 'referring to
all the Greeks because they have set up markets and practise buying
and selling, whereas the Persians themselves do not have a custom of
using markets; nor do they have any marketplace whatsoever'. The
Persians' disdain for marketplaces became a truism, repeated by later
Greek historians. Herodotus linked it to the first warning from a Persian king that one day he hoped to invade the Greek world, the place
where men met and cheated one another in markets.

It is one thing to disdain the ethics of business, another to renounce
business itself. No doubt Thasian nobles expected their own slaves

and agents to take each wine harvest from their personal estates and sell it on their behalf in the Market of Lies. Nonetheless, there was more than mere trade to their way of life. In his survey of his third year on Thasos, the doctor refers to the two 'brothers of Epigenes' who fell sick at the very same hour. They both reached a final crisis and survived on the seventeenth day, but only after relapsing and recovering at different times. They 'were lying' near what all but one of our manuscripts represent as 'the theatre'.[63]

If correctly transmitted, this landmark is of exceptional interest: it becomes the first textual reference to a theatre in the entire Greek world. Was it present in the doctor's original text? Galen's comments on the passage are known most fully in the text of his commentary translated into Arabic: he remarked that the word *theatron* indeed stood in the old copies of the *Epidemics* text, but a recent commentator, Artemidorus Capito, active *c.* AD 120, preferred a different word, evidently *theretron*.[64] One of the early medieval Greek manuscripts of the *Epidemics* even contains this very rare word.[65] Galen understood it to mean a 'place for summer', as if second homes for the summer were owned by Thasos's upper class away from the heat of the city centre. However, he dismissed the word as a footling change to the text, irrelevant to the topic of medicine: he scorned Capito for trying to explain it as a threshing floor.[66] It is irrelevant to what the doctor had actually written. He never left the city for any case which he describes in his text, let alone to visit summer country homes or threshing floors. He had indeed mentioned a theatre, present in the old manuscripts of his text, as Galen attests.

The world's first theatre is associated, rightly, with Athens, where tragedies were invented in the late sixth century BC. Like Athens, Thasos had very strong links with Dionysus, one of its protectors, the god who was honoured with theatre festivals and choral songs. Archaeologists date the earliest excavated Greek theatre to *c.* 460 BC, found in Attica outside Athens city, but the doctor's text guarantees an earlier one much further away.[67] A theatre existed on Thasos by *c.* 470 BC.

Perhaps in the 470s this theatre was a venue for choral singing and dances which honoured Dionysus. However, tragic dramas were already famous in Athens. Plays by the great Aeschylus and his contemporaries sometimes travelled to be restaged outside Athens, where audiences or benefactors were willing to pay their main actors. In 472 BC, Aeschylus had delighted the Athenians with his play *The Persians*, in which he presented the Persian king, Xerxes, returning humiliated after his defeat in Greece. The play is then said to have been taken west and performed in Sicily at the invitation of Hieron, ruler of Syracuse and owner of the great racehorse Pherenicus.[68] In 472, Hieron was still suffering from those pains which Pindar had hoped to alleviate by an ode. Perhaps *The Persians* gave him some relief.

I like to think that the play had another foreign performance. In 481 BC the Thasians had entertained at vast expense that very same Xerxes when his huge army travelled west to Greece. There was no apter play for them to restage in their theatre than Aeschylus' *Persians* in which the shattered Xerxes comes back to his palace at Susa, having lost his entire force. If they restaged it in 472/1 BC, after its first performance in Athens, the doctor might even have been in the theatre to see it.

The site of Thasos's later theatre is known: it stands on the hillside overlooking the main city, from which paths lead on up to the acropolis and the temples of Apollo and Athena.[69] It is superbly sited so as to give its audiences a view out over the sea, the glistening backdrop to performances. It was surely already the site of the theatre which the doctor visited in the city. In the 470s and early 460s Thasos was ruled by a noble oligarchy. It had an active theatre culture, but that culture had nothing locally to do with democracy.

15

Patients of Quality

I

From the two brothers near the theatre to the man from Paros above the sanctuary of Artemis, the doctor's text and modern archaeology throw fascinating, but flickering, light on the city of Thasos and its citizens' behaviour. Archaeology fixes some of the locations he mentions and helps us to track him in the city, but his text mentions much which archaeology has not yet been able to find. Only once might he seem to have gone further afield than the city's perimeter wall. He treated that unnamed woman who had just given birth and was lying sick by 'the cold water': three days after giving birth to a daughter, she began to shiver, having been feverish and off her food for a long while before going into labour.[1] On the eighth day she was struck with delirium, but it did not last. On the fortieth day coughing began, but stopped on the sixtieth. On the eightieth day she died. In general, he remarks, she had been 'off her food, dispirited; bouts of anger and of taking things badly'. 'In respect of [her] thinking, melancholic', he notes most strikingly.

During nearly twelve weeks the doctor and his assistants attended and observed this woman's case, spending time on it which was far more prolonged than the rapid throughput imposed on a modern general practitioner. Where did he go for this observational marathon? The Thasians had a revered source of cold water in the hills beyond their city, but there were springs in the city area too. One of them was noted near the shore in the south-west section of the city by Alexander Conze, a visiting German student, when he landed on the island in May 1858.[2] Perhaps it was the very spring by which the doctor had attended this poor lady. If so, it too was within easy reach of his urban base. He stayed firmly within the city, never straying out into the countryside beyond, let alone to the rough landscapes of the island's centre or parts of the south, to the mines on the north-east coast or the marble quarries, manned by slaves.

He is also silent about one important class of patients: the elderly. Only once, in his survey of the third year, does he happen to mention a specific condition of the over-sixties, as he calls them: during the red-rash fever they showed little scars initially on their heads. Otherwise, he never describes any of his patients as aged or old. Almost all of his female patients, twenty-four of them, were no such thing. Fourteen of the seventeen individual females in his case histories were, or had just been, pregnant or still had periods: another one was still a young girl. Of the other two, one was surely of childbearing age, but the other, the 'wife of Dēialkos', might perhaps have been old, maybe about fifty, as we will see. However, her husband was exceptionally eminent and the doctor already attended a patient in his gardens: she would be an understandable exception to the usual pattern. In the third constitution he remarks that 'many' women were ill, though fewer than men and fewer of them died. He then remarks their difficulties in childbirth and their periods: once again he is not thinking of old women at all. When he refers to males as 'older' (*presbyteroi*), he always uses the term relatively, to contrast them with boys and children. He does not use it to mean 'elderly' as a class. Their absence from almost all forty-two case histories and the twenty-six references to named individuals in the third survey is very striking. He and his helpers might note a particular effect of a disease on the elderly in general (those over-sixties with red-rash fever), but, when he himself chose whom to see, he never went out day after day to observe and attend an old person. Did he reckon, like the goddess Aphrodite in her hymn, that they were not worth the effort of striving to keep alive?

Inside the city there is still the problem of patients who were 'lying down', even if they were not old and near death. Some of them were 'lying down' near the sanctuary of Heracles, say, or of Dionysus or the theatre, sites central to civic life: as big spaces, they had room enough for patients to bed down. Were there external reasons why they were doing so? In his masterly account of the great plague in Athens, the historian Thucydides remarks how Athenians came into the city from the country and were obliged in the heat of summer to live in 'stifling shacks' and even in the sanctuaries.[3] However, the reason why the country-dwellers had crowded into the city in 430 BC was that the Attic countryside was being ravaged by Spartan invaders. Even if the doctor had been at work in 411–407 BC, there was no comparable annual ravaging of the countryside around Thasos's city by hostile troops. In the late 470s there was no war or siege at all. His patients had not crowded into the city in a crisis and bedded down wherever they could find space.

To put them into sharper focus, a third source of light is needed, the inscriptions, once again, which list Thasos's senior magistrates and their fathers' names. About 50 per cent of the names in Thasos in *Epidemics* 1 and 3 also occur in the big inscribed lists of the city's magistrates: are they the same people, members of the city's small governing class? Some of these names need to be considered with caution: they are quite common in the Greek world. Others are attested only on Thasos, some predominantly on Thasos, although some of these may have also been current in lower social levels, ignored by the lists of magistrates. Others, however, like Cleanactides or Cratistonax, are very likely to have been names in the upper class only, because of their hints of clannishness (-*ides*) and social pre-eminence (-*anax*, or 'king').[4]

In the list of the city's *theoroi*, in the eleven years after Antiphon son of Critobulus, five names match names of the doctor's patients: four of them are the fathers of magistrates, one a magistrate himself.[5] The overlap may seem notable, but as three magistrates were inscribed each year together with their fathers' names, there were sixty-six names in these years alone: the proportion is not so striking and four of the names are not unduly rare. At least it implies that several more people whom the doctor went out and treated may have been from families or households in the city's highest society. Can this implication be strengthened?

Two earlier names in the theoric list have been recognized as particularly relevant to this question. The doctor visited a garden whose owner's name was transmitted as 'Dealkes' or 'Dealdes' and attended someone 'lying' there.[6] This patient was presumably a slave. It was in his fourth year that he also treated the 'wife of Dealkes', who lived by the 'plain' (*leiou*) or, in my view, by the sanctuary of Heracles ([*Herak-*]*leiou*). An acute fever 'seized' her after 'grief': she died on the twenty-first day.[7] In each case manuscripts transmitted the name in various ways, but rather than follow their perplexity it is sounder for historians to follow solidly attested names known independently in Thasian families. Main sources of them are the magistrates' lists, though they modernized names' original spellings. The name Dēialkos recurs prominently there, probably in one and the same family, across nearly two centuries, including four in the big theoric list before 400 BC.[8] Both the name and the high status of the doctor's patron can therefore be restored with some confidence: he was one of the city's very top people, whose garden was very sizeable. Possibly he was Dēialkos, the father of Pythonax, a *theoros* three years after Antiphon. If so, his wife was fifty or so when the doctor attended her, uniquely old among his female patients.[9]

The doctor also attended a man whose name the manuscripts transmit as 'Daitharses', 'Daiphrases' or 'Dēithrases'. Dēithrases is a name specific to Thasos and, so far, very rare even there: in the doctor's clientele it should be corrected to its historically attested spelling.[10] In 1893, this client was acutely connected to a Dēithrases in the list of *theoroi*. The doctor refers to this upper-class family twice. He attended Dēithrases' son and also his virgin daughter, both in the third year of his time in Thasos. He observed the girl at an exact moment in her life: she had a nosebleed and at the same time a period, her first ever.[11] This double expulsion of blood helped her to survive the year's outbreak of burning fever, whereas pregnant women, unable to bleed as she did, succumbed to it, the doctor noted, and died.

Her first period is an important clue: she was in her fourteenth year, the usual age for this physical change in this era.[12] Her double bleeding occurred in the doctor's third season on Thasos. On my dating of his work, that year was somewhere between, say, 473/2 and 469/8 BC: the girl, bleeding at both ends, had therefore been born some time between 487/6 and 483/2. On the traditional chronology of the list, a Dēithrases son of Heragoras was a *theoros* in 453/2.[13] If *theoroi* were elderly persons, as that chronology supposes, he cannot possibly have fathered a daughter as late as 422/1, ready to be thirteen and bleed in 409/8, the date considered to be the doctor's third year on the island. A like-named nephew has to be postulated so as to bridge the gap. On my higher chronology of the list, there may be no need to invent one. Dēithrases son of Heragoras was a *theoros* in 506. If he was born *c.* 536, he could indeed have fathered a daughter when in his early fifties, one born between 487/6 and 483/2 and ready to start bleeding in her fourteenth year between 473/2 and 469/8. The doctor noted that Dēithrases also had a son who survived the same burning fever.[14] The boy might have been a little older than his sister, born perhaps *c.* 490.

Whatever these patients' exact family relationships to the like-named *theoroi*, they are highly likely to belong in the *theoroi*'s family clusters. The list of *theoroi* places them firmly in the city's exclusive governing class. Though 'lying down' by the sanctuary of Heracles, the wife of grand Dēialkos was certainly not sleeping rough when the doctor attended her. Her family had a proper house, evidently near the sanctuary, and if she 'lay down', she lay down in it as a resident, especially if, as I suggest, she was aged about fifty. The doctor's word for lying down, therefore, does not entail that someone was sleeping rough or bedding down in a public space. It usually has a Shakespear-ean sense of 'lying': residing or lodging.[15]

Other females whom he treated reinforce this social context. Out of the thirteen he identifies by a personal name, he refers to eleven indirectly, by the name of their nearest male kin, the 'wife of Epic-rates', or, if unmarried, 'daughter of Euryanax'.[16] This oblique mode of address has been definitively studied in the lawcourt speeches of orators in fourth-century BC Athens.[17] There, too, orators referred to females by elaborate family periphrases, sometimes using the name of a woman's brother to identify her. Of course, a husband might refer in public to his wife by her personal name, but a polite outsider would not. Orators called a female by her own name only if she was dead or belonged in their opponent's circle, or was, by implication, a prostitute. She was named only on her tombstone, if she had one, or in an inscription recording a dedication by her to a god. This general rule became less prevalent in texts after the classical age, but the doc-tor on Thasos is cardinal evidence for its observance outside Athens as early as the 470s BC.

The female patients named in this way were married women or daughters in respectable households: some of their fathers' or husbands' names are among those attested in the upper reaches of Thasian society.[18] When they were 'lying down', these women of distinction, like the wife of Dēialkos, were not bedded down in the open air. They too were residing in households, often beside or near a landmark which the doctor specifies.

Nonetheless, the doctor-author was not a doctor for the rich only. He also attended slaves in their households, precious assets whom the houseowners would want to keep alive.[19] He calls most of the other eleven females in his text simply 'the young girl who . . .' or 'the woman who . . .'[20] One of them is specifically a slave, and as three others are identified as being in another man's household, they are probably slaves too. Five of the unnamed eleven were pregnant and they too are likely to be slaves, rather than daughters of citizen parentage who had become pregnant after premarital sex. Free-born girls, even when shamefully pregnant, would still have been identified by a family member's name.

Three times, at most, the doctor's text refers to a woman as if by her own name. One may be the patient who had had headaches for a long time, but the manuscripts differ over the personal name, either Philistis, a female and perhaps a sex worker, or, with Galen, Philistes, a man (the reading I prefer). Another is Kratie, but as she was living 'by Xenophanes' she was probably a household slave.[21] The other is Melidia, who lived near Hera's sanctuary.[22] She 'began to be afflicted strongly with pain of the head, neck and chest'. On the sixth day she was observed to be 'comatose, nauseous and shivering; redness around the jaws; she was slightly out of her mind'. On the seventh day, she sweated and the fever left, only to reappear in a relapse. Again she sweated and on the eleventh day the fever reached a crisis: this time she survived. She was not, then, named personally because she was dead: in fact, the doctor never refers to a female patient by her own name in any case which proved fatal. Melidia had no husband and perhaps no kin either, so she too might have been a sex worker whom it was permissible to call by her personal name. However, her name may have been incorrectly transmitted in manuscripts of the text. 'Melidia' already stood in the copy used by Galen in the second century AD, but error could begin early.[23] There is no support in the manuscripts, but perhaps we should read 'Mendaia', meaning the 'woman from Mende', a wine-producing city to the west with which Thasos had frequent contact. Elsewhere in the Epidemic books, individual foreigners are identified by their ethnic, the adjective which denotes their city of origin.[24] On any view this female is evidence that the doctor sometimes treated people outside the households of the upper class.

Five other Thasian patients are left unnamed, implying they too were of lowly status, and three of them are related to a place. One was the woman who was 'lying down' on the shore, where she had given birth; another, the young boy, perhaps a slave, who was lying down 'on', explicitly, the Market Place of Lies, and the third, an unnamed woman who, on the likeliest reading of the text, was doing likewise in the same place.[25] As these people were lowly people, they were probably sleeping for free by the sea or on that hotbed of ungentlemanly behaviour, the Market Place of Lies. The centre of the city was quite another matter. Unlike modern Havana or Los Angeles, it was not a city in whose open spaces people slept rough wherever they could. Thasos's magistrates would have had them removed, like disagreeable dung in the main street.

This understanding of the phrase 'lying down' reflects yet again on archaeologists' continuing attempts to illumine the city. Residents with grand names like Cleanactides 'up beyond the sanctuary of Her-acles' were not living in humble houses, but, so far, traces of only one major house occupied in the fifth century BC have been excavated and published, down in the south of the city, just south of the temple of Heracles, where it was part of a bigger block of dwellings.[26] The doctor's text implies that others existed, apparently as free-standing addresses, but it has yet to be amplified by well-placed finds on site.

His women, at least, can be linked to finds in the city's sanctuaries. Those to whom he refers by the names of their male kin relate to female figurines which were vowed to the goddess Artemis in the early fifth century BC. These votive items have been convincingly interpreted as representations of the very types of female who offered them.[27] They show women wearing the usual tunic and over-mantle of respectable, body-muffling Greek female dress. On top, their head-dresses, veils and diadems have been plausibly understood as markers of their status: unmarried girls wear diadems only, but married ladies wear veils too. The doctor's 'wife of Dēialkos' or 'wife of Dromiades' were evidently used to wearing a veil.

Like the doctor's style of address, these figurines reflect a hierarchy of gender in Thasian families. A fine marble relief from Thasos expresses it well.[28] It shows a male figure, tidily bearded, reclining on a couch and holding out a shallow cup in his right hand. On the left side a smaller male is shown standing naked, evidently a slave, as he is drawing wine from a big bowl on a stand to the left of the couch. Behind the reclining man, to his right, the much smaller figure of a woman, evidently his wife, opens with her right hand a bottle of scent. She holds it upright in a container which resembles the alabaster containers found in Greek females' graves. She is represented as about a third of her husband's size and is seated on a chair with her feet on a stool. A mirror is hanging on the wall above her: she is evidently shown in a separate women's room in the household, not being able to attend her husband's drinking on a couch.

Such 'banquet reliefs' are familiar as a general type in Thasos's sculpture. They are generally interpreted in relation to a cult of a hero, as if the male shown centrally is a hero, perhaps an ancestor of the family honouring him.[29] This relief was found in the centre of the city and may have once stood in a hero-shrine, but no worshippers are shown, and the items evoke an earthly reality which the family concerned lived out in their own lives. A war helmet hangs on the wall above him, as does a distinctive crescent-shaped shield: it is a clear example of a *pelte*, a type of shield which Greeks copied from Thracian warriors.[30] As a Thasian, the dead man had lived, and no doubt fought, among Thracians who were a constant presence on the mainland and on Thasos itself. Beneath his couch and his wife's chair are other everyday items: for him, a well-sculpted hunting dog, its nose to the ground, and for her, a partridge, a frequent pet in Greek households. Warfare and hunting evoke his way of life, whereas scent and a pet bird evoke hers. The relief is a well-cut work of the Severe style, to be dated *c.* 470 BC, the very time when the doctor was active in such households, working for upper-class clients like Antiphon the feverish son of Critobulus or Dēithrases, father of the daughter who bled for the first time from both ends of her body at once. Traces of paint on various items in it show that its details, including the partridge, were once painted. It may well be the work of pupils of the great Polygnotos, who left Thasos for Athens in the mid-470s BC.

II

Why did the doctor record so many of his patient's names with the addresses where they 'lay down' and resided? He was not compiling a medical version of satnav for the benefit of himself and his followers.[31] Nor was he pinpointing them because he considered their exact location to be relevant to their condition. Seasons and the weather, not streets and hills, were the factors which his text related to illness. Were the patients' names included as an *aide-mémoire* for students? They would help them to remember particular cases, but not every patient or place is named. I think the details were recorded to help the doctor himself. They enabled him to arrange in alphabetic order the tablets on which his notes originally stood. They were the database which he had to integrate into a text.

One very important person was not named and identified: the doctor-author himself. He uses the first person singular, but gives no clue to his identity: on the original roll of his text, he was probably not named, either. In antiquity, scholars assumed, surely rightly, that he was from Cos. This text turns out to be the earliest datable text which we have by a Coan doctor. Might its author be Hippocrates himself?

In antiquity this attribution was widely endorsed, and until relatively recently accepted by most modern scholars.[32] In the second century AD Galen accepted it, but did not extend it indiscriminately to the other Epidemic books. The problem is that Hippocrates' name became attached to many other texts in the Corpus and accepted there in antiquity, even though most, perhaps all, of them are not by him at all. Hence moderns refer now to the 'Hippocrates problem'. It has attracted many attempts at an answer, but as there is no external evidence to resolve it, none has succeeded in convincing its critics.[33] The result has been general scholarly scepticism that the problem can ever be resolved. Why are *Epidemics* 1 and 3 any more likely than other texts now in the Corpus to be the real Hippocrates' work?

One apparent obstacle needs to be removed. As we have seen, an ancient biographer of Hippocrates alleged that he was born in the year 460 BC, but this date and the claim that it had been preserved in city archives on Cos are not historically founded. As they are worthless, Hippocrates could perfectly well have been born in *c.* 500 BC and lived on into the 430s, the dramatic date of Plato's references to him as an authoritative doctor, an Asclepiad from Cos. During a lifespan from *c.* 500 to *c.* 430 BC, Hippocrates could have written the Epidemic text in his early thirties and lived on to be the person Plato evokes.

The result of our study of his biographies and so-called letters has been to endorse the reduction of facts known about him to almost none. There is nothing in them to support or challenge his author-ship of any medical text of fifth-century BC date, *Epidemics* 1 and 3 among them. However, from Plato onwards, references to Hippocrates' opinions occur in ancient texts, seeming to offer a control. In his fine dialogue, the *Phaedrus*, Plato makes Socrates discuss the method to be followed in a well-founded inquiry, but since antiquity the mean-ing of his words has been disputed.[34] Socrates puts to Phaedrus that understanding of the soul requires understanding of 'the nature of the whole', which in context is best understood as the whole soul. Phaedrus assents, remarking that Hippocrates considers that under-standing of the body requires understanding of 'the nature of the whole', the whole body. This rather general principle can fit several works in the Hippocratic Corpus, including *Epidemics* 1 and 3. If Plato based it on reading, not on hearsay, it has some relevance for Hippocrates' possible authorship as it helps to disqualify several texts supposedly by him. However, being so general, it does much less to qualify any particular text as one by Hippocrates himself.

A more promising reference is that one attributed to Ctesias, the doctor from Cnidos who was active at the Persian court at the end of the fifth century BC. According to Galen, as we have seen, he crit-icized Hippocrates' advice on how best to reset a thigh dislocated at the hip. Such advice can be found in the text *On Fractures*, surviving in the Hippocratic Corpus and linked to the one *On Joints*. However, Galen, as elsewhere, may have supplied the mention of Hippocrates, whereas Ctesias may only have been criticizing the procedure, one which other doctors also used. Sceptics remain unconvinced about Hippocrates' authorship of these two texts.[35]

Other references in antiquity to what Hippocrates said are even less compelling, especially when read in their context. They are not exact quotations of his own words and are not made in the fifth century BC. In 1890, a fragmentary papyrus text was discovered and aroused scholarly excitement because it claimed direct awareness both of what Aristotle, c. 330 BC, ascribed to Hippocrates and also of Hippocrates' very words: it even expounded 'what Hippocrates says'.[36] However, almost none of what it presents about him as ascribed to Aristotle matches any surviving Hippocratic text, except parts of an unlikely candidate, On Breaths. Its own presentation of 'what Hippocrates says' survives fragmentarily, but at least it mentions bile, phlegm and heat. It could fit Epidemics 1 and 3, but it also fits other Hippocratic texts: there is the further problem that the author of the papyrus text was writing as late as the first century AD.[37] His view of 'what Hippocrates says' and what Aristotle also ascribed to him was based on whatever texts had been attributed by then to Aristotle and to Hippocrates. He quotes them confidently, choosing them out of potentially many others, but sceptics can well contend that he had no adequate idea of what Hippocrates himself had really written.

Why insist that he wrote any medical text at all? Here, it is worth comparing antiquity's most famous healer, Jesus. Like our Epidemic doctor, he aimed to help, but unlike our doctor he did not always 'do no harm' (those poor Gadarene swine . . .). He also differed in his output. At least in the Gospels he never tells his followers to write down anything he says. He was not credited with writing a text himself. Correctly, therefore, a 'Jesus Corpus' was never ascribed to him. Hippocrates, by contrast, became credited with an ever-growing one, presumably because he was considered to have written a text or texts of his own. At least one by him ought, then, to survive among all the texts now gathered in the Corpus, and my favoured candidate is Epidemics 1 and 3. It has turned out to be the earliest datable one there, others being placed on grounds of thought or language no earlier than c. 450–420 BC. Unlike many of them it was not enlarged by anonymous additions over time. Its original order has been slightly dislocated: what is now its ending was probably transposed from a slightly earlier position in the text and the work, originally one, became divided rather clumsily into the two separate books we now have.[38] This division and dislocation were consequences of later copyists' errors, not of actual additions and insertions by later readers, updating what we now read. The words in the text remained one author's work. It also inspired heirs to write separate Epidemic texts of their own. No other 'Hippocratic' text had this sort of impact, and one reason might be that its direct heirs were convinced that it was indeed Hippocrates's own. Without decisive contemporary evidence, the case obviously remains open, but meanwhile I suspect, no more, that in its terse style and exact language, in its observations of people like the young Meliboean or the young man of Abdera, heated from sex and drinking, or of the many Thasians, from the Market Place of Lies to the hillside theatre, from the unnamed woman near the 'property of Pylades' to the virgin daughter of Euryanax, both of them despondent in the course of illness, Hippocrates may be addressing us directly.

Whoever the author is, his case histories and year-surveys treat the city of Thasos as a self-contained item. He names one or two individuals from elsewhere, the Parian or the Clazomenian, but they are residing at fixed points. He never comments on the arrivals of traders or other temporary visitors, let alone on their timing, origins and peaks and troughs. He does not even comment on arrivals to the city from the rest of the island. He never notes its contacts with a wider world, even though, as we have learned to our cost, they can be crucial for a city's public health. Consideration of his thinking will help to explain his silence.

PART THREE

The Doctor's Mind

Old men mostly are less ill than young men, but such ill-
nesses as they have, mostly die with them.

Hippocratic *Aphorisms* 2.39

If you want to stop a monthly period in a woman, apply the
biggest possible cupping glass to her breasts.

Hippocratic *Aphorisms* 5.50

The attempt to recreate Thucydides' experience should (and
will) never be dropped.

H. T. Wade-Gery, *Oxford Classical Dictionary s.v.*
Thucydides (Oxford, 1949, and subsequent editions)

16

By the Bedside

If the travelling doctor-author of *Epidemics* 1 and 3 was indeed Hippocrates, he was, as Plato testifies, an Asclepiad. That honourable name might be borne by associates of the medical clan, but it also belonged to descendants by birth of the clan itself, people who were well-born citizens of many centuries' standing. Hippocrates' name, with its classy connection with horses (*hippoi*), implies that he was upper class by birth, and could, then, be a true Asclepiad by descent, not one by association as a pupil. As tradition assumed, he was a citizen of Cos.

In the Greek world, prophets and diviners also included members of family clans, people well categorized by Robert Parker as 'wandering aristocratic opportunists of hereditary skills'. A true-born Asclepiad doctor was a similar opportunist, but better skilled.[1] As inheritances in Greek families were shared equally between sons, he was not a dispossessed younger son, forced to acquire a skill as a livelihood and then wander abroad for a living. The doctor-author of the Epidemic text had no reason to mention payments for his services but he surely expected them without scruple, whatever his social status. Meanwhile, how and why did he observe his patients?

Even resolute sceptics about his identity accept that this question is discussable: whatever the name of the author, *Epidemics* 1 and 3 exist as texts. I will refer to him as the 'Epidemic doctor' or the 'doctor-author', meaning the author of books 1 and 3. For comparison and contrast I will sometimes refer to 'Hippocratic doctors', meaning, as before, authors who wrote later but whose works were regarded at least by 300 BC as works by Hippocrates and eventually found a place in that hospitable construct, the Hippocratic Corpus.

As *Epidemics* 1 and 3 is now the earliest datable text in it, it strikes me as the likeliest candidate to be Hippocrates' own. However, dating alone cannot establish the point: I will allude occasionally to the possibility, considering it to be alive but not established.

In his Epidemic text, the doctor-author attends, usually at the bedside, the individual: the sick person is central, whereas in much modern hospital and laboratory medicine he or she simply disappears. However, he alludes only eight times to treatments. They are mentioned incidentally and are mild and entirely traditional: laxatives and bathing and, once only, that he drew blood from a patient's arm.[2] He says nothing about prescribing drugs or remedies. He never claims credit for a patient's recovery, itself a minority outcome. Twenty-six of his forty-two case histories end by simply noting the patient's death. In 1836, therefore, a French doctor characterized him as 'an interested but callous spectator who looks on unmoved while his patient dies'.[3] The *Epidemics'* great English translator, W. H. S. Jones, disagreed, but nonetheless considered the author 'to be for a time a spectator looking down on the arena, exercising that *theoria* [contemplation] which a Greek held to be the highest human activity' (Jones was thinking here of Aristotle and his personal ideal of *theoria*, but it was the ideal of very few Greeks indeed).[4] When Dēithrases or Dēialkos on Thasos paid him to come and attend their daughter or their wife, did they receive nothing more than a sequence of observations day by day until the case concluded with death or recovery? If so, why did they continue to let such an ineffective contemplator into the house? Why did the patients put up with him for so long?

Twice he hints at more. At the end of his third year-survey he writes more generally about the sources of 'our' knowledge about diseases and gives a long list of items which have to be considered. He discusses varieties of feverish heat and then comments that by 'examining these things, regimens must be applied'.[5] 'These things' are not just items relevant to fevers: they are items relevant to diseases of every kind. So, therefore, are regimens. The doctor adds that there are many other signs closely akin to the ones he has just specified, 'some of which have been mentioned in writing, others of which will be written about'. In the light of this written material, he states, a doctor will reason, scrutinize and examine 'to whom something must be administered or not, and when and how much and what the substance applied shall be'. The 'written material' is his own text. It will guide future treatment.

These remarks are excellent evidence for his intended audience: they are indeed doctors and medical pupils who will read his work and draw conclusions which will affect their treatment of a case. The role of literacy in fifth-century BC Greek culture has been much discussed in modern scholarship, even with an extreme claim that a culture of reading took root only in the late fifth century, but here, *c*. 470 BC, it is already assumed to be essential for best practice in the medical craft. At, or near, the very end of his text, the doctor-author returns to the point. He states that he considers it a 'great part of the craft' to be able to 'examine rightly what has been written'. Anyone who 'realizes this and uses it seems to me to be unlikely to go far wrong in the craft'.[6] His emphasis implies that consulting the written was not generally best practice for others: in Herodotus' story even Democedes is never said to have taken written material with him on his many travels or to have consulted it before treating Atossa's breast.

The first two authors known to have written prose texts in Greek on a technical subject are both architects. However, doctors' texts are the first prose texts written on a general subject, not a specific example of it. Cnidian doctors' *Cnidian Sentences* was presumably a text of short statements about what to do, but the *Epidemics*' Coan author wrote something more intricate, subtle and interrelated. Thanks to *Epidemics* 1 and 3, Coans could study statements about ethics and best practice in a general setting of yearly weather, diseases at large and case histories. They had a much richer resource than anything the Cnidians had yet produced.

Near the end of his work, the Epidemic author again states a range of items which need to be considered, including the state of the seasons: his point is that it 'will be easy to examine the order of critical days from these [signs] and to predict from them [the critical days]. For someone who knows about these things, it is possible to know for whom and when and how a regimen must be established.'[7] Again, he has treatments clearly in mind. They are the note on which he ends either his entire text, following the order in the manuscripts of later copyists, or, following the order endorsed by Galen and others, his general survey of the fourth year.

Given this emphasis, why does he almost never mention treatments? It was not that he believed in faith healing. He never advises that the patient must believe in the doctor in order to be restored to health, let alone, like Jesus in the Gospels, that it is sufficient for the patient to do so. Nor was he a committed believer in spontaneous 'natural' healing left to its own course. Famously the author of 2, 4 and 6 was to write: 'Of diseases natures are healers', adding that Nature 'finds a way for herself not from thought: for instance, blinking; and the tongue assists and other such things'. He adds that Nature 'does what is required but not as a result of her having learned it'. He was not excluding treatments here: he meant that the body's natures work to heal and will do so, implicitly, after help from a doctor too.[8] The doctor-author of books 1 and 3 would agree.

In antiquity, his silence about treatment was already a matter of debate, as a recently published text on papyrus has revealed. It shows that the admired doctor Asclepiades, active c. 100–80 BC, added treatments to his commentary on these Epidemic books, but others criticized him for trying to supply from insufficient evidence what the author, in their view Hippocrates, had left out.[9] In practice, did he prescribe and apply treatments more often than he records? There was a strong reason for him not to have done so. One of his primary aims was to enable informed prognosis about the course of a disease and to allow an appropriate regimen to be founded on it in future. If he was prescribing remedies, but not including them in his case histories, he was omitting important factors in each case's development. These omissions would affect his text's usefulness when similar cases recurred: their doctors needed to know how they were likely to develop and when best, therefore, to intervene. I assume that his patients were under observation, but not on trial. They were specimens, not guinea pigs.

Trialling would have distorted the necessary first step, understanding of how and when a sickness would naturally evolve. That understanding was invaluable for future doctors and future cases: meanwhile, he offered cautious predictions to his patients on the basis of what he was seeing. They may not seem much use to our age of pills and interventions, but patients and families would welcome them: they helped to allay fears about what would happen and when. With justice, he could also explain that treatments favoured by lesser practitioners would be worthless. He would not cut up patients or burn them with hot irons. Even when they raved and became delirious, he never intervened to restrain them forcibly. Four times he uses the phrase 'he/she could not keep control', but it refers to the patient, not to the doctor or his aides.[10] There was nothing to fear from letting him come in.

Presumably, he instructed his assistants to follow this same routine, because their reports were to shape his surveys of each year. He had, he implies, a reasoned long-term aim. First, he would record how a sickness developed, usually without any intervention from him personally: then, he would write it up and preserve it so that the next case of a similar illness could be treated with a better idea of its course. Patients like Heropythos in Abdera, observed for a span of 120 days, or Chaerion, observed for twenty days by Thasos's temple of Artemis the Delian, were not volunteers for a trial. They were being observed in order to improve the treatment, not of themselves there and then, but of similar patients in future. Perhaps they did not fully realize why they were being observed with such care.

Befitting his aims, their doctor-author was admirably concise. His style and language convey a highly distinctive sort of 'voiceprint'. Unlike many Greek prose authors it does not emerge from his use of small expressive particle words: on the contrary, it emerges from their absence. He sets one observation after another without connecting them with any adverbs or particles. He is crisp, clear and exact. His use of compound verbs and adjectives is particularly fine. His choice of prepositions to compound them is the choice of someone who is exercising sharp, detailed discrimination. He is sparing in his use of metaphor and he avoids similes. He fits words to things as tersely as possible. He is never rhetorical.[11]

He does not hide behind impersonality, as if his text is 'the' text, not his own text. He uses the first person 'I' or 'me' seventeen times in his work: 'I myself know' or 'as it seems to me' or even 'I knew / know none who . . .' These personal remarks are integral to his discussions and are in no way insertions by later readers into what they were reading for study: the text we have has not grown by accretion.

Whereas the philosopher Heracleitus (c. 500–490 BC) and the geographer Hecataeus of Miletus (active by c. 510 BC) had used the first person, boldly and programmatically, in the very first sentence of their texts, the doctor started with concise, objective remarks about the weather: 'on Thasos during autumn around the equinox and at the time of the Pleiades many rains, continuous, gently, among southerly winds . . .' He used 'I' and 'me', but only incidentally at later points and always in connection with his evidence or lack of it. He was not strident or composing for a general lecture audience. He was not promoting himself: he was addressing his evidence with a sense of its limits and qualifications. This voiceprint is indeed new in surviving Greek prose-writing.

He uses a revealing word for his observations: *akribōs*, not 'precisely', but 'accurately'. The word had been applied once in a post-Homeric poem, to Podaleirios' exceptional god-given knowledge of the inner man, but the Epidemic doctor's use of it is the first to survive in Greek prose.[12] Whereas Podaleirios' knowledge of 'accurate things' was granted to him alone by Poseidon, the doctor's accuracy was the result of bedside visits and questioning, an ideal which anyone could, and should, emulate.

His practice exemplifies what has been praised in previous Greek authors, the 'instinctive exactness of early Greek observation': with him the instinctive becomes explicit. For how long did he personally observe a patient? By the sanctuary of Heracles on Thasos he records details about Pythion during ten days until his death.[13] On the fourth day, he notes: 'early on, calm; towards the middle of the day everything became more acute; chilling; could not speak, no voice; got worse; warmed up after a time; black urine, with cloudiness; at night, calm; was asleep'. On the eighth day he records changes between early morning and nightfall. Did he stay with Pythion twenty-four hours at a time, waiting to note these significant changes even during the night? It seems more likely that he came in from time to time by day and relied on questioning others about events meanwhile, asking family members and slaves and the patients themselves, so long as they were still able to speak and think straight. Certainly he relied on them and those around them for facts before he attended a case: two days before the 'wife of Epicrates' gave birth, she had not become heated, 'so others were saying', he records, presumably family and slaves telling him what he had not been present to see.[14] On the tenth day of her burning fever, eventually fatal, the virgin daughter of Euryanax lost her wits and 'they were saying' that it was after eating a grape. Again, he had elicited this information from onlookers, presumably as he had been absent when it happened.

His patients, when not delirious, helped him to know where a pain was being felt or what had passed during a night. Otherwise, the judgements of their moods ('angry', 'taking things badly') are the doctor's own, as are almost all the descriptions of physical and mental signs. Little in the record results from a two-way dialogue between doctor and patient. Much of it rests on the doctor's own judgement, sometimes explicitly 'what seemed to me'. Yet, so far from being callous, he is explicit about the ethics of his craft, that feature which is prominent in later Hippocratic texts. Famously, he states two principles concerning the treatment of diseases, both of which are admirably humane: 'to help or not to do any harm'.[15] They may seem banal nowadays, but even if they were already current in other areas of Greek life, 'doing no harm' was a most striking aim for a doctor, especially when other healers were making big claims for their exotic treatments in order to impress potential clients and be hired. Six centuries later, Galen testifies to the impact this aim made on him when he first read it.[16] It is still one which modern doctors cite prominently, often believing, wrongly, that it is stated in the Hippocratic Oath.

The Epidemic doctor also lists the protagonists in a disease: 'the disease, the sick person, the doctor'. The doctor, he states, is 'servant of the art', medicine: the sick person is to 'oppose the disease with the doctor's help'. The sick person is not strictly a patient: he has an active role, fighting against the disease.[17] This metaphor of fighting a disease is still very popular, but it was already in this doctor's mind c. 470 BC. Some six centuries later, Galen approved this advice, but made a revealing change to its order: he promoted the doctor to second place and put the sick person in third place.[18] In fact the doctor-author had given the doctor a less prominent role. He was to be a collaborator, he wrote, and a 'servant'. He should not see himself as taking charge of a disease and as triumphing imperiously whatever a sick person may tell him.

Before this statement he remarks in passing what a doctor must practise: 'state what has already happened, understand what is present, foretell the future'.[19] Again, he does not emphasize treatment or healing. His stress on prediction is not at the forefront of modern medicine, but it is present in later fifth-century BC Hippocratic texts, including one which eventually received the title *On Prognostication*. Why?

The author lived in a world in which indications of the future were frequently sought. They were requested from the gods at oracular shrines. They were detected by experts in divination who saw signs from the gods in the flight of birds, in weather signs, the entrails of sacrificial animals and so forth. Dreams too were widely believed to hint at the future, as other doctors, whose texts are now in the Hippocratic Corpus, accepted.[20] Like doctors, diviners claimed to practise a craft: was the Epidemic author's stress on foreknowledge a spin-off from it?

He wrote of the need to assess the 'past, present and future', words which Homer had used, but in a different order, when presenting the diviner Calchas.[21] However, his own technique was quite different. Diviners assumed that 'signs' in the world about them were indications of outcomes in what are, to us, wholly unrelated fields of human activity. Those signs, they assumed, were sent intermittently by the gods. Doctors, by contrast, studied signs shown by the human body, their subject, which indicated the body's course. They considered that those signs were given off by an underlying nature, the body's own. Their predictive medicine was not a branch of divination at all.

Astronomers were a more relevant model, as they too observed daily natural phenomena to predict what might come next, in their case in the skies, but in the early fifth century BC their astronomical prediction was probably still quite reliant on guesswork.[22] The doctors' likeliest model was simpler and practical. As the fifth-century Greek historians show, foresight was prized in daily life as a valuable human knack, whether in politics or in the rapid choices everyone faced.[23] His interest in it relates to that way of thinking.

Unlike some of his later followers he applies the verb 'foretell' only to the future. Doctors, he considers, should not divine what has happened, but 'state' it, presumably only after dialogues with their patients and with those around them. As for foretelling the future, he refers admirably to the need to consider generalities and specifics and to take similarities and dissimilarities into account.[24] He also has a guiding principle. About disease in general he states that 'our thorough study' involved 'learning from the common nature of all and the particular nature of each'. He presents this method in the past tense ('involved'), as a method, therefore, which he and his helpers had already followed. Its study of 'the common nature of all' was not a general natural philosophy about the world, the 'all', but of 'the common nature of all *diseases*'.[25] He was looking for general patterns in them, but, as he well recognized, the 'particular nature' of each disease complicated the search. In his year-surveys he often states what was 'in general' or 'for the most part' likely, language which anticipates Aristotle's admirable caution about animal and human behaviour many years later.[26] Sometimes, a particular case history fails to conform to the general patterns of sickness which his survey of its year has just set out.[27] These discordant cases are not a betrayal of his principles. Deviation, he accepted, was always possible.

His concern with prediction answers an otherwise puzzling question: why did he record so many cases which ended in death? They were hardly the most compelling advertisement for his craft, but they were recorded because they would help doctors in future whenever they confronted a similar sequence of signs and sufferings: they would know what was likely to come next, including death, and would be able to intervene or not accordingly. A crucial concept for him was the 'critical day'. It was not a day of grave crisis, in our sense of the word, but one of decision, favourable or unfavourable.[28] The idea of such a day was his innovation, Galen believed, but even if it was not, the insistence that it can be rationally estimated may well be, driving his entire text. Knowledge of a critical day's likely timing would guide when to apply treatment in future or prescribe a helpful regimen.[29]

This notion is far removed from the activities of healers in Homeric or post-Homeric poetry. It was a great advance, but nonetheless the doctor-author qualified it in many ways. The analogy of a modern professional investor may help here, someone who also acts on a theory, a price-to-earnings ratio, say, or a three-stage sequence in a market's direction. He too has a basic insight: over time a range of equity investments will indeed eventually make money. The doctor-author had such an insight too: in medicine, similar illnesses will often come to a conclusion at similar times. Prediction, therefore, is possible: neither the doctor nor, contrary to popular wisdom, the investment expert would be better off picking outcomes by the spin of a coin. His aim and insight were a breakthrough, but, in practice, unqualified theories of prediction would sometimes fail. When they did, they risked undermining a doctor's reputation and his claims to have a genuine craft. So, he listed ever more qualifying conditions, like the admonitory small print beneath an investment adviser's recommendations. The occurrence of a critical day in one season of a year, he accepts, may differ from the timing of a critical day in the same sort of fever in another season.[30] Weather patterns are essential to the patterns of disease in each year, but they are noted in such detail that many years would never conform exactly to the examples given.

Such differences gave predictors a let-out. Nonetheless, about the timing of critical days the author was explicit. An exacerbation of an illness on an even-numbered day, he states, will be followed by a crisis on an even day. An exacerbation on an odd-numbered day will be followed by a crisis on an odd day. He then gives the even and odd numbers which are indicative: the fourth, sixth, eighth, tenth, fourteenth, twentieth and so forth, but numbers between fourteen and twenty are missing.[31] A crisis may occur on another day, he also states, but if it does its significance will be different: it will be followed by a relapse. Prediction is also complicated by the existence of fevers of a specified length, 'tertians' or 'quartans'. They may not exacerbate at all.[32]

How would these principles work out in practice? The case of Meton on Thasos is a simple example.[33] He was 'seized' by 'fiery heat' and had a painful heaviness in the loins. After drinking water on the second day, his excretions were a favourable sign, but those on the third day were 'thin, bilious and somewhat red'. On the fourth day the condition 'was exacerbated', but he had slight nosebleeds from the right nostril, usually a promising sign. His urine was rather black and had a cloudy deposit in it, elsewhere a bad sign. On the fifth day, he had a 'violent flow of unmixed blood from the left nostril; sweats'. It was the crisis. He was sleepless after it and his mind wandered. He had more nosebleeds. His head was bathed and he recovered. Elsewhere, rather black urine with a cloud in it is an unfavourable sign, but in other cases, nosebleeds are a very favourable sign, so it seems from Meton's recovery that the doctor presupposed a trade-off between good and bad signs. He did not spell it out for his medical readers. They shared it already, but it cannot always have been easy for later readers to assess.

How could a student of the rest of the text be sure of predicting the outcome of Meton's case? It departed from the odd-and-even rule, itself a postulate which gave too much weight to this illusory aspect of number. On the third day, nothing hinted at an exacerbation on the next day, the fourth. Its exacerbation on an even-numbered day should have been followed by a crisis on another even-numbered day, but the crisis occurred on an odd day, the fifth. It was not a relapse, as the author's odd-and-even theory had specified. It was a complete cure.

Meton's case exemplifies a general problem. With so many qualifications and variables at issue, how could a doctor correctly predict a patient's death by comparing his condition with the evidence of other cases in the text? It was not that the very moment or definition of death, as nowadays, was in dispute. Rather, half-way through the day-by-day symptoms, much in a case could still vary and so a doctor needed to wait and see in order to be sure of its future direction. The doctor-author had some invaluable insights, that a disease could be a process, that cases could be similar, that they could come to a foreseeable crisis, but he hedged their predictive power because cases did not always follow the same pattern. He should not be assessed as a philosopher: he was no such thing. He had some big ideas and in practice tried to save their credibility. To do so, he covered his tracks, but by covering them he weakened others' chances of following them with certainty.

17

Filtered Reality

I

'As it seems to me . . .', 'I know none who . . .': what did this author study and what does it imply about his view of the body and its workings? As he never records cases of fracture, wounding or bodily damage, his subject matter is quite different from the cases presented in Homer's Iliad. It is also irrelevant to a major part of the Hippocratic Corpus, its texts about traumatic and non-traumatic surgery. Instead he observes and records malfunctions of the body, whose perceptible signs hint at processes occurring inside it.

In the survey of his third year on Thasos he gives his readers a long list of what to consider: it includes urine, excretions, spitting, vomit, coughing, flows of blood, piles, shivering, breathing and so forth.[1] It never occurred to him to consider the pulse, which was not recognized as a sign in Greek medicine until nearly 200 years later. However, he realized the value of psychological signs, including silences, dreams – 'what sort and when' – and, most strikingly, thoughts. He considered it relevant to observe patients' pulling of their hair or scratching or weeping. He laid emphasis, wisely, on patterns of sleep, recorded over several days. He also paid careful attention to the voice and to loss of it.[2]

Doctors before him had surely considered some of these signs. Their practice of bloodletting implied a belief that there was something about a patient's blood which meant that, in many cases, less of it would be better. His novelty probably lies elsewhere, in the range of what he considered relevant. First-class medics still recommend to their students 'topical examination', the alert study of externals in a patient, to be conducted even before the results of laboratory tests are consulted. The doctor-author exemplifies topical observation in his case histories. Dēialkos' wife was 'always silent . . . scratching, pluck-ing [her hair] . . . gathering her hair up'.[3] During more than two weeks he records the nightly sleep, or its absence, of his patient 'lying' in Dēialkos' garden. As for dreams, he notes them on the fourth day of the fever which gripped Erasinus 'by the watercourse of Bootes': when Erasinus regained his wits on the following day, he presumably recalled for the doctor that he had been dreaming during the night.[4]

When he notes sweating, he discriminates very finely between cold and hot sweats. He notes chilling, especially of the body's extrem-ities. He records shivering, as in the case of the wife of Epicrates 'by Archēgetes'.[5] He reports the colour, quantity and texture of faeces and urine with particular care, showing that they had been collected assiduously and sometimes left for a while before being fully described.[6] Once, he notes that sufferers from stomach sicknesses passed abundant urine, 'not resulting from the drinks that were administered, but far in excess', a rare hint of observation with a quantifying eye.[7] He also notes deafness.[8] However, he never records close examination of the colour and qualities of his patients' eyes. Once, in the fourth year, he comments on eye conditions as 'watery and long lasting', but he notes only the accompanying pain and the presence of growths on both the outside and inside of the eyelid, 'what they call "figs"'.[9] Conversely, he pays some attention to the tongue. He notes when its surface is rough, especially when a patient has become thirsty.[10]

His noting of mental conditions is most striking: he records them on a par with physical signs.[11] He is not concerned with the inner workings of the mind or that great enigma, its relation to other bod-ily elements. He never even refers to the soul, or *psyche*, a crucial component of bipartite Pindaric or Platonic man. His notion of the person is also quite different from Homer's.[12] He does not refer to the mind, or *phrēn*, although he refers to cases of phrenitis, a mentally related condition which his medical readers were expected to recog-nize. Nonetheless, he assumes a connection between mental and physical symptoms. He notes when patients begin to ramble and talk nonsense, to go out of their wits, as we still say. He uses a vivid range of verbs to express an implicit progression in derangement, from 'being struck and going astray' (*parakrouein*, used fifty-four times, three times more than in all the other texts in the Hippocratic Corpus combined) to 'going completely mad' (*ekmainesthai*, a verb used by other authors only for going mad with passion for somebody). He never assumes, as most Greeks did, that madness is an affliction sent by an outside power.

His range of observation is admirable, but observation is guided by presuppositions: there is no such thing as the virgin mind. He is also selective. The most striking example of his selectivity relates to women. He records six cases of women who have just given birth, but never once does he refer to their milk or any problems with it. As the milk in a woman's breasts was considered to be displaced blood, its state was highly relevant to his theory of the body's functioning and its emphasis on expulsions of excess blood as a step to recovery. There are other absentees too. In the general year-surveys he gives details of sufferers from *phthisis*, to us consumption, as the verb *phthinein* means to wane or waste away. If these details are compared with discussions of *phthisis* in later fifth-century BC Hippocratic texts, they omit some major items, including pain in the back and a whistling and dryness in the windpipe.[13] He can hardly have failed to notice such a conspicuous feature of the condition.

Was he presupposing that his readers knew it anyway? One way to save his credit is to argue that he presupposed knowledge of any obvious sign he failed to mention. However, it is not obvious that this defence of him is correct. A related defence is to regard whatever he recorded in a case history as being a significant sign with a hinterland which explained its relevance, but as readers would know this hinterland, he left it unstated. 'Nothing irrelevant is mentioned', his great translator W. H. S. Jones believed, 'everything relevant is included.'[14] Yet when surveying the incidence of *phthisis*, or consumption, in the fourth of his year-surveys, the doctor notes that the 'type' included people with a slightly white skin and people with a slightly reddish skin and so forth.[15] He does not say why these signs were related to consumption. It is more likely that his general surveys sometimes omitted a feature by accident and that his case histories recorded anything striking which he observed or learned, even when he did not know exactly why it was significant. It was better put in than left out.

No doubt a Sombrotidas or a Democedes had considered some of these signs in sick patients, but it is most unlikely that they considered such a range so carefully. What the doctor-author looked for was guided by his general view of the body and its afflictions: what did he think was happening behind the visible signs?

He uses the word 'nature', *physis*, only three times, once for those 'for whom their nature inclined to the consumptive', once for a woman's nature after giving birth and once for the 'common nature of all', where he means 'all diseases'.[16] He never refers to what we would spell as Nature or to the natural world, though those concepts were fundamental to other Hippocratic texts and indeed to their very invention of medicine. Yet he was not distanced from their thinking. His entire text presupposed those concepts, not least by giving detailed records of the seasonal weather as a pertinent factor in diseases which were at large: he did not need to spell out Nature's relevance.

The concepts of a disease and a symptom, a diagnosis and a cause, are standard assumptions in modern medicine, but they cannot be taken for granted in an early doctor's thinking. The Epidemic author certainly has the idea of a disease, one and the same condition which affects different people simultaneously and over time: his description of mumps is an immediate example and his description of *phthisis*, or consumption, is another. He puts 'the disease' first in his list of the protagonists in an illness. He has a simple range of names for diseases, *peripneumonia*, *phthisis*, *phrenitis* and *dysenterie* being examples. He also assumes that a disease can change into another disease during an illness, as indeed it can.[17] He is no Cnidian, splitting illnesses into many sub-divisions, each with a separate name.

How concerned is he to diagnose? The word 'diagnosis' occurs nowhere in the *Epidemics* and only once in the entire Hippocratic Corpus. Although he uses the verb *diagignōskein* twice, he means by it to study or know thoroughly.[18] Five cases conclude in the manuscripts with a single identifying word like 'burning heat', but those conclusions are rightly regarded as a later reader's additions to the original text.[19] He gives quite general names to diseases in his survey or at the start of his case histories, but he does not explain the grounds for his identifications. Perhaps he expected his medical readers to be able to supply them anyway, whereas his main aim in his text was to record signs and assist prognosis in future. The first case history makes the point clearly. An 'acute fieriness', it records, seized Philiscos, who resided 'by the wall' and died about noon on the sixth day. On the fifth day, Philiscos had a 'little' nosebleed.[20] The same case happens to have been cited previously in the survey of its year, the doctor's third: only there does he remark that Philiscos was a sufferer from the burning heat which was at large in Thasos and that a 'small' nosebleed was not sufficient to bring on a crisis and help him to recover.[21] In the case history he did not relate the case to this widespread disease. It is by chance, not design, that the year-survey happens to make Philiscos' disease more intelligible.

When he notes 'signs', he is not regarding them as symptoms of a particular condition. He looks on them as signs in a developing process. Does he strive, nonetheless, to identify causes? The very definition of a cause is still controversial among philosophers, whether it is an event, whether it admits exceptions or even, on one view, whether it can follow its effect. He never uses the most familiar Greek word for one, *aition*.[22] Whenever he uses the word 'for' (*gar*), he uses it to introduce an amplification, not a causal explanation. After a crisis on the sixth day, 'among the bouts of fieriness and after the crisis', a girl in Larissa had a period for the first time: 'for', he adds, she was a young virgin. Neither her age nor her virginity caused this first period.[23] Of course, he accepted that the conditions he described resulted 'from' other observable factors, but only once does he use the usual preposition meaning 'because of'.[24] He did not specify the cause of each condition in his text: causality was not uppermost in his approach.

In four contexts he uses the distinctive word *prophasis*, which Galen later took to mean 'cause' or 'visible cause'. However, a better translation, certainly in three of the contexts, perhaps in all four, is 'first manifestation'.[25] Once, he notes red rashes with a *prophasis* and relates them to blisters visible on the body, evidently a first manifestation of the complaint. He also mentions red rashes without a *prophasis*, but he is not meaning that these rashes had no cause at all. In cases of mumps, he notes that the swellings by the ears did not rot like swellings resulting

from other *prophaseis*. Again, the word means 'other first manifesta-
tions'. His usage of it is telling because the word *prophasis* had already
been used by Homer and post-Homeric poets in a causal or purposive
sense. In the Iliad, Homer's King Agamemnon swears that he has not
used the captive girl Briseis 'for the *prophasis*' (purpose) of sex.[26] In the
later Greek historians, Herodotus or Thucydides, the word has wider
senses too, including 'pretext'.[27] By using it to mean a prior manifestation, this doctor-author, *c.* 470 BC, was restricting its scope. The
restriction fits his emphasis on the observation of visible signs.

Underlying the signs which he observed were processes inside the
body: he mentions them, but does not stop to explain his view of how
they work. His pupils were expected to know, whereas we can only
infer them. One such process is *pepasmos*, or ripening, usually translated as concoction; another is *apostasis*, a process of setting aside or
depositing.[28] They read like terms with a technical use, ones which
are never mentioned in evidence for doctors in action from Homer to
Pindar. They may well be new to him and his associates. They had
not transferred them into medicine from natural philosophy or from
theories of the formation of the universe, as if the body was a micro-
cosm of that bigger macrocosm. *Pepasmos* was readily observed in
the ripening of fruit, but more pertinently it related to the everyday
skill of cooking, a source, too, of Hippocratic doctors' thinking about
pregnancy: seed, they considered, 'cooked' and 'matured' inside the
womb, an oven-like receptacle, just as our phrase a 'bun in the oven'
still assumes.[29] The closely related word *pepsis* was also applied in
the study of scent: the scents of various leaves and types of bark were
considered to be 'concocted' by the plant's own heat.[30] As for an
apostasis, it was a separation and depositing. In political contexts the
word was used to mean rebellion (our 'apostasy'). The doctor used it
as a technical term of his craft, perhaps one of his own invention.

How did these technically named processes lead to a crisis? Again,
his thinking has to be inferred, as he is not wholly explicit. One way
of amplifying it is to look ahead to later fifth- and fourth-century BC
texts in the Hippocratic Corpus and interpret this author, perhaps
Hippocrates, by the 'Hippocratic'. As I ascribe only *Epidemics* 1 and
3 to this one author and as I put at least forty years between them and
the Hippocratic texts which are considered to be most helpful for his

thinking, I will concentrate on what his *Epidemics* say and use later Hippocratic material very sparingly.

II

All but one of his case histories begin with, or very soon record, their patient being 'seized' by fire or by acute fieriness. Heat was a constant element in the body, in health as in sickness, but it could become *puretos*, fiery heat, or 'acute' fiery heat. It might be provoked by extreme seasonal weather or it might result from exercising and what is called, in the case of Pythion beyond Heracles' sanctuary, a 'careless lifestyle'.[31] Excessive sex was also thought to dry the male body and therefore heat it. Later Hippocratic authors remark how excessive exercising can cause the flesh of untrained bodies to heat up, but in this case they think it becomes moist and raises the levels of fluid inside them.[32] Probably, the Epidemic doctor shared this interpretation, implied by visible sweat. Fiery heat could also appear in women who had given birth a few days before, an exertion indeed and certainly one by which body fluids were increased.[33]

Worst of all, fieriness might become *kausos*, or 'burning heat', a severe condition especially remarked during the spring to autumn season of the third year on Thasos. It was then that it afflicted Dëithrases' young daughter, the one who had a period for the first time, the doctor remarked, as did a young fellow-sufferer in Abdera in the following year, also a virgin.[34] Both of these girls had slight nosebleeds at the same time and then recovered. The assumption in all these cases is that an expulsion of blood leads to recovery from burning heat. The author never refers to bad blood or explains that there was something in the blood in cases of fieriness which meant that it needed to be expelled.[35] He seems merely to think that the less fiery blood there is in a body, the more the heat inside it will subside and allow the patient to recover. He was the heir, after all, to the long-practised art of blood-letting, exemplified by the practice of cupping and its instruments.

How did concoction work? Food, he considers, is generally turned into blood, and in women into menstrual blood: he has no idea of ovulation, of course, or why periods occur. Parts of an intake of food, what later texts call 'residues', would be separated, worked on and excreted.[36] Examination of urine and other excreta could therefore show whether concoction, a good sign, had occurred or not. So could examination of phlegm, whether as spittle or catarrh. Why was concoction significant? He states crisply that concoctions signify that a crisis will come swiftly and that there will be 'healthy safety'.[37] There might also be rotting, or *empyema*, another significant sign. Rotting might show in flesh near the skin, in what we would call ulcers, or in general decay. In the fourth year he describes as a consequence of a red-rash fever the rotting of flesh from patients' arms and legs, but comments that it was 'more frightening than bad' as a sign.[38] Horrible though it seems, he appears to interpret rotting as an indication of concoction inside the body, a good sign.

As for *apostasis*, it means a setting aside, usually as a deposit or displacement of a fluid from one place in the body to another.[39] It could also be the displacement of a disease into a different disease during a period of illness. Deposits were of two sorts. They were bad ones if the fluid transferred was too much for its new location in the body. The fluid would then move back and prompt a relapse. Otherwise, deposits were good signs. In one of the burning heats, the blood was displaced to settle beside the ears, where it caused swellings. When the swellings disappeared, the blood was considered to have been displaced down to the left flank and hip. After pain and a crisis, slight excretions of urine occurred, probably a result of 'strangury', painful and limited urinating. Then, a better sign occurred: blood began to flow slightly from the nose.[40] This burning fever was the one which 'seized' Antiphon son of Critobulus and Dēithrases' daughter. After forty days, the fieriness ceased in Antiphon and a final crisis came. The daughter of Dēithrases reached the crisis more quickly because she had her first period as well, expelling blood twice over.

The case of Pythion who lived by the temple of Earth shows how rotting, strangury and displacement could interrelate.[41] Pythion's hands began to tremble; on the first day an 'acute fieriness' seized him; he lost his wits; he continued to tremble; eventually on the eleventh day he had sweats and slightly concocted spittle. 'It was a crisis', but he had to wait another forty days before localized 'rotting' began and then an *apostasis* occurred involving painful slight urination.

Presumably it led to the final complete crisis: he survived. Rotting was not itself a displacement, but it was a significant prelude to one.

These cases, beginning with fiery heat, very seldom lead the doctor-author to comment on the state of the body's inner organs: unlike many of his followers, he never infers that in a distressed woman the womb may be moving around and wandering. He is much more interested by a few basic fluids than by inner bodily structures and their interrelations. How or why does fiery heat arise? He does not specify, but his careful presentation of the seasonal weather patterns in each year implies that he shared the old view, that it was intensified by seasonal weather. He does not, however, consider, as Hesiod did, that it was provoked specifically by the Dog Star or by any one constellation.

When fiery heat becomes a fieriness or a burning fever, patients are never said to catch it: instead, it catches or 'seizes' them. Here he has what strike us as a massive blind spot. He never considers that diseases are contagious, let alone infectious by other means. He never considers that another person, insect or animal might be transmitting them. He remarks how the disease we identify as mumps was especially common among young men in the gyms and wrestling rings, and not among girls, but he never infers that physical contact passed it on. He may well have thought that this bodily exercise heated up the young men's bodies and so they were 'seized' by fiery heat more readily: it would be wrong to think that he implies an idea of contagion here and does not need to spell it out for his medical readers.[42] He never considers in any context that disease might be provoked by bad entities, to us viruses or germs, which invade a body from outside.

Germs, let alone viruses, require a microbiology which the ancients never formulated. For all their later skill in optics, they never even invented a microscope. Contagion seems an obvious inference to us, but awareness of the effect does not appear in our evidence until witnesses of the major plague in Athens in the 420s BC. It might have been recognized even earlier by other silent witnesses, herdsmen and shepherds in charge of animals. As flocks were individuals' valuable property, there was a pressing need to find a way of limiting a widespread sickness. One obvious way was to kill diseased stock in order to dispose of it and separate it from healthy others. 'Herd immunity' was not a notion farmers would ever have formulated, but, without knowing why, they might have found that individual culling minimized casualties.

In neither case did awareness originate with doctors. In the 420s, Athenians simply noticed how birds and animals which ate plague victims' dead bodies died themselves. As for culling by farmers, if practised, it was not a treatment doctors could apply to their human patients. Why did the Epidemic doctor not think of transmission by contact or invisible entities? As later authors pointed out, the passage of an invisible influence from one object to another was not unknown: the obvious example was what we call magnetism. However, the most widely credited examples were two popular misapprehensions: invisible pollution or *miasma*, and effects of the 'evil eye'. They were examples of the sort of superstition which Hippocratic doctors consciously rejected. Right on these points, they may have been blinded to another, the transmission by contact of something else they could not see.[43]

The Epidemic author was not alone in his blindness. Later Hippocratics shared it. Their preferred explanation was bad air, but the Epidemic author, earlier than them, never states as they do that disease is related to the air we breathe and that in an epidemic we should try to breathe as little of it as possible. He never even dwells on bad water or his cities' orientation, factors emphasized by one of his fifth-century successors, the author of *Airs, Waters and Places*. He seems content to relate his diseases at large to details of the weather and season without being more specific. He leaves unstated what exactly the link between the climate and bodily heating is. Hot sun would obviously provoke sweating, and cold weather would provoke shivering, but the 'fire' or 'acute fieriness' which seizes patients is more profound. Acute fieriness sometimes coexists with shivering, as he well observes. It could also take place in winter, a cold season.

Why did diseases attack some but not all at the same time in a particular type of season? Once, he refers to people of a consumptive 'nature', as if there is a propensity in some people to a particular condition, but he does not analyse it or try to fit it into a coherent theory.[44] He appears to think that particular natures, or propensities, are worked on differently by weather patterns and lifestyles and that, together, they explain how a disease can appear in many people at once, but not in everybody. Even now, the particular incidence of an epidemic is related to individuals' natures, their genetic makeup, in order to explain it. However, the Epidemic author's 'natures' are quite different. At the end of his third year-survey, he sums up that 'most' of those who suffered the diseases, apparently all of them, and died were 'young adolescents, young men, those in their prime, smooth[-skinned] with white skins, straight-haired, black-haired, dark-eyed, those who had lived with a random and lazy lifestyle', and also those who 'stuttered or were irascible: and most of the women who died were of this type [probably some or all of the features given]'. He does not explain why straight hair or stuttering might be relevant, and his emphasis on excess blood applies only to some of the features named. The discrimination of physical detail is fine, but much here has gone unsaid, if indeed it fully could be.[45]

The most remarkable account in the Hippocratic Corpus of the relationship between hot weather, inner heat and disease is given by the author of what is now called *Diseases*, book 4, a Cnidian writing c. 420–400 BC. He considers that hot weather 'disturbs and heats' all the wetness in the body, an effect which he compares, vividly, to the Scythians' practice of churning milk and causing it to separate into different levels.[46] Indeed, they churn it still in central Asia, where girls pour mares' milk into cans and turn the handles of their churns to separate its cream, a dietary challenge for foreign visitors to their yurts. The author argues that, likewise, heat churns the fluid in the body, bringing bile to the top, then blood, then phlegm and, lowest of all, the heavy 'dropsical' element. If the Epidemic author, a Coan, had ever heard this fascinating comparison of the over-heated body to a Scythian churn, a *bishkek* in the Kyrgyz language (from which their capital city takes its name), he would have relished it. However, his own thinking seems simpler, that the body has a natural heat that sometimes, like fire, intensifies and seizes someone and then burns more intensely.

Without an explicit comparison to churning, he at least implies a link between excessive inner heat and excessive bile. The normal heat of a body, he assumes, acts on food or drink to produce bile, phlegm and other juices. They are not mere postulates: bile is visible if someone vomits, phlegm if they sneeze or cough. Once, he calls blood, too, a juice, or *chumos*, his only use of the word which moderns translate rather too precisely as a 'humour'.[47] From Galen onwards, humours and the balance between them came to be considered the very hallmark of Hippocratic medicine. The Epidemic doctor never uses the word *krasis*, the word for such blending.

He accepts that women too have bile, though blood is a particularly prominent fluid in them, as their monthly periods attest. Does he think there are several distinct biles in the body? He implies bile can be yellow, because he repeatedly refers to patients' vomit being 'bilious, yellow': anyone who has vomited repeatedly will indeed recognize what he meant. He also implies that bilious vomit can become rusty green, like verdigris: that, thankfully, is rarer in modern experience.[48] He has no idea of the active role of the gall bladder in its production and is probably thinking of one and the same bile, changing colour according to circumstances. Sometimes it even becomes black, because he refers twice to the 'melancholic', or black-biled.[49] Once, he distinguishes three 'types' of people who fell sick with particular types of disease: one type are the 'melancholic and somewhat sanguine'.[50] Only in one fifth-century BC Hippocratic text are there specifically four humours, to be related explicitly to four temperaments and even to the four seasons and the four ages of human life.[51] No such system is stated in *Epidemics* 1 and 3, although their melancholic 'type' is a first step towards one. The author also remarks on a female patient, the one by the cold water in Thasos, whom he summarizes after her eighty-day illness as 'off her food, dispirited, sleepless; bouts of anger; taking things badly; in respect of [her] thinking, melancholic things [*melancholika*]'.[52] Here, he assumes that someone's mood, as we would say, can be melancholic, though it is not in itself what the word melancholic always refers to. The origins of the idea of melancholia and the condition we associate with it have been endlessly discussed, but this female patient, around 470 BC, is the first person known to have been described as melancholic. From the eleventh day until her death six weeks later, the doctor-author notes that her urine was persistently black: presumably he called her melancholic because she had an excess of blackened bile, to us, not him, a sign of a malarial fever.[53]

*

She is a significant presence, but throughout *Epidemics* 1 and 3, there is an even more significant absence. Nowhere in any of his year-surveys or his case histories does the author ever mention the gods or their possible intervention in a cure or sickness, not even in the attack of mumps which spread so swiftly and painfully among young men in the gymnasium. Never once does he mention that a prayer or vow was made by a patient or by himself. His summary statement is most revealing: the craft involves three things, 'the disease, the patient and the doctor'. The gods are not so much as mentioned. In the 470s and early 460s BC, he was inhabiting a different mental universe to Homer's and Hesiod's. Unlike them he presents misfortune without any reference to random actions by a divine power.

His exclusion of the divine is even more remarkable because his patients, their friends and families are immensely unlikely to have followed suit in their own lives. When struck by pain and sickness they surely made vows and prayers, dedications and offerings, all of which were staple Greek responses to distress. Not only was the city of Thasos well supplied with sanctuaries, niches and altars at which to do so: in each month of the year, there were official festivals for specific divinities. Gods were protectors of the city, as their representations on each of its gateways testify: they were embedded in its civic structure, as their role as heads of the citizens' brotherhoods, or *patrai*, shows. They were the immediate recourse whenever help was needed: hence the many figurines and other items which individuals dedicated in thanks for, or hope of, divine favour in the city's sanctuaries. The gods were especially manifest in times of personal or collective crisis. One of the regulations for best practice in the sanctuary of Delian Artemis is beautiful evidence. Anyone who transgressed its rules about not keeping animals in the sanctuary was warned they would have an *enthymiston*, or 'scruple', on their minds.[54] It was not a penalty imposed by a human judge. Sent by a divinity, it would weigh on them later when things went wrong with their lives.

In the doctor's case histories, patients often lose their wits, and talk nonsense, and sometimes even go mad. Never does he imply that these fearful symptoms were caused by a god or resulted from a transgression by a patient or a family member against a god or a sanctuary. He never explains them as arising from a scruple, or *enthymiston*, sent by a divinity. He was not an atheist: atheists in our sense of the word are unattested in the Greek world in his lifetime. He probably accepted that the gods were the ultimate source of whatever happened, including the craft of medicine, but as he assumed, without needing to state it, that the body had its own nature and was part of the natural world, he explained it and its diseases in natural terms.

The presumptions of a Homer or Hesiod were excluded, that a plague would be sent by angry Apollo and a long-lasting sickness by Zeus or a *daimon*. The doctor-author's craft was not reliant on making offerings to divine powers. The gods' interventions were unpredictable, whereas his explicit aim was to detect similarities and, if possible, predict patterns. Even though he never had the ideas of contagion, transmission and infection, he never once considered that a disease was, or might be, heaven sent.

In the 470s BC he shines out for this concerted rationalism. He was not a philosopher. He was not an experimentalist. He had no idea of germs or genes, the circulation of the blood or the ovaries or the role of the brain. He had, however, grasped one great truth: diseases should be observed and treated and their course predicted without explanations in terms of the gods. Working on Thasos *c.* 470 BC, he is the first such practitioner known to us in non-philosophical history.

18

Retrospective Diagnosis

I

Magnificently excluding divine interventions, the doctor-author claims accuracy, not just precision, for his observations. Can they be taken further: can any of the conditions he records be diagnosed retrospectively and identified with diseases which are still current? Since the late nineteenth century, doctors have made many attempts, not always in agreement. Obviously, they face practical and intellectual obstacles. Greek words which the doctor-author uses, like *typhos* or *anthrax*, *emphysema* or *herpes*, are not to be translated by their English medical equivalents simply because they are spelled alike. Their meanings may be quite different. He notes a range of signs in his cases and year-surveys, but he is not thinking in modern clinical terms and is unaware of germs, infections and, therefore, contagion. What if practical limits and these presuppositions caused him to omit signs which would turn a modern diagnosis in a very different direction? He could not give blood tests. He did not have a stethoscope and he never checked pulse rates. To be fair to him, a modern diagnosis may not even be possible. Over time, many diseases die out or mutate: what he describes may be a disease which is, mercifully, extinct. Modern names for diseases still group different conditions under one convenient word. They too may not apply helpfully to his.

What, indeed, is the gain for historians in a retrospective diagnosis? Of the seventeen female patients in the case histories, nine were pregnant or had just given birth. What does it add to the understanding of them if cases of 'fiery heat' a few days after birth are identified as 'puerperal fever, with sepsis', except to reiterate that the author

236

had no notion of infection? Nonetheless, there is more here than a guessing-game for medical experts. If correct, retrospective diagnoses validate the author's claims to practise accurate observation. They can even, on one optimistic view, allow editors to improve the manuscripts' text, as in that case of poor Parmeniscus in *Epidemics* 5 and 7. He has been diagnosed retrospectively as suffering from a melancholic attack, but as we have seen it requires him to be without fever (a-*puretos*) rather than having a fever (*puretos*), as the manuscripts transmit. The word *puretos* has been changed here in modern editions to fit the rest of his symptoms, rather too boldly.[1]

At a more general level, retrospective diagnoses relate a patient's sufferings directly to ours, tightening links in that chain which Chekhov once powerfully imagined as connecting our present experience to that of the distant past, a chain which we can still pull on and activate by empathetic understanding. Like his weather surveys, his case histories may also validate a long continuity and show that diseases, like Thasos's weather, have not changed in human experience since the 470s BC. Correct diagnoses can also enhance our sense of the hazards of Greek city life in the fifth century BC. If they confirm the presence of killer diseases, they strengthen the structural studies of 'risk and mortality' which historians now try to extrapolate from other evidence, especially for ancient Rome.[2] Better than the ancients, we know what sort of damage these still-active diseases would cause and spread through a population and why.

Extreme sceptics, doubting if retrospective diagnosis is ever secure, are refuted by the very beginning of the doctor's text. He does not name what he describes, but manifestly the Thasians suffered an outbreak of mumps during his first year in their city. After so many centuries, its symptoms have not changed, least of all for young men.

What about *phthisis*, the 'consumptive' fever which he describes in the same year, beginning early in the summer and persisting into winter? He is widely believed to be describing TB.[3] However, his survey says nothing about coughing up blood, the best-known symptom of pulmonary TB to outsiders. He mentions only the coughing up of bits which were 'small, frequent, concocted' and coughed up with difficulty, whereas those most violently afflicted 'continued to cough up raw', that is unconcocted, material. Perhaps they were coughing the cheesy 'necrotic tissue often teeming with bacilli' which is also expelled by TB patients, but the coughing of blood was remarked by later doctors in antiquity and is much too distinctive for the onlooking doctor to have missed it.[4]

He gives a case history, the daughter of Euryanax, which he relates to consumption and which almost certainly belongs to this first year's outbreak.[5] Again, he notes no coughed-up blood. So far from having the surge of hope which occurs in many terminal cases of TB, this poor girl, no Traviata, remained 'low spirited; despairing of herself'. He even presumed that consumption might be hereditary, an interpretation he applies in passing to her case. TB may perhaps relate to a genetic predisposition but it spreads by contagion, through coughing and sneezing. He and his helpers never imagined such a thing.

In his fourth and final year, consumption was evident again, beginning in winter and causing chills, fever, sweats and a loss of appetite for food and drink. His description of it at that point is his most detailed, differing on several minor, but finely observed, details from his survey of it in the first year.[6] He notes persistent coughs, in some cases lasting for months without interruption, and he also notes that they brought up 'ripe and wet' matter without any real pain. In the first year's survey he had noted raw matter being coughed up too, and then only with difficulty. He and his helpers had clearly paid careful attention, but there is still no mention of the coughing of any blood. Instead he notes that much that was 'sticky and white and wet and foaming' kept coming down from the head, presumably because it was then coughed up and made visible. He also remarks that 'all that was around the lungs' was emitted 'below', being excreted, therefore. This remark is evidence of his lack of understanding of hidden interconnections inside the body.

The duration of this *phthisis* is duly noted. It became evident in winter, and most of the patients, but not all, took to their beds. Most of those in bed died in the spring, but the rest of them continued coughing. They had a remission in summer but in autumn all took to their beds again and died.[7] These observations are indeed compatible with TB, which is often a protracted condition and can occur at any time of year. The author concludes with notes on what he calls the 'type' disposed to it. They were 'the smooth[-skinned], the white-ish [TB, the 'white death', does indeed bring about anaemia], the lentil-coloured, the reddish, the clear-eyed, the white-phlegmed [sufferers from swelling ascribed to white phlegm], the winged [explained by Galen as people with protruding shoulder blades]'.[8] The clarity of the patients' eyes probably lay in their colouring, not in the ethereal look well known in advanced cases of tuberculosis. Simply from this 'type', no doctor would diagnose TB.

Phthisis, the Greek word for consumption, relates primarily to wasting away, a state observed in patients' externals: sufferers from it were noted as no longer wishing to eat. Among them, probably, were cases of what we know as TB, but they were not the only consumptives covered by the Epidemic doctor's use of the term. This partial, but not total, overlap between a general Greek term and a modern condition is one which can be traced elsewhere.

II

In the winter of the third year, many were beset with *paraplegie*: the doctor notes it as 'endemic' at that time in the city and that some of the patients died swiftly.[9] It was a widespread affliction, affecting 'many'. He had already related *paraplegie* to people whose inner heat is low, especially the elderly.[10] Its appearance in winter is consistent with his interpretation.

Paraplegie here has been diagnosed as polio, rather tenuously, especially as it was noted only for one season in one year. Polio had long existed in antiquity, as skeletons recovered from early Egypt attest, but outbreaks of it in an acute form are not otherwise in evidence.[11] Like consumption, the word *paraplegie* seems to have had a wider reference. By Thasos's seashore the doctor attended an unnamed woman, three months pregnant, who was seized by 'fire'. On the third day her right arm was paralysed with a spasm, he records, 'in a paraplegic way'.[12] This case, at least, was not polio: her arm was loosened on the very next day.

Here, too, one modern diagnosis does not fit all the cases subsumed under a single Greek word. Precise modern diagnoses proliferate nonetheless. By that new wall, Hermocrates, seized by 'fire', has been read as a case of Weil's disease, accompanied by the typical yellow jaundice and the typical rise and fall of fever: it intensified, retreated, then recurred until Hermocrates' death on the twenty-seventh day.[13] The doctor-author also records upset stomachs and watery excretions from them in what he calls *dysenterie*. They allow a secure identification of this condition with modern dysentery.[14]

In the fourth year, he records something curious, 'mouth ulcers and sores. Fluxes round the genitals in many cases, sores, swellings inside and outside around the area of the groin. Watery eye conditions, long lasting, chronic, with pains, growths on the outside [and] inside of the eyelids, destroying the sight of many sufferers, which are called "figs" [styes]. Many growths on the other ulcers, even on the genitals.'[15] Initially, modern diagnosers considered that this condition included venereal disease and cited it as crucial evidence for VD's existence in classical antiquity. However, in 1937 a Turkish skin specialist, Hulusi Behçet, identified a particular condition, Behçet's syndrome, which combines ulceration of the mouth, inflammation on a layer of the eye and ulceration of the genitals. Further study of cases in the 1950s, including one on the Greek island of Rhodes, added more symptoms and seemed to strengthen its diagnosis in the Epidemic doctor's patients.[16] The syndrome has continued to attract specialist attention, but the more it is studied, the more it poses problems, some of which are still unsolved. One modern name for it is the Silk Route disease because it is well attested in Japan, also in China and along the route to Turkey and even in north Africa.[17] That route was irrelevant to Thasos.

Inflammation of the eye is indeed part of the syndrome ('iridocyclitis'), but growths on the eye have yet to be verified in a modern case, let alone growths which deprive sufferers of their sight. Behçet's syndrome damages the eye more intimately, including the retina, but the effects are only detectable with modern equipment. When the Epidemic doctor notes 'wateriness of the eyes, long lasting', he does not specifically notice inflammation, but Behçet's syndrome can certainly cause pain to the eyes, a factor he also remarks. He may be wrong that it was the outer growths, or 'figs', which destroyed many sufferers' sight, but correct that sight was often lost. However, its loss from Behçet's syndrome is usually the result of episodic outbreaks in an individual over a long time: the Epidemic doctor seems to have noted quite rapid destruction among 'many' at once.[18]

He does not class the condition among 'fiery heats', to us 'fevers': these, he says, will be written up later.[19] Fever can occur in Behçet's syndrome, but it is rare. Pain in the joints, especially in the knees, is more common, but the doctor-author does not mention any. The major problem in this retrospective diagnosis is that he describes the condition as affecting 'many'. Behçet's syndrome has a genetic element and, so far, has not been identified as infectious. It cannot, then, be the disease which the doctor describes as particular to the fourth year, but no other, and as affecting 'many' at that one time. Despite the seductive similarities, it is a separate condition, perhaps closely related to the bowel fluxes and bloody flows which the doctor also notes in the year, though apparently believing that they were distinct from the ulcerations of the mouth and genitals. Some of them may, in fact, have been interconnected.

Two cases in Abdera are more revealing. In one, Apollonios (perhaps the magistrate 'Apol-' on the city's coins) had an enlarged spleen and 'around the liver, continual pain for a long time'.[20] He remained upright, nonetheless, for a long time: he then had jaundice and suffered from wind. He then 'heated up a little and went to bed'. The doctor relates this change to his moving from a diet of sheep's products to one of cows'. He even emphasizes that previously Apollonios consumed boiled and fresh goats' and sheep's milk in great quantity and followed a 'bad lifestyle': these factors caused him 'great harm everywhere'; his fever intensified, his stomach could not excrete much; his urine was thin and slight; he had wind; he could not sleep, but he was comatose and very thirsty. 'Swelling up of the right side of his abdomen, with pain.'

The long phase of liver pain and a swollen spleen, then the jaundice, the drowsiness and the pain in the right side of the abdomen fit well with cirrhosis of the liver, or, even more likely, liver cancer.[21] On the thirty-fourth day poor Apollonios died. He had had some shivering meanwhile, followed by heat and a loss of his wits, suggesting a secondary fever had begun on top of the cancer. The doctor's careful and distinct observations allow a confident diagnosis, although he had no idea of it or its underlying causes. His own explanations in terms of changes and excesses of diet were wonderfully misplaced. Though wrong, he allows us to be right.

Also in Abdera, Anaxion, 'who was lying by the Thracian gate', was seized by acute heat and had 'continuous pain in his right side'.[22] Among other signs, he had a troubling cough which persisted and became dry. His breathing was difficult for a while but eventually, during a relapse into fever, he coughed up much 'ripe', or concocted, matter and recovered on the thirty-fourth day. Noting the pain in his right side, Galen interpreted the case as one of *pleuritis*, or pleurisy, but that condition is nowhere mentioned in *Epidemics* 1 and 3. The modern retrospective diagnosis of streptococcal pneumonia is preferable.[23]

Back in Thasos city, the brief case of the woman described as suffering from 'throttling' [literally, 'dog throttling', observed in dogs] well illustrates the complexities.[24]

> The throttling began from the tongue; unclear voice; tongue red; it dried out on the surface. On the first day, shivery; she heated up. On the third day, shivering, acute fieriness, rather red swelling, and hard, of the neck and of the chest on both sides; extremities cold, livid; breathing from high up [only]; drinking through the nostrils; she could not swallow; excretions and urine stopped. On the fourth day, everything intensified. On the fifth she died.

Retrospectively she has been diagnosed with Ludwig's angina (hence the throttling) or scarlet fever (hence the red tongue and red swelling, but scarlet fever is not attested in the ancient world), or diphtheria, a favourite modern candidate as it affects the throat and the voice (hers was unclear), paralysing the palate (the throttling) and spreading out to affect the body elsewhere (the swelling on either side of the chest). Bubonic plague has also been proposed, but the swellings were only on her chest: if it had been present it would surely have attacked many more people and become a lethal epidemic. My amateur guess is indeed diphtheria.

III

The strongest of these diagnoses, mumps, TB and liver cancer, confirm the author's insistence on accuracy. They validate his careful observation, a skill which is best seen in his case history of Philiscos, recorded on Thasos. He concludes it, unusually, with a brief summing-up: 'in this patient, right through to the end, breathing as if he was calling it up: for it was spaced out, loud'.[25] This rather literal translation shows that he was observing the effect known as Cheyne–Stokes breathing, named after the two Irish doctors who singled it out in the nineteenth century, 'a deep and noisy respiration', a modern expert remarks, 'that gradually diminishes and gives way to an apnea [absence of breath] that can last as long as 20–30 seconds. During it one has the impression that the patient has forgotten how to breathe; then he catches his breath, as though by conscious effort.' It often occurs near death. At death's approach, the 'Hippocratic face' has also remained famous, with sunken eyes, stretched skin and so forth, but it is not mentioned in *Epidemics* 1 or 3. It is first described in the text *On Prognostication* in the later fifth century BC.[26] If Hippocrates was indeed the author of the Epidemic text, he never mentioned the face which now bears his name.

Inevitably there were misunderstandings. One of the most influential was the disease phrenitis. When the doctor-author first mentioned it in his year-surveys, he assumed that his medical readers were already familiar with it: it was not, then, his own invention. Sufferers from it shiver and have a fever; they have heavy pains in their head and neck: they vomit rusty-green matter: they suffer spasms. Some of them die quickly, but in the third year-survey they are said to have reached their crisis on the eleventh or the twentieth day.[27] They generally lose their wits and go mad, though once, in the fourth year, none of them did so. Instead they suffered from sloth and became comatose, another frequent symptom.

A case history, probably on Thasos, sets out how this disease progressed.

The phrenitic patient: on the first day took to his bed, vomited up rusty-green matter, lots of it, thin: fever with shivering; much sweat continuously throughout; weight in the head and neck with pain; urine thin, small matter floating in it, which did not settle; from his stomach, excreted matter in a mass; much delirium: did not sleep at all. On the second day, voiceless early on; acute fieriness; sweated; did not leave off; palpitations all over the body; during the night, spasms. On the third day everything exacerbated. On the fourth day, died. [28]

These symptoms are confirmed and amplified by later Hippocratic texts. In the seventh book of Epidemics, in a case observed between c. 360 and 350 BC, the author vividly records Hippis's sister, who was 'phrenitic; erring with her hands and busying them, she lacerated herself on the fifth day; on the sixth, in the night voiceless, comatose, blowing into her cheeks and lips like people who are asleep; died about the seventh day.'[29] The Hippocratic text On Diseases is even more explicit, probably setting out what the author of Epidemics 1 and 3 and his readers already took for granted. The blood, it states, 'contributes the greatest part to the intelligence in man, or some say, all of it'. Phrenitis begins when bile starts to move, stirs the blood and therefore brings about fever: on Thasos the doctor-author, too, linked phrenitis to people of the 'rather blood-red and black-biled type'.[30] Such was the prestige of these descriptions, all believed to be by Hippocrates himself, that on into the nineteenth century people were characterized as suffering from brain fever, although the Epidemic author did not believe that intelligence, or the wits, reside in the brain. Phrenitis, in his view, affected the mind, but it was not located in the diaphragm, either (a usual meaning of phrenes). The mental and physical connection was left unspecified. According to On Diseases, sufferers become deranged and therefore refuse food and drink. They waste away, become cold at the extremities and then die. More specifically,

something like a thorn seems to be in their innards and to prick them; loathing grips the patient and he shuns the light and other people and he loves the dark and fear seizes him. His phrenes swell out [probably the diaphragm in this context] and he feels pain when touched. He is afraid and sees frightening things and terrifying dreams and sometimes, dead people.[31]

Here, the condition has taken on even more graphic contours than those which the Epidemic doctor-author specified. It 'grips most people during the spring', whereas on Thasos it was present in winter too. Sufferers even 'gaze steadfastly' with a fixed upward stare, the very word which was later to be used in the Acts of the Apostles for the heavenward gaze of the apostles at the Ascension and of the first Christian martyr, Stephen, when claiming to see the glory of God.[32]

Fever, shivering, delirium and refusal of food are the common symptoms which the Epidemic author, too, records in this affliction. Phrenitis has been diagnosed retrospectively as a 'toxic infectious attack on the central nervous system', but that reading does not pin it down. Perhaps, like '*phthisis*/consumption', the term covered several conditions and one accurate modern diagnosis is impossible. Or perhaps a phrenitic reader of this book can be more specific.

IV

The most challenging condition for modern diagnosers is burning heat. The doctor-author remarks that it shared signs with phrenitis, but unlike phrenitis, it was not related to bile nor were the excreta of its sufferers noted to be bilious. Nosebleeds distinguished it from phrenitis, he remarks, and were a good sign as they expelled blood and reduced heat. In the third of the year-surveys on Thasos, sufferers from burning heat had a typical shivering and some of them went on to develop jaundice on the sixth day. If they did not have a nosebleed, they often had swellings by their ears. Urine was thin and pale. Burning heat struck in spring and early summer, but few died of it. It also attacked in the winter.[33] Women suffered from it too, but less often than men: it was particularly dangerous for those who were pregnant.[34]

In this same year some of the cases followed a pattern which led to recovery. The final crisis occurred on the seventeenth day, but was preceded by a preliminary crisis and then a relapse. This pattern was exemplified by the brothers of Epigenes, who resided by Thasos's theatre.[35] Here moderns have diagnosed relapsing fever, transmitted by ticks or lice.[36] The swellings which sometimes developed by the ears of sufferers from burning fever are indeed an occasional complication of this relapsing disease, but cases of relapsing fever involve very heavy sweating followed by a brief bout of very high temperature, whereas the doctor-author records only slight sweats. The diagnosis remains open.

There is a good reason why it remains unresolved. The case histories of burning-feverish patients show that this term, too, does not refer to one disease which we individualize. Two of the cases suffered for as long as 120 days, one of them dying, one not. One of them went deaf on the fourteenth day, another one, a virgin girl also in Abdera, went deaf on the eighth day; both had pains in their body, one in the right hip, the other in her feet.[37] One modern condition has nonetheless been diagnosed: typhus.[38] This fearsome disease is transmitted by lice or fleas. As jail or camp fever, it was rampant in German concentration camps during the 1940s and rampant again in downtown Los Angeles in 2018. It is typified by a rash all over the body, but no such rash is reported in these cases. A likelier sufferer is young Silenus, after his drinking bouts and 'untimely exercise in the gym'.[39] He had an acute fieriness from the second day onwards, intensifying on the third when he lost his wits, kept talking at length, laughed and sang: delirium can indeed accompany typhus. On the sixth day his bodily extremities became cold and turned a livid colour, certainly one symptom of typhus, and on the eighth, as well as continuing to sweat, he developed little round red spots on his body 'like buttons: they remained'. They too fit the profile of typhus. On the eleventh day he died. Drink and over-exercise were not, then, the causes.

Fleas were no doubt active in fifth-century Thasos, but other carriers were even more important. The doctor-author distinguishes burning heat from another category, fierinesses (fevers) which were time-specific, tertian and quartan (two and three days long), and others, which were 'continuous'.[40] In the first year he also notes prolonged, but not fatal, fierinesses which had crises on the twentieth, most often the fortieth, but often, also, the eightieth day: if a relapse followed this crisis, another crisis occurred, usually after the same interval of time. Moderns have suggested brucellosis here, the so-called 'undulant fever', but brucellosis's typical spikes of high temperature are not noted and the sweating was only very slight.[41]

Among the 'continuous' fierinesses, typhoid, or enteric fever, is a convincing candidate, medically endorsed by specialist readers.[42] It spreads from water polluted by human excreta, all too readily available in a Greek city, despite those rules in Thasos city against leaving dung out on the main street. It is highly likely to be one of the continuous fierinesses presented in the third year. Among individual cases, experts consider it to have affected the man from Clazomenae who suffered for forty days, or the virgin by the Sacred Way in Abdera, who shivered acutely, went delirious and deaf, and had slight nosebleeds, as typhoid sufferers sometimes do, and pain in her feet.[43] In Thasos the 'wife of Dēialkos' wrapped herself up in her bedclothes, scratched and pulled her hair, became insensate on the seventh day and then died on the twenty-first. She too had Cheyne–Stokes breathing, persisting 'right until the end': her erratic hand gestures also suggest typhoid.[44]

The time-specific fierinesses point to a different culprit: malaria.[45] It fits very well with the author's tertians and quartans, the forty-eight hour or seventy-two hour types of fever. In some of the cases yellow jaundice is mentioned too, as is the excretion of black urine, both highly compatible with malarial fever. Malaria is caused by three types of parasite, transmitted into the bloodstream by bites from mosquitoes. Each type is now attested in antiquity, including the most dangerous one, which would indeed cause the doctor-author's 'malignant tertian', a fieriness recurring every forty-eight hours.[46]

In Thasos city, the case of Philiscos, in the third year, is the most compelling evidence.[47] On the third day, his fieriness intensified. From then on he passed black urine; he sweated intensely; he became delirious; his extremities became cold and then livid; he lost his voice; his spleen became hard and rounded; he fell into Cheyne–Stokes breathing; on the sixth day he died. The major absentee here is jaundice, a frequent accompaniment of malaria, but 'in a population whose skin was tanned by the sun and sea air', a modern medical scholar has claimed, 'a slight case of jaundice could easily escape notice'.[48] This explanation of its absence is rather evasive, but if it holds good, then the other symptoms, especially the hardened and rounded spleen, fit well with a diagnosis of malarial haemoglobinuria, or blackwater fever.

The case histories of women are also relevant here. Most of them relate to pregnancy and childbirth, times when a woman's immune system is lowered: the 'woman by the cold water' in Thasos, an eminently suitable breeding ground for mosquitoes, is noted as ill before she gave birth. A series of signs then follow, including black urine, which fit well with a malarial diagnosis.[49] Malaria may also underlie as many as eight of the nine case histories of pregnant women, where previous scholars merely diagnosed puerperal fever.[50] There is also the possibility of malaria as a secondary condition. Quite often, it develops in patients weakened by other diseases. It may, therefore, be a secondary presence in other Hippocratic cases, complicating a modern diagnosis of only one disease.

Greeks were well aware of mosquitoes' presence in damp and marshy areas, a profile which fits Thasos city. The area beyond the main *agora*, leading down to the sea, is now considered to have been flooded for part of the city's earlier existence.[51] The water table remained high there and elsewhere in the city plan, making mosquitoes a likely hazard. Mosquitoes are also sensitive to wind, affecting their range of flight: at Abdera, especially, the prevailing northerly wind blew directly into the city area.[52] The doctor surely knew the insect, but he had no idea of its role as a carrier of parasites. He continued to observe the sufferings of his specimen cases and to note the relation of their critical days to sweating and nosebleeds. Their very regularity supported his conviction that the critical days in illnesses could, within limits, be predicted. He had no idea of the reason why. Once he gave a malarial patient, Philiscos, an enema.[53] He never imagined he needed bug spray, let alone that faraway wonder, quinine.

V

If the 'three great killers', TB, malaria and typhoid, were indeed present in Thasos, they bear on the city's health and on the doctor's own achievement. They relate to his blind spots, contagion (airborne, for TB), infection (typhoid) and transmission by carriers (malaria). However, he had not gone to Thasos because epidemics had already broken out and needed urgent medical study and advice. In the first year, *phthisis*/consumption began to be prominent only in early summer, nine months after his arrival. It then accounted for 'many' lives, persisting into winter, but it had not been evident at the start of the year. Otherwise, diseases in his first year were mild and not fatal.

In his second year there were uncomfortable sicknesses, but still not fatal ones. The really dangerous ones began only in autumn, persisting into winter and including time-specific and continuous fevers, some of which recurred for several months. Malaria and typhoid were almost certainly prominent among them. Many were affected, but deaths were not the usual consequence, he noted, although they especially befell children. In the third year, however, mortality intensified. *Paraplegie* affected many in winter, but only some of the victims died. The bigger problem was burning heat, which began in spring. Malaria probably underlay much of it, peaking on the fourth and sixth days, but even so it did not prove fatal. Then in November and on into winter, phrenitis predominated, and in this bout 'many died'.[54]

In the fourth year, fatalities were even more frequent. Beginning in winter there was *phthisis*, probably quite often TB, which killed people either rapidly or later on, in autumn. A red-rash fever, or *erysipelas*, started in early spring and 'was killing many'.[55] Also in early spring, ardent fevers and phrenitis began and were joined by the ulcerating disease which affected eyes and private parts. Stomach sicknesses affected people too and 'to put it summarily, all of them, both those who were ill for a long time and those who had a short sharp illness, in each from stomach problems, kept on dying'. There were also many fevers, including the tertians and quartans and continuous ones.

These years are a unique glimpse of life in a Greek city. Whereas historians now labour to compile estimates of mortality and its structure in ancient Rome, the Epidemic author has left four years of observations in Greek Thasos, differentiated by the sufferers' sex, age and outcome. He does not refine them numerically: how many were his 'many'? The impact of the three great killers would vary according to density and distribution of population, local immunity from previous infections and so forth, none of which we can now compute. However, he is not describing only a few cases. If his years had been the years from 411 to 407 BC, 'many' Thasians would have been debilitated by sicknesses while they were fighting to regain their mainland dependencies and resisting Athenian military attempts to reassert control of their island. In *c.* 470 BC, the correct date of his presence, during that *belle époque*, life, we now see, was not so *belle* for the many who had one of the major fevers. Even so, from 465 to 463 enough Thasians were sufficiently robust to resist nearly three years of intense siege by the Athenians and their allies.

In early Christian Rome, historians now discern a 'mortality spike' in late summer, deduced from dated Christian epitaphs in the city.[56] In Thasos, eight centuries earlier, the doctor-author gives a more nuanced picture from his own directly observed evidence. In the second year, continuous fierinesses and time-specific ones occurred during autumn and on into winter. In the third year, burning heats were killers especially in autumn and on through winter. In the fourth year, consumption began in winter and had already killed many by the end of spring. The red-rash fever proliferated in early spring too, when burning heats and phrenitis also began: all three persisted significantly into winter. The doctor concludes that 'in all the conditions described, spring was the hardest time, whereas summer was the easiest', exactly the opposite pattern to the mortality structure which moderns postulate for ancient Rome.[57]

In ancient Rome, the late-summer spike has been explained by malarial fever. The most malignant of the parasites tends not to incubate in humans over the winter. It is transmitted only when the weather warms, making a surge in deaths during late summer more likely.[58] However, neither in Thasos nor in Rome was malaria the only killer at large. TB was present throughout the year, a disease transmitted by infection, especially inhaling: it is incubated for varying lengths of time, whatever the season. If typhus existed in Thasos, it too is not specific to one season, though warm summer weather might exacerbate the poor hygiene and the bad food and water which help to transmit it. In Thasos, the pattern inferred for Rome did not apply: disease did not spike in a late-summer surge. One reason may be the range of serious diseases at large, keeping people ill throughout the year; another may be variations in climate. In the fourth year's summer, there were periods of 'great stifling heat' but the usual Etesian winds from the north-east were only faint and intermittent.[59] In that summer, ardent fevers were relatively few, perhaps because the lack of wind affected the mosquitoes' flying zone and the consequent diffusion of the parasites which cause malaria.

VI

These variations are traceable only thanks to the doctor's detailed noting and discrimination. There was even more to his achievement than careful observation: he and his helpers were admirably courageous. Fearlessly, they observed lethal fevers at close quarters over long periods of time, recording cases whose symptoms, like the rotting away of flesh, were extremely distressing. As soldiers, Greeks saw frightful physical damage and suffering in battle, but they did not choose to spend months among it, returning voluntarily to record what they saw. The doctor and his helpers were selflessly courageous, under-appreciated heroes of the early fifth century BC, acting without an awareness of infection by contagion but well aware of the constant risk of death. They also had ethical ideals for the proper exercise of their craft.

They were acting to a long-term plan, the collection of data to help future hypotheses and predictions of a case's course when it recurred. The Epidemic author never records an experiment. He never attempted a randomized trial and never used measurement. His theory of patterns in odd- and evenly numbered days harked back to older notions of the significance of numbers. He had no theory of probability, no statistical method, no primary interest in causation. Nonetheless, his approach and medical style are unprecedented in our evidence for the Greek world before 470 BC.

His text recorded no recipes or treatments, items which sometimes reveal the low level of 'science' in other Hippocratic texts.[60] We do not know if he too endorsed treatments with exotic, mostly useless, ingredients or recipes based on a misplaced belief that 'opposites cure opposites' or 'like cures like'. In the absence of such prescriptions his reasoned thinking seems particularly impressive: how best should we characterize it?

The beguiling modern title, 'the first scientist', has been widely applied, whether to his Ionian predecessor Anaximander, a natural philosopher, or to Aristotle a century and a half later, or to Roger Bacon or Leonardo da Vinci or to protagonists of the seventeenth century's 'revolution'.[61] In his fine recent study of Aristotle's natural science, the biologist Armand Leroi isolates two styles of empirical investigation, one in 'which causal hypotheses are tested by deliberate, critical experiment', the other 'less familiar but hardly less important ... in which data are amassed, patterns sought and causal explanations inferred from those patterns'.[62] He sees the latter exemplified in the modern emphasis on big data and its ordering, a principle of science which has become central in the age of computer studies. He credits Aristotle with the second style and as a result credits him with the 'invention of science'. The Epidemic doctor did not approach his case studies with a research hypothesis in mind; nor did he use any word for 'discovery'; but his aim, too, was to collect, interpret and present big data, from which he and others could apply inferences fruitfully in future.[63] He was not necessarily the first to collect in order to predict, but he is the first person known to us to have done so. His notion of causality was not explicitly stated and defined. He did not bring about a paradigm shift in thinking, a broadbrush term which has become controversial among historians.[64] He was aspiring to be the master of a craft, not a philosophy.

'There was no science as we know it today in ancient civilisations', its eminent analyst, Geoffrey Lloyd, has justly concluded, 'Yet there were analogous ambitions – in relation to understanding, explaining, predicting a wide variety of phenomena.'[65] It is on those ambitions, the first to be fully attested for us, that the doctor-author's claim to scientific primacy rests.

19

Philosophers and Dramatists

The Epidemic author's thinking is terse, scientific and admirably free of gods. In the 470s BC, how does it relate to other surviving texts, both from neighbouring non-Greek cultures and from Greek philosophers, dramatists and historians who were active in the same period and its immediate aftermath?

In the East, in Babylonia, the old *Diagnostic Handbook* was still being amplified and copied in the fifth century BC.[1] Like the doctor on Thasos, it presumed that prognosis was a central aim of a healer's skill. Unlike *Epidemics* 1 and 3, it gave diagnoses of individual diseases, but they were not related to individual case histories developing over time. Nowhere does it use the Epidemic author's first person 'I', which would have shown that its contents were matters of personal judgement and discovery. Instead, anonymity gave an impression of objectivity, as if it was 'the' result, not someone's view of it. Its diseases were named by interventions of a god, or his 'hand', as if those intrusions accounted for them.[2]

Away in the West, among the Etruscans in Italy, interpretations of disease and misfortune were also most unlike those of our Epidemic doctor. They have taken on new life thanks to Jean MacIntosh Turfa's recent study of an old Etruscan calendar (the Brontoscopic Calendar).[3] Preserved in a late Greek translation, it notes the significance of thunder on each day of the year and makes frequent reference to warnings of sickness. As an authoritative reference work it too had no use for the first person 'I'. Among many examples, if there is thunder on 7 August, it 'signals harsh winds and sicknesses at the same time'. If thunder occurs on 14 September, it 'threatens diseases'. On 21 October, it means there 'will be coughing sicknesses and oppressions of the heart'. On 29 January, thunder means the 'condition of the air will be oppressive and disease-bearing for all'.[4] The Epidemic author, though aiming at prediction, never cited thunder as an omen sent by the gods: for him, omens were irrelevant. He discussed weather patterns in each season, but only because they related naturally to the diseases which prevailed in them.

Among Etruscans, divinely sent thunder threatened even more. On 5 August, it signified, ominously, that 'the women are more sagacious' and on 6 September 'if it should in any way thunder, there will be greater power among women than what is appropriate to their natures'.[5] On 25 January, it means there will be 'slave-fighting' and on 19 August that 'the women and the servile class will dare to carry out murders'.[6] These recurrent fears in the Calendar match the Greeks' impression that Etruscan society was based on a huge servile class and allowed far too much licence to women.[7] There were slaves on Thasos too, but they were never predicted to cause trouble, no more than were 'wives of' Thasian citizens, unmentioned by their own names. In antiquity the fears of the Etruscan calendar are unique.

In the Greek world, instead, there were philosophers. Their general interest in nature had led them to an interest in astronomy and to the formulation of theories to explain it: the Epidemic author, too, used astronomical observations. He marked turning points in the year by the rising or setting of the Pleiads or of Arcturus, points which had been known to farmers and herdsmen for many centuries and had already been cited in Hesiod's traditional poetry.[8] His reference to the 'turnings' of the sun, the winter and summer solstices, was also traditional: Hesiod too had used them. However, in the very first line of his text he referred without qualification to the 'equal-day', our equinox, the autumn one: he goes on to refer to it again, this time the spring one.[9] In a late fifth-century BC Hippocratic text, the famous *Airs, Waters and Places*, the author refers to solstices without qualification, but to equinoxes as 'what are called equinoxes': this qualification has been taken to show that, for him, the equinox was still something new and rather technical.[10] However, c. 470 BC the doctor on Thasos had been using the word without any such qualification. The date at which Greeks did, or could, calculate an equinox has been keenly disputed, but the high dating of *Epidemics* 1 and 3 changes the argument. Later antiquarians credited Anaximander of Miletus (*c.* 500 BC) with the concept, one which was related to the pointer he devised for a sundial, but their claim is controversial.[11] When using the term, the Epidemic author was not being innovative: he was more at home with a natural philosopher's discovery than was one of his successors in the later fifth century.

Among philosophers, the obvious forerunner for his medical thinking is Alcmaeon of Croton, settled in south Italy *c.* 500 BC. Alcmaeon is reported to have considered that:

> on the one hand, disease comes about through an excess of heat or cold; on the other hand, through an excess or lack of nutriment: its location is the blood, marrow or brain. Sometimes disease may also result from external causes, from the quality of the water, the local environment, exhaustion, hardship or something similar. Health, by contrast, is a harmonious blending of opposites.[12]

The Epidemic doctor would agree with most of these views, except for the role of the brain, which he did not endorse, and the emphasis on health as a harmony or balance, two words which his Epidemic text never uses. He never adopts Alcmaeon's notion of health and disease as related to pairs of opposite qualities, let alone as 'equality' or 'monarchy'. Unlike Alcmaeon, he never gives any hint of believing that his patients have a soul which is immortal: they simply 'end' and die. Alcmaeon's expert philosophic expositor, Jonathan Barnes, even considers there is no known 'argument for the immortality of the soul one half as clever as Alcmaeon's, the very first argument in the field'.[13] Admirably, Hippocrates considered there was nothing even to argue about.

The most famous of his philosophic contemporaries is Empedocles, active in Sicily from *c.* 470 BC onwards. In verse, Empedocles describes most touchingly how he was followed by big crowds wherever he went, 'some wanting oracles, whereas others, long pierced by grievous pain, demand to hear the word of healing for all sorts of illnesses'.[14] These verses are a vivid reminder of the scarcity of public doctoring in Greek cities and its inability to cope with the acute and widespread conditions it confronted. They also imply that Empedocles was more of a guru than a methodical doctor: he did not diligently collect cases and try to generalize from their outcomes. He even composed a medical work in verse as well as one in prose, a reversion to the poetic communication which had prevailed in an earlier age but was eschewed by our doctor-author and his followers.[15] Nonetheless, he was a natural philosopher and so medicine was important to his interests. He postulated four basic elements, earth, air, fire and water, and believed that differing proportions between them account for the differences between substances, including blood and bone. He even believed that they account for differing natures in individual humans: the Epidemic doctor, too, implies that individual natures exist in individual patients.[16] Like him, Empedocles gave a major role to heat in the body. He too accepted that food becomes blood and that excess blood is the source of the milk in a mother's breasts. However, he considered that food becomes blood by a rotting process, not by one of concoction or cooking. Here, as elsewhere, he differed fundamentally from the Epidemic doctor's views.[17]

In one poem he offered to teach its recipient 'all the drugs that are a defence against sickness and old age'.[18] Then, as now, anti-ageing prescriptions were not medicine: they were mumbo-jumbo. Recent discoveries on papyrus have made plain that in Empedocles' writings mumbo-jumbo and medical thinking co-existed, even in one and the same poem.[19] By contrast, the Epidemic doctor ignored Empedocles' underlying elements, fire and so forth. He had nothing of the guru or wonder-worker about him. In a later anecdote Empedocles is said to have diverted two rivers into the main river beside the city of Selinus in south-west Sicily in order to rid it of an epidemic.[20] The Epidemic doctor might have been impressed by this concern with the water supply as a natural cause of sickness, but he would never have endorsed Empedocles' more grandiose claims of a personal influence over the weather and the heavens.

Empedocles was credited by Galen with medical pupils, men of Sicily.[21] In due course Empedocles' four elements were to be vigorously denied by another author in the Hippocratic Corpus, but there is no evidence that on Thasos the Epidemic doctor had ever read a word of him.[22] It was with another philosopher, Democritus, that later fiction associated Hippocrates himself.

Democritus was a man of Abdera, visible just across the sea from north-east Thasos, a place which the Epidemic doctor indeed visited, treating patients there, three of whom were holders, I have suggested, of a top magistracy attested by the city's coins. However, Democritus was not even born before 460 BC: in real life the doctor-author of the Epidemics, perhaps Hippocrates, never met him before or during his work in the city.

A more tantalizing near-contemporary is Hippon, born on the island of Samos, though perhaps, like the doctor Sombrotidas, he migrated later to Italy. He is only dimly known to us but is described as *atheos*, godless, by later authors. They are Christians, however, who applied this term quite freely. It seems to have arisen from Hippon's views about the universe and his theory that it originated from nothing but water.[23] To assert it, Hippon did not need to have been an atheist in our modern sense, someone who denied the existence of all gods. He may only have denied their interventions in human affairs: if so, he is an interesting parallel for the Epidemic doctor's exclusion of divine intervention from his medical cases. However, a similar exclusion had already been evident in Alcmaeon: he is the most pertinent philosophic predecessor for the author of *Epidemics* 1 and 3, at least in evidence which survives, but in no way had he formed him or given him all his views.

In the 470s and 460s BC the Epidemic author also coincided with great changes in art and tragic drama. In sculpture, the new style of these decades is classed by many moderns as the 'severe style', detectable in such marble sculptures as the wondrous 'Kritian boy' in Athens, or in those carved for the temple of Zeus at Olympia *c.* 460 BC. Many of its figures present a stern expression, but the style's expert, Brunilde Ridgway, has picked out, too, an increase in 'characterisation', so that 'one should no longer speak of *types*, but more correctly of subjects', and an 'interest in the mechanics of expression. The range goes from quiet brooding to worried forethought to physical distress, and ends with the uncontrollable muscular distortions of death.'[24] Many of the Epidemic case histories also end with death, but they never dwell on its last-minute distortions. They too note brooding, worry and physical distress, but they do so as signs in a process of sickness, one whose episodes no sculptor tried to represent. In Thasos, the severe style was also present, whether in sculpted reliefs of gods, nymphs and Graces in the so-called 'passage of the *theoroi*' or that relief of the male banqueter with his wife and hunting dog. The severe style in this art, still somewhat archaic in Thasos, was not shaped by the invention of medicine, contemporary with it; nor was the Epidemic doctor reflecting some wider mood of the times when he combined brief, often austere, observation of individuals with a noting of accompanying emotional changes.

Statues do, however, add to our mental image of the men among the doctor's patients. He has no reason to mention it, but they were likely to have had a distinctive type of hair-styling. Some of them on Thasos may have begun to adopt a short haircut, in contrast to the older style of long flowing hair, the mark of many previous aristocrats. More strikingly, they will also have had their pubic hair stylized into neatly trimmed shapes. Visible on statues even in the age of the severe style, after 480 BC these razored male trims retained the 'standard shapes of a horizontal bar or neatly cut diamond', as their expert, R. R. R. Smith, has carefully itemized, with a fuller growth of hair inside this defined area, 'carved or engraved with small tight curls'. Around 450 BC these artificial patterns 'were abruptly abandoned in favour of an unstyled natural growth'.[25] On Thasos, the doctor's upper-class clients will still have had their pubic hair trimmed into shapes, just before the age of the untrimmed 'democratic bush'.

From the 470s BC onwards, tragic drama, like male pubic hair, was also undergoing remarkable changes. They are immortalized by the three surviving Athenian playwrights, few though their surviving plays are out of their far larger dramatic output, now mostly lost. The oldest of them, Aeschylus, is the Epidemic doctor's direct contemporary, active from the late 470s to the 450s. Indeed he sometimes uses the imagery of disease, drugs and healing: he even refers to distant Etruria as a land of 'many drugs'.[26] He also uses words whose basic meaning derived from a medical context. The most striking are words, used once, for a 'devouring' ulcer and for a 'licking' skin condition: they are known otherwise in the fifth century only in medical texts.[27] Was he aware of the new thinking and writings of the Epidemic doctor and his helpers, already active in the late 470s?

As the doctor's text was written for fellow-doctors and medical pupils and probably enjoyed a limited circulation, it would be most surprising if Aeschylus knew it directly. However, many other doctors were practising all around him and some of them might perhaps have exchanged ideas with him. One of them might even have passed on the Epidemic author's views second-hand. Much modern comment relates Aeschylus' use of these particular words to their occurrences in later Hippocratic texts, as if they are yet more examples of the great dramatist's awareness, even endorsement, of progressive thinking in his lifetime. However, words are one thing, their underlying theory quite another: words alone do not mean that Aeschylus was a champion of the rational, natural and ethical medicine to which the Epidemic doctor and his like adhered. They do not even entail that he had read or studied medicine in any way. Aeschylus may have been using the words at several removes, much as non-medics now talk freely of 'coronavirus'. He uses no word which is used technically in the Epidemics. He presents man's innards and emotions very vividly, but he links emotion and the liver as the Epidemic doctor never does.[28] He refers once to the effects of a skin disease as 'blossoming' on a sufferer, as does the Epidemic doctor, but this vivid image may have been in wide popular use.[29] He also writes of bile. When the chorus in his *Agamemnon* confront the crazed Cassandra, prophesying Agamemnon's murder, they refer to a 'yellow-tinged drop' which has 'run to their heart', as it does to men 'falling to the spear' in battle: perhaps bile is what Aeschylus means here too.[30] He even refers to the plugs which were inserted into open wounds.[31] Again, he did not need to have studied medicine in order to know about these items. The yellow drop, he states, was known to warriors, and the plugs, too, were familiar to battle-hardened soldiers. Famously, Aeschylus had been one himself.

So much for his supposedly up-to-date medical allusions: the plots of his greatest surviving plays are even more telling. They are in direct opposition to the Epidemic doctor's way of thinking. Admittedly, the doctor never had to attend a crazed matricide like Aeschylus' Orestes, but he would never have ascribed his distressed mental state to an ancestral curse or fault, let alone to hounding by the Furies. Imagine if he had gone down to Athens about ten years after his

observations on Thasos and attended in spring 458 BC the first performance of Aeschylus' great trilogy, the Oresteia. Like the dramatist
and his audience, he had long been exposed to the spectacle of human
suffering but, unlike many of them, he had not encountered it primarily as fiction in the theatre. He had faced it day after day in the
real world, in the agonizing cases which confronted him repeatedly in
his working life. The Oresteia's language and staging might have fascinated him, and like other attentive listeners he would surely have
responded to the cardinal remarks of the old men of Argos in the very
first chorus of the first play, the *Agamemnon*, that the god Zeus has
established that 'By suffering man gains understanding.'[32] This much-
admired statement had a specific scope in its context, referring not to
a general widening and deepening of our humanity but to a lesson
not to repeat what had hurt so badly. Once bitten, twice shy: 'instead
of sleep, the pain of remembering [his] suffering drips before man's
heart', constantly working on it, and so 'even to the unwilling, dis-
cretion comes'. In his medical work the doctor had repeatedly seen
pain in action, but he had not described its victims' responses in the
aftermath. He had no medical reason to note whether someone like
Nicodemus on Abdera, who had suffered after sex and drink but
recovered after twenty-two days of intermittent pain, had been
brought by remembrance of his suffering to self-control and dis-
cretion thereafter.

The Oresteia's plot would have seemed to the doctor to be from a
different mental universe. Set in early Athens it was not that it was
located in a different 'galaxy, far far away', but its presentation of guilt
and retribution indeed belonged very long ago. The central agents in
the drama, the Furies, Apollo, Athena and *miasma* (pollution), had no
place in his own way of describing diseases and derangement. Aeschylus even described Apollo as an *iatromantis*. The word means 'a seer
who is at the same time a physician'.[33] It was at home in the Homeric
and post-Homeric world with the likes of Calchas or Melampus,
but it was wholly at odds with the Epidemic doctor's conception of
the craft of medicine and its human capacity to predict by natural
observation.

As for Sophocles (dramatically active, *c.* 468–406 BC), forty or
more years after *Epidemics* 1 and 3, he was still presenting an

epidemic in Thebes as sent by the 'fire-bearing god' Apollo and caused by Oedipus' criminal incest and parricide.[34] His description of this pestilence and its dire effects on plants, cattle and people owed a debt to older poetic representations, but none to medical observation.[35] His play presents the cause of Thebes' pestilence as beginning to be understood only by a consultation of the oracle of Apollo. The old Homeric explanation of disease was still dominant. If Sophocles had been confronted with mumps on Thasos, he would probably have traced its predominance among young men to the anger of Aphrodite, goddess of sex.

Like Aeschylus he was highly unlikely to have encountered the Epidemic text, written for fellow-medics, not for general circulation. During his adult lifetime yet more texts on medicine began to be composed, several of which related to the context of a public lecture. If Sophocles ever heard or read one of them, he flatly ignored it in his dramatic writings. Later in life he presented the hero Philoctetes as incurably infected in his foot with a rotting condition which was only to be healed by attendance on the god Asclepius. This recourse to Asclepius was Sophocles' own addition to the play's plot: it befitted his well-attested welcome for the god into Athens and his composition of a poetic paean in his honour.[36] Hippocrates himself was an Asclepiad, one of the extended family or grouping on Cos which traced itself back to Asclepius, but if he was indeed the author of *Epidemics* 1 and 3, he gave his divine ancestor none of the role which Sophocles portrayed. When Criton on Thasos began to have violent pain in his foot, starting from the big toe, and when swelling spread all over it and little black blisters appeared, the doctor merely observed the case until its rapid conclusion, death on the second day.[37] He never considered that Asclepius would be any help to it whatsoever.

Sophocles' younger contemporary, Euripides (active *c.* 455–406 BC), might seem a more promising comparison for the doctor-author's medical understanding. Immortalized by the comedian Aristophanes for his trendy thinking, he seems the likeliest dramatist to have drawn on the doctors' new craft when presenting sickness and mental derangement on the stage. In his long lifetime, ever more medical theorizing was being presented orally and in texts, not all of which was for informed students only. Euripides was not wary of using anachronism in his play's speeches and he was well able to exploit the fashionable philosophic contrast between 'nature' and 'convention' in debates between his protagonists. Surely the new medicine would appeal to him too? It has even been considered crucial for the very making of his drama.[38] More precisely, Hippocratic medicine has been upheld as a source of enrichment and extra layering in his presentations of mental and physical suffering on stage. The onlookers and sufferers alternate, on this view, between old and new views of the condition, thereby adding 'uncertainty about which story to attach to symptoms'. Their differing interpretations, some being pre-Hippocratic, others not, are considered to engage the audience in interpretation of these powerful scenes, thereby adding dramatic complexity.[39] As Euripides' most famous plays were put on at least forty years after the cases in *Epidemics* 1 and 3, he had had plenty of time to grasp the insights which they presupposed, even if he too never encountered the text directly. He could also draw on Hippocratic thinking which had been advanced in the intervening years.

Certainly, Euripidean characters sometimes express progressive views on sickness and health. In lines preserved from an otherwise-lost play, one of them distinguishes diseases brought on by man himself and diseases sent, it seems, by the gods.[40] However, the former are probably wounds, and the wording and implication of these lines and their context are not at all clear. In another lost play the view was expressed that 'whoever knows how to doctor well must look to the lifestyles of those who inhabit a city and to their land when examining diseases'.[41] This stress on *diaita*, or lifestyle, and environment was indeed a hallmark of the new craft of fifth-century medicine.

Like Aeschylus' plots, the plots of Euripides' tragedies did not endorse that medicine's central aspect. In his *Hippolytus*, the chorus of women wonder what is causing the sickness of Phaedra, unaware that she is pining with desire for her stepson. As a first idea they propose that she is possessed by a goddess, possibly by Hecate or Cybele or Artemis. Only then do they wonder if her disease is caused by her 'distress' or perhaps by her 'woman's nature'.[42] These latter two suggestions can be matched in case histories of women described on Thasos, but it is the chorus's first suggestion, intervention by a goddess, that turns out to be correct in the context of the play.[43] It is also one which Hippocrates never endorsed: a later Hippocratic text was to dismiss it as the sort of explanation which charlatans typically offer.[44] If Phaedra had been a patient of the Epidemic doctor, he would have studied her urine and diet, her sex life and dreams, and reached the correct diagnosis, that she was suffering from rampant lust.

In 1929, the great classicist E. R. Dodds remarked on the 'fascinated precision with which [Euripides] explored those dark tracts of the spirit that lie outside the narrowly illuminated field of rational thought'.[45] The Epidemic doctor too observed madness and derangement with fascinated precision. No fewer than twenty-two of the twenty-six case histories in his first three years on Thasos and thirteen of the sixteen in his fourth year include symptoms of delirium, loss of reason, nonsense-talk and so forth.[46] However, they contrast sharply with presentations of the 'dark tracts of the spirit' in Euripidean drama. The doctor-author never adduces a god as their cause. He presents madness as arising internally in the human body, whereas Euripidean madness arises from divine agents, from vengeful Hera, her agent Lyssa (Frenzy), the Furies or cruel Dionysus. It is also much more extreme. It leads to matricide or parents' crazed killing of children. There are no such cases in any Hippocratic text.

In his *Iphigeneia in Tauris*, composed between 414 and 412 BC, Euripides presents Orestes' deranged state in notable detail. He stands stock still; he falls silent; he groans aloud; his limbs shake; foam collects around his mouth; his eyes roll.[47] These symptoms are symptoms of epilepsy and are noted exactly by the Hippocratic texts *Airs, Waters and Places*, of late-fifth-century BC date, and *On Breaths*, now considered to be a text of the late fifth century too.[48] However, there is no hint of these texts' thinking in Euripides' presentation. Madness comes over Orestes in what Euripides calls a *pitulos*, an onslaught, a word never used in the Hippocratic texts. Unlike the Hippocratic cases of epilepsy, Orestes' madness is sent by the gods. Euripides' text shows no interest in contemporary scientific thinking about epilepsy's relation to excess moisture, to the brain, blockages of blood and so forth.[49] The symptoms the play cites were everyday knowledge to anyone who had seen or heard about an epileptic fit,

There is one apparent exception: in Euripides' *Orestes*, composed in 408 BC, Orestes regains sanity and says '*Aluo*', I am distraught, or at my wits' end.[50] The very same word is used in *Epidemics* 1 and 3 for the wife of Dromiades, who had just given birth to a girl on Thasos, and for that young virgin at Abdera who was seized by acute fever, probably malarial had the doctor but known.[51] '*Aluo*,' Orestes says, 'releasing *pneuma* [air] from my lungs,' using the same word for lungs as in medical texts. 'From the waves I see calm again,' he then says, a line which became notorious in antiquity for being mispronounced by its actor. The Hippocratic *On Breaths* (*c.* 420–400 BC, probably) relates epileptic fits to the piling up of air, *pneuma*, in the veins, thereby blocking blood, until the air is released, partly as foam, and is expelled or diverted. 'Calm', the author notes, then occurs in the body.[52] However, this passing of air and foam occurs when its sufferer is still mad: Euripides' Orestes had just regained sanity. The doctor goes on to connect recovery to a warming of the blood and dispersion of the inner air, partly by breathing but also, a very different point, by its diversion into phlegm.[53] Euripides has no such theory in mind. He merely links Orestes' recovery to his release of breath from his lungs, a readily observed occurrence in epileptics who breathe deeply again after shallow panting. Although he and the doctor share the imagery of a storm and calm, it was surely current, as nowadays, in everyday speech about epileptic seizures.

In his *Heracles*, perhaps ten years earlier, Euripides had already presented detailed and awesome signs of Heracles' madness. 'See,' says its divine agent, Lyssa. 'He is shaking his head and rolling his eyes. His breathing is uncontrolled and he is roaring like a bull.'[54] The chorus, old men of Thebes, immediately present Lyssa as coming to madden Heracles with her 'head gleaming with the open mouths of a hundred snakes'. The mortal onlookers are in no way unsure of the agent, nor do they introduce dramatic uncertainty into the symptoms' interpretation. A messenger comes out of the palace and reveals that Heracles 'is casting out' bloody veins in his eyes; he is dripping foam from his bushy beard and has been speaking with 'skewed laughter'.[55] On Thasos the doctor had described young Silenus, who showed on the third day of his condition 'much talking, laughing, singing; he was unable to control himself'.[56] However, Euripides did not need to read a doctor's text to find these symptoms. He could draw on precedents in Aeschylus and in Sophocles' *Ajax*, including a specific mention of crazy laughter.[57] Next, his Heracles begins to hallucinate. He sets about killing his wife and children while imagining he is competing in the Isthmian games and able to break down the very walls of Mycenae single-handedly.[58] His madness here has been diagnosed 'as clearly . . . belonging to the manic-depressive type', but delusive mania and an epileptic seizure are quite different conditions.[59] For a dramatist, however, this unmedical combination had a great advantage. It could be readily acted and visualized, much more so than the prolonged cases of madness which are described so patiently day after day in the Epidemic doctor's text.

Heracles then recovers and, like Orestes, reports more about his breathing. He is now exhaling hot breaths, *metarsia* ones, he says, but not firm ones yet from his lungs.[60] The word *metarsia* occurs nowhere in a Hippocratic text, but a very similar phrase does, *meteoron pneuma* in *Epidemics* 3: Galen explained it as air released by breathing from the upper part of the chest only.[61] On Thasos the doctor attended that woman afflicted by a sort of 'throttling' and on the third day observed that there was shivering and *pneuma meteoron* and that drink came out of her nostrils: as she was constricted by angina, she could not swallow and could only breathe shallowly.[62] Like her, Euripides' Heracles was recovering, but was not yet fully recovered, as his constricted breath showed. Once again, Euripides pinpoints a particular type of breathing, but it is an obvious one in a case of epilepsy. Either he inherited it from general knowledge or he himself may have seen it in an epileptic case. It did not depend on any up-to-date medical study.

These scenes of transformation, from madness back to sanity, are dramatically brilliant, but they differ from Euripides' ultimate masterpiece, composed in 407 BC, the scene in the *Bacchae* with Agave, driven mad by Dionysus. Here too her symptoms begin from an epileptic template.[63] Agave foams and then her eyes stare. A fine analysis by George Devereux, himself a professional therapist, showed the clinical exactness of the sequel: Agave passes from initial 'hypomanic exaltation' to a full exit therapy by what Devereux well calls 'insight and recall', prompted by skilful questioning from the elderly Cadmus at her side.[64] At first she is evasive and still in inner denial, but then she passes to resocialization, Devereux notes, to the recognition of personal responsibility and promptly to the exoneration of herself from the dreadful deed she has done in her madness, the killing and rending of her own son. Unlike Heracles, she has been led to this realization by a third party, Cadmus.

Why has Euripides chosen to present a transformation in this new way? In Macedon, where the *Bacchae* was composed and first performed, there were living Dionysiac women, people who were linked in explanatory anecdotes to rites and dancing for Dionysus in the wild.[65] Real dancing in the mountains is attested among them, driving young girls beyond the limits of everyday health and safety.[66] In Macedon, Euripides may indeed have seen exit therapy being practised on a dazed maenad by 'some psychologically perceptive' person, Devereux suggested, one who had not 'been wholly superseded by the psychologically somewhat short-sighted methods of the Hippocratics'.[67]

Was the Epidemic author really so short-sighted ? Thirty-five of his forty-two cases record patients who lose their reason, talk nonsense and are mad and out of control. Did he just sit by the bedside during these intensely upsetting scenes, like those with the unnamed woman beside the cold water in Thasos who lost her reason and talked nonsense, regained it, then lost it and showed fits of anger, taking things badly and being *melancholikos*? [68] During her ramblings might he have tried a strategy of insight and recall in the hope of bringing her back to her senses? He shows no evidence of having visited Macedon, but Dionysus was a prominent presence in Thasos, accompanied by maenads in the city's public sculptures and in that imagery of a maenad stamped on its coins: maenadism among Thasos's girls is eminently likely, though unattested before a verse inscription of the first century AD.[69] The Epidemic doctor may, then, have been directly aware of this therapy's value in cases of mental disturbance. Yet, nowhere in the Hippocratic Corpus is it mentioned: it was not to Hippocratic doctors that Euripides owed his remarkable scene.

Unlike the woman by the cold water, Agave had been driven mad by the intervention of a god. Here, the doctor-author was not 'psychologically somewhat short-sighted'. The plots, myths and genre of tragedy required their dramatists to make gods the causes of their protagonists' sufferings. The Epidemic doctor, by contrast, looked for natural explanations and natural remedies. Hippocratic medicine was not important for the 'making' of Euripidean tragedy. Euripides ignored its approach, its theorizing and its central god-free core. He used a repeated template of symptoms related to an epileptic fit. Their origin was not always evident to their victim, but there was no inherent 'uncertainty about which story to attach to [them]', as if new medical thinking had complicated understanding of the predicament and thereby enhanced the drama.[70] In these scenes, as elsewhere, it is enhanced by something quite different, by pathos and irony, effects as old as their master Homer: the victim fails to understand the extreme cause and course of his derangement, whereas onlookers already do. Nor was tragedy important for the Epidemic doctor, perhaps Hippocrates himself. His case histories noted the preludes to, and arrivals of, times of crisis, but drama had nothing to do with how he presented them and the days of progress towards them.[71] The gods, essential to tragic drama, were kept out of it.

20

Epidemics and History

I

The doctor's understanding of the human body and illness was quite different from the presentations of the great Athenian dramatists of his lifetime. Does it overlap with the presentations of two other geniuses of the later fifth century BC, the historians Herodotus and Thucydides? They present a fascinating contrast between old and new ways of thinking about disease.

In *c.* 471–467 BC, the doctor-author on Thasos had preceded them both. In modern terms he had coincided with the passage from the archaic to classical, terms which apply in different ways and at differing times to art and literary history. In sculpture, the transition is typified by that new severe style with its interest in individuals, not types, and its concern with 'the mechanics of expression' and so forth, although its dating is disputed. In places where it appeared, especially in Athenian sculpture, it was certainly in evidence in the 470s; it was evident then too in Thasos, just when Hippocrates was active there in the north Aegean.[1] It was not, however, indebted to the new medicine which he represented; nor was he indebted to it.

In literature the transition from archaic to classical is also controversial, but I equate it with two manifest changes, the break with an old and pervasive acceptance of hereditary punishment of individuals and communities by the gods, the 'ancestral fault' whereby descendants might suffer for their ancestors' misdeeds, and the sharpening of language into a more exact relation with its items of reference.[2] This sharpening is seen in the move from the pervasive metaphor of an Aeschylus (*c.* 472–455 BC) and the 'linguistic density' of the philosopher Heracleitus (*c.* 500 BC) to the lucidity of Thucydides' narrative and the search for exact definitions by Socrates, both active from the 430s BC onwards.[3] The transition tends to be placed in the 440s–430s, although even then it was not universal. Four decades earlier, however, the doctor-author of the Epidemics was already not in the least archaic on either count. His language eschewed cloudy metaphor and made the most finely judged use of verbs and adjectives, exactly correlated with what he had observed and wished to convey. The old archaic bogies of pollution by misdeeds and divine retribution for earlier generations' wrongs had completely disappeared from his understanding of misfortune.

The shift from archaic to classical among the Greeks cannot be pinned to a single decade in every form of expression, let alone to everyone who was active at one and the same time as a sculptor, thinker or writer, as if they all shared one 'spirit of the age'.[4] It is, however, most manifest in the two great historians, Herodotus (at work from *c.* 455 to *c.* 425 BC) and Thucydides (writing *c.* 433–399 BC). There is nothing archaic about Thucydides' history, nor about Herodotus' language, which is in no way 'dense' like the philosopher Heracleitus' or riddled with cloudy metaphor. How do they relate, or not, to what the doctor-author of the Epidemics had exemplified about thirty years before they were at work?

Both of them included medical material in their histories. Both of them related it, albeit differently, to the craft, as Herodotus indeed calls it, of the new medicine. As their inquiries relied mainly on oral evidence, they too worked by questioning, listening and making judgements about people with whom they talked. Remarkably, both of them visited Thasos. Herodotus went there, probably on the journey which took him into Macedon, apparently to the court of King Alexander I.[5] As Alexander died in *c.* 454 BC, this journey probably occurred in the earlier 450s. Thucydides' links to Thasos went much deeper. In 424 BC he visited it briefly during his short-lived career as an Athenian general, but his family already owned property over on the mainland, including the right to work a gold mine. This mine's whereabouts have been contested, but Thucydides himself states it was in the part of 'Thrace around these places', these places, in context, being Thasos.[6] As it was a gold mine, it is highly likely to have been on Skapte Hyle, the Dug-Out Forest, which Thasos had lost to the Athenians in 463 BC: it was at that time that Thucydides' father received this valuable concession, helped, no doubt, by his family's connections with leading Thracians.[7] After his failure as a general in 424 BC, Thucydides withdrew from Athenian public life. While writing his history, he is highly likely to have spent time at the property near to Dug-Out Forest and the family's slave-worked mine on the southern slope of Mount Lekani, just inland from Kavala with a daily view across to Thasos itself. A late Byzantine life of him even refers to him writing there 'under a plane tree', but the author for this intriguing detail is not cited and may, alas, be fanciful.[8]

II

Herodotus' Histories have a wide curiosity and range, one of their abiding distinctions. In them he makes use of the contrast between custom and nature which fifth-century BC thinkers also exploit. He endorses the relations between climate and human character and climate and social organization which other fifth-century authors explore too.[9] His engaging curiosity about what he sees and how it is to be explained may suggest he is an up-to-date thinker, moving with the intellectual changes of his later years. However, the relationship between climate and human society was not a new discovery in the last decades of his life. He gives the impression that his mind was mostly formed earlier, *c.* 470–460 BC, when he was still a young man. His political outlook fits neatly into that era.[10] So does his religious outlook. He declares himself utterly against any 'speaking against oracles as if they are not true'. Denials of truth in oracles fit most readily with new thinking being advanced in the later years of his life. They were too radical for Herodotus.[11]

His Histories allude quite often to the health and medical treatments of people they cite. Much of this knowledge is likely to have been widely current, not as 'folk wisdom' but as the general knowledge of educated landowners like himself and his family, people who were in daily contact with women and children and with animals, sick or healthy, who were known in inner detail when dissected and offered to the gods.[12] As the author of *On Regimen* later reminds us, there was a general public with a broad interest in matters of health: Herodotus is an example. In the 460s, he was in his twenties while the doctor-author of *Epidemics* 1 and 3 was observing and writing his detailed text, that very model of the new medicine. He shows no sign of having read something whose assumed readers were specialized doctors. Like Aeschylus', Herodotus' medical allusions were not based on a text or learned from study with a medical specialist. Nor was his view that diseases come about from changes, especially changes in the seasons.[13] That general belief had been current at least since the time of Hesiod, who remarked on the direct effects of the Dog Star (the hot 'canicule' still dreaded in modern Parisian life), or of Pindar, who assumed a link between seasons and weather.

From the outlook of the doctor-author on Thasos, Herodotus, like Aeschylus, is separated by a profound gulf. When sicknesses and epidemics befall individuals and communities at random, he repeatedly ascribes them to divine retribution and intervention.[14] The most memorable involves the feminization of men who were well known as riders on horseback. Herodotus mentions that in the distant past, to us *c.* 680 BC, Scythians invaded Syria, where a rearguard stayed behind and ravaged the famous temple of Aphrodite in Ascalon.[15] As a result, he observes, they and all their descendants were afflicted by the goddess with the 'female sickness'. He adds that the Scythians agree on this divine cause for the condition and that visitors to Scythian lands can still see descendants who suffer from it. When he cites here what the Scythians say, he is not intending to distance himself from the explanation: the Scythians confirm what he has just stated in his own person.[16]

A Hippocratic text of late fifth-century date, the *Airs, Waters and Places*, also remarks that the Scythians ascribe this female disease to a god.[17] However, quite unlike Herodotus, its author disagrees with the Scythians' own explanation. He considers, memorably, that as Scythian males spend so much of their life riding horses, they are 'bumped' by them, presumably by their backs, making them 'feeble for sexual intercourse', a wonderfully incorrect belief, as life-long male riders well know. Like Herodotus, he also explains why 'most' men among the Scythians become womanish, speak with high squeaky voices and cease to be able to have sex with a woman. Their prolonged riding makes them lame and damages their hips, so they try to cure themselves by cutting the vein behind each ear and restricting the blood supply. When cut, these veins make them impotent.[18] The author ascribes their condition to natural causes, whereas Herodotus ascribes it to the vengeance of Aphrodite, the goddess of sex and love. His distance from enlightened Hippocratic thinking is manifest.

The most-cited counter-example in his history is more apparent than real. When Herodotus discusses the death of the Persian king Cambyses, he has seemed to some of his readers to come very close to a Hippocratic explanation.[19] He observes that troubles beset men for many reasons and then gives two possibilities for Cambyses' mad behaviour toward members of his household. Either he behaved insanely because of the god Apis, whom he had affronted in Egypt, or because 'from birth' he had the 'disease which some call sacred', what we now call epilepsy. Herodotus considers it 'not unlikely' that 'when the body is sick from a major disease the wits should not be healthy, either'.

The Epidemic author never described a case of epilepsy on Thasos, but later in the fifth century a Hippocratic doctor in what is now entitled *On the Sacred Disease*, probably composed *c.* 420 BC, denied that this disease was any more divine than any other: he described it as 'the so-called sacred disease'.[20] However, he considered that epilepsy might occur hereditarily, whereas Herodotus considered that Cambyses might have been afflicted from birth onwards, a different point.[21] When Herodotus called it the 'disease which some call sacred', unlike the doctor he was not distancing himself from this name for it. He was not deliberately contrasting a divine cause, the god Apis' revenge, with a natural one, epilepsy. His alternative explanations illustrate his previous point, that troubles beset men for many reasons. In the passage in which he has sometimes been considered to come closest to advanced Hippocratic thinking, he is not close at all.

He is no closer to it in his notions of foreknowledge. The word *prognosis* occurs nowhere in his Histories. In them, the guides to the future are oracles, though often ambiguous, and omens. In a brilliant chapter of his *The Greeks and the Irrational*, E. R. Dodds set out how the mass of traditional beliefs which passed down unthinkingly to each generation about gods and inherited guilt, what he and Gilbert Murray called the 'inherited conglomerate', began to break up in the late fifth century BC.[22] In tragedy it was very slow to break and in Herodotus' Histories, completed *c.* 425 BC, it was still intact. However, back in the 460s, the Epidemic doctor, perhaps Hippocrates himself, had already made the break before Herodotus wrote a single word. He had exemplified it in a text which was not the text of a showy sophist or an 'adversarial Greek'.[23] It was the text of a patient practitioner who wished to make the minds of students into quiet rationalists, in a decade which modern scholars place for other reasons on the cusp between the archaic and the classical age.

III

Herodotus' later years of writing his Histories overlapped with the early years of Thucydides writing his. To read Herodotus' younger contemporary is to engage with a markedly different thought-world. Like the doctor-author on Thasos, Thucydides is explicitly concerned with the accuracy, *akribeia*, of what he reports.[24] The word and its relations are used nowhere by Herodotus. Like the doctor, Thucydides stresses the value of foreknowledge and links it to the usefulness of study of his text.[25] He is quintessentially an author of the post-archaic age.

Some of his modern scholars have read him as if he approached his entire subject, the Peloponnesian War (431–404 BC), in the manner of a Hippocratic doctor. A 'medical template' has been discerned in several of his extended discussions, not just in his famous discussion of a major plague but in his extended account of factional political strife, though its presence there is far from obvious, and, even less persuasively, in a brief digression on the Athenians' finances, their changes in tribute and their, supposedly, 'impassioned and diseased behaviour' in connection with their disastrous expedition to Sicily.[26] No one model will account for Thucydides' approach and aims, but there is certainly still force in the contentions that he would have compared his procedure to the 'practice of any honest doctor' and that 'the physician studied [man's] clinical behaviour, Thucydides his political behaviour'. [27] He presents what participants did and suffered as a guide to what might happen again in future: on Thasos the Epidemic doctor had done much the same. Like him, Thucydides was not affirming cast-iron rules: he too was aware that particular circumstances might vary in future times.[28] Like him, he wished his text to be useful for future examples of similar predicaments, in his case those of a war and the motives, decisions and conflicts it would involve. Just as the doctor gave no specific advice on treatments in his text, so Thucydides never gave advice in his own person on how citizens should behave and decide in comparable circumstances in future. Above all, like the Epidemic doctor's text, his Histories never cite intervention by the gods as an explanation of events. Without any intervening technological change, Herodotus' thought-world has disappeared. There is not a word in Thucydides about fate, personified retribution or divine punishment.

Did the new medical thinking give Thucydides the impetus to write history in this novel way? One problem with postulating a direct medical influence on his undertaking has been uncertainty about the dates of the Hippocratic texts. When Thucydides began writing in the late 430s, which, if any, of them existed? Now, a major medical text turns out to have been in existence, *Epidemics* 1 and 3, based on observations as early as *c.* 470 BC. It was compiled and composed before Thucydides was even born.

At first sight Thucydides is less engaged than Herodotus with medicine. He does not dwell on individuals' illnesses and has no reason to describe the health and types of healing among non-Greek peoples. He never even refers to doctors as present with an army. However, he makes two references to them, both of importance. One is that tantalizing remark in the fateful debate to reopen the Athenians' decision to send an expedition against Sicily, the ill-judged result of a people's vote.[29] Thucydides presents the reluctant general Nicias as advising the official who is presiding over the assembly to put the matter to a second vote and ignore allegations of illegal behaviour. He must act 'like the doctor of the city which has counselled badly' and he must 'help it as much as possible, or do it no willing harm'. Exactly this ideal, 'to help or do no harm', had been stated by the author of *Epidemics* 1 as the aim of a good doctor.[30] The parallel is striking but, of course, it does not prove that Thucydides had read and remembered the medical text. The words may have been, or become, a widely held view of a doctor's ideal, possibly known to Thucydides from general discussions in the world around him.

Even more famous is Thucydides' long presentation of the fearful plague which was to devastate the city of Athens from summer 430 BC onwards, killing perhaps as much as a quarter of the population.[31] He was explicit that nothing like it had occurred anywhere before. Exactly observed and profoundly moving, his account remains a high point of history-writing, one which was to be imitated, at least in part, by Greek historians on into the sixth century AD.[32]

Thucydides' account of this plague has many points of contact with *Epidemics* 1 and 3. Like their author, he states that his presentation of the illness will help anyone confronted with a future outbreak if one ever occurs again: 'having some foreknowledge the reader will not be ignorant'. What he means here, as his modern connoisseur Geoffrey de Sainte Croix well understood, is not just that the reader will know that it is the same disease again, the one which attacked the Athenians: he will know how it will tend to develop and so he will be able to add that knowledge to his reactions to it.[33] This eminently Hippocratic aim is the basic reason why Thucydides devotes such length and detail to the plague's description.

Like the Epidemic doctor on Thasos, Thucydides does not prescribe any treatment for the disease he describes. He too prefaces his account with remarks about the season's weather, though in his case in order to note that there was nothing exceptional about it: he is not deliberately dismissing the weather's possible relevance in other cases, but he is implicitly emphasizing a major theme, that this particular disease was unlike any before.[34] Like the doctor-author of the Epidemics, he too refrains, in his case explicitly, from proposing a specific cause. He too uses the word *prophasis*, and in the context of the plague uses it like the Epidemic doctor, to mean a 'first manifestation' of what became the disease in full.[35]

These similarities are striking but Thucydides also observes that doctors had no understanding of why the plague had arisen or how to treat it.[36] Was he implying that medicine, after all, is not a well-founded craft? Like the remedies which were attempted, he notes, details of the disease's attack varied from one individual to another. He also notes that lifestyle, *diaita*, made no difference.[37] The word *diaita* is the word widely used for a helpful regimen in doctors' texts. Thucydides also refers to 'purgings of bile, all those such as have been named by doctors'. However, in none of these remarks is he belittling medical expertise in general.[38] Authors of the Epidemic texts, from books 1 to 7, are also dismissive of other doctors' mistakes without in any way dismissing medicine itself. When Thucydides remarks on doctors' inability to cure or explain and on *diaita*'s irrelevance, his intention, yet again, is to emphasize the disease's exceptional nature. He too is not wishing to discredit medicine as a whole. Far from it: when he describes the disease's main features, he uses language which is often the sort of language a trained doctor would use. 'The overall impression . . .', Rosalind Thomas has well concluded, 'is of a sustained display of medical terms, sophisticated and unusual, if not actually "technical" '.[39]

His use of such terms proceeds from the head downwards. He begins by describing the plague's effects on the head and then follows its progress down the body until it 'fixed itself' in the stomach. This 'fixing' is an item in Hippocratic texts, but Hippocratic authors specify that what moves downwards is a specific flux or humour.[40] Conversely, Thucydides mentions the downward movement only of general pain. He also notes blistering and red rashes, items noted by the Epidemic doctor in many cases, but they were surely too obvious for any description of the plague to have omitted them. More strikingly, Thucydides too characterizes cases of the disease by their duration in days, 'the seventh or ninth day' being often fatal.[41] He states that sufferers died from 'the internal burning heat', that repeated item in the Epidemic doctor's physiology.[42] Like him, Thucydides also notes sleeplessness and attacks on the body's extremities. However, these attacks were not signs of a critical day, as they often are in the Epidemic doctor's text: Thucydides remarks on them for a different reason, the fact that they caused permanent damage to people who had already begun to survive the disease.[43] Like the Epidemic doctor, he notes mental derangement, but in his presentation it too occurs only after the disease had subsided, when some of the survivors lost all memory of themselves and their families.[44] The Epidemics' cases note the coming and going of mental derangement only *during* a disease's active course. They agree on the harmful effect of 'despondency', but it is Thucydides, not them, who gives an exact account of the beliefs from which despondency most often arose.

'I had the disease myself and saw others suffering,' Thucydides remarks with exemplary brevity at the beginning of his account, something the doctor never says of himself on Thasos or elsewhere. So precise is Thucydides' description that, like the doctor-author's cases, the plague has attracted frequent attempts at a retrospective diagnosis. Over the centuries many candidates have been advanced, smallpox, measles, typhus, glanders, ebola, SARS and even a toxic influenza labelled 'Thucydides syndrome'.[45] On one or other point each of the candidates fails to match what Thucydides described. Is his presentation of it therefore deficient?

There is no other contemporary textual evidence to validate his description or help to identify its subject, but, recently, archaeology seemed to offer a control: a mass grave was excavated in Athens's Kerameikos cemetery, dating to *c.* 430–420 BC, and was taken to support Thucydides' memorable remarks about the throwing of sufferers' corpses into heaps instead of giving each a proper burial. When DNA was extracted from three sample teeth, it was argued to show the presence of typhoid fever.[46] Not only has this conclusion been contested: the plague which Thucydides describes cannot be typhoid anyway. It infected birds and animals who came into close contact with human corpses: Thucydides singles out birds of prey and dogs who died or avoided the dead bodies, having seen, presumably, the results of close contact. Typhoid fever never transfers from humans to animals.[47]

As the Epidemic doctor's case histories have already exemplified, retrospective diagnosis is a fragile business, beset with difficulties at both its ancient and modern end. Bacteria evolve over time and diseases die out: Thucydides' description fits no disease known nowadays, not because it is inaccurate, let alone a literary fantasy, but because its plague no longer exists. At most, a descendant of it, much milder in form, may still be at large, measles being a favoured candidate, but a descendant is not the same as a parent.

IV

As a historian, Thucydides was especially interested in the cohesion of communities and the values and actions of individuals. Unlike the Epidemic doctor, therefore, he dwells memorably on the plague's impact on morality and morale. None of the Epidemic authors ever included such observations: they were irrelevant to their medical focus. He remarks that no fear of the gods nor human law restrained people. Considering that they would die anyway, they lived for the moment, he comments, putting pleasure first.[48] They saw how abrupt were the changes of fortune of those who were prosperous and then suddenly died, and of those who had had nothing but then took the dead's property. They considered both their bodies and their possessions to be ephemeral, Thucydides states, and acted accordingly. He alludes here to a sexual free-for-all (the bodies) and a redistribution of property (the possessions), neither of which is fully allowed for in modern social histories of Athens in this decade.

He does not mention the plague's physical impact among women, a point which the Epidemic doctor regularly noted in his surveys of a year's diseases. He does, however, bring out the 'most dreadful aspect of the entire evil': the despondency which it caused in victims and the fact that those who took to nursing one another became 'filled up' with the disease 'and died like flocks [of sheep]'.[49] He writes poignantly of those who, out of 'shame', went to help their friends, although family and kin had given up on them. These selfless helpers then fell sick and died themselves: so, at times, did attendant doctors. Survivors, Thucydides remarks, had noticed that the disease never attacked them fatally if it returned a second time. He even notes how, in their joyous relief, they expected that no disease of any kind would ever attack them thereafter.[50]

Among this masterly social and psychological analysis, Thucydides has been credited personally with two novel medical insights: contagion and acquired specific immunity.[51] However, he does not single out either with an abstract name and he was not the only person to notice examples of them. Acquired immunity was implicit in the survivors' belief that the disease would never attack them fatally again. Contagion was implicit in the observation, also not confined to Thucydides, that carnivorous birds and animals who came into contact with the corpses caught the disease, and above all that those who went to care for the sick fell ill themselves and often died. 'One from another,' he wrote, 'in their attentive care' they were 'filled up'. Maybe the 'filling' is a metaphor, but it shows a striking awareness that the disease passed 'from' one person to the next.[52] Survivors, friends and families of the carers and even dog-owners surely noticed these facts, but Thucydides is the first extant author to give the underlying observations. Made by eye witnesses, not specifically by doctors, they then featured, as never before, in how people were said to regard the risks of attending a very sick patient: carers, it was realized, might become sick too. In the first century BC a notion of invisible entities in bad air was then postulated, probably by the doctor Asclepiades, from whom it passed into Latin and featured in a vivid description of plague by the poet Lucretius.[53] Thucydides had never mentioned it.

Even so, the most striking aspect of his account is something else. Like the Epidemic doctor, he nowhere ascribes this epidemic disease or its origin to divine intervention. At the end of his account he notes two oracles which contemporaries applied to it.[54] One, he explains, involved them citing an oracle with a wording which supported its apparent relevance: men 'were adapting their memory to what they were suffering'. The other oracle is the subject of his last words about the disease. It was known only to some, he says. It was the oracle which Apollo at Delphi had given to the Spartans when they asked if they should go to war: 'if they fight with all their might,' the god's oracle declared, 'victory will be theirs and he himself will assist.' Thucydides then notes that 'people were surmising that what was happening was similar: the disease began when the Spartans and their allies invaded and it did not extend south to the Peloponnese to any notable degree.' Is he himself implying that Apollo, the god who traditionally sent plague, was indeed behind the epidemic and was active, as promised, on the Spartan side?[55] He does not say so. He reports only what people 'surmised'. He had already insisted that, for him, the cause of the disease remained open: 'let anyone, whether a doctor or a layman, speak about it as he thinks'. Thucydides was not an atheist. He assumed that gods existed, but he did not think that their interventions could be ascribed with certainty to human events. So he left them out of his historical narrative.

Illuminatingly, his reticence here was beyond the capacity of his imitators. The closest was about thirty years younger, the Sicilian political figure and historian Philistus, known to Cicero as 'almost a mini-Thucydides'.[56] Philistus is the ultimate source of another account of a devastating plague, one which afflicted a big Carthaginian army in Sicily as it attacked Syracuse in the 390s BC.[57] Following Thucydides, he described the epidemic and its effects, both medical and social. He too remarked that the troops were crowded together and that the summer was unusually warm. He even considered that the environment was relevant, as it was marshy and had a previous history of disease. He noted that the mornings began by being chilly as a breeze blew over the water, but that the middle of the day was stiflingly hot. The 'stench of the unburied bodies' and the 'rotting from the marsh', he considered, caused the disease to begin with catarrh. Yet, beyond these natural causes and their link with bad air, he stressed something else: the Carthaginians had begun by looting the temple of Demeter and Persephone in Syracuse's suburbs, thereby bringing divine punishment on themselves.[58] Mini-Thucydides explained the plague on two levels, one divine, one natural. If Thucydides had wished to do so, he too would have been explicit. He did not so wish, and even his closest imitator found the godless account of his master too reticent for him to follow.

V

Why was Thucydides able to give such an account whereas Herodotus, his older contemporary, could not? Part of the answer is that they belonged to different generations, Thucydides maturing in the generation of the 440s BC, among hard-headed realists in Athens, active and articulate in the years of Empire and increasingly exposed to the intellectual lecture culture and accompanying questioning of the travelling thinkers, whom Plato categorized as mere 'sophists'. There was also, I believe, a difference in their awareness of a crucial medical text.

If anyone on Thasos had offered to show the visiting Herodotus the Epidemic doctor's text about the sicknesses which had beset them and their families in the late 470s and early 460s, he would have excused himself from taking the offer up: the sicknesses belonged a decade after the concluding point of his Histories and, as usual, he had no inclination to wade through such dense and specific information in a text. For Thucydides, that text was much more pertinent. His family, owning property nearby, had friends on the island and a long familiarity with its main city. In the 440s BC, when Thucydides was young, the text of *Epidemics* 1 and 3, now some twenty years old, was certainly available in principle, although it was a text aimed at pupils and medical followers. It was close in both space and milieu to young Thucydides as it was not to an Aeschylus or Euripides. Parts of his formative years were surely spent in the family home near Dug-Out Forest, looking across to the very island on which most of the text was set. He could have had access to a copy owned by an aspiring young doctor on Thasos whom he knew personally or a copy belonging to one of the upper-class families there whose very members the text named. Written as a teaching text, it did not circulate widely, but it was not kept hidden from outsiders. It had good reason to interest Thucydides with his local connections and emerging intellectual interests: it was set on one of his family's doorsteps.

If it came his way, it was, I believe, of much more than local interest: it was a revelation to his thoughtful mind. In the 440s it was the one example he could have found of year-by-year, day-by-day narration which intermittently used the first person 'I', and which made no reference whatsoever to divine involvement. It was not just giving wise advice, like a speaker in Herodotus' Histories, or saying that the past is a guide to the future, a platitude which Thucydides could have heard very often in tragedies performed in his native Athens. It rested on a different realization, that the nature of the human mind and body will respond in similar ways to similar stimuli, so much so that its responses can be predicted, at least for the most part, from cases observed and recorded.[59] It was, I think, a revelation he never forgot.

There were to be many other formative influences on him, including many debates and conversations now lost to us among his social peers in Athens. There is no proof, because Thucydides never cites any such readings, but one major influence, I believe, was simply this: he read or heard our *Epidemics* 1 and 3, a work, I also believe, of the great Hippocrates, written in the 460s BC. He did not try to imitate its terse, disconnected style. As a historian he had to be more expansive. It fascinated him with something else, its exact observation of day-by-day cases and seasonal changes in order to enable prognosis about possible future recurrences. Thucydides 'was not such a fool as to think that "history repeats itself" ', Geoffrey de Sainte Croix has again well observed, 'it is above all *patterns of behaviour* which are likely to be repeated, although even then there will always be different factors involved'. In his Histories, Thucydides provided a 'series of political case histories' and 'like the author of the Hippocratic *Epidemics*,' de Sainte Croix, a connoisseur of Greek science, concludes, '[he] intended the knowledge gained from them to issue in informed and intelligent *action*.' [60] Above all, no unpredictable god was involved in the narration and its analysis.

Thucydides' history moves in a different thought-world to Homer's poetry. It has nothing to do with Pindar. It has nothing to do with tragedy, either, whose divine framework it excluded: even its most poignant catastrophe, the Athenian army's retreat from Syracuse, is told without the inevitability that a tragedian would have imposed on the plot. Into its making went something very different, an awareness of the Epidemic doctor's thinking, probably his very text. By *c.* 433 BC, when Thucydides began to work on his Histories, there was no other medical text which combined case narratives and such lapidary observation. Medicine was not the one and only 'model' for Thucydides' histories, but it was certainly one of importance.[61] It lies behind the length and detail which he chose to give to his account of the Athenian plague. It presented a breakthrough in accuracy (*akribeia*), in qualified prognosis and in godless narrative. They were breakthroughs which the *Epidemic* author exemplified and which Thucydides sustained in the next generation, but which passed Herodotus, his older contemporary, by.

21

Hippocratic Impact

So far as we know, the Epidemic doctor wrote up only four years of observations in cities from Thessaly to Thasos and beyond. If he was indeed Hippocrates, he was not forestalled by death, not even by one of the fevers which he attended and recorded day by day without any idea of infection or contagion. As Plato's references to him imply, Hippocrates lived on for more than thirty years into the 430s BC. No other text survives with a vocabulary close to *Epidemics* 1 and 3. He may have written one, but only if his style, language and thinking changed drastically might it be a text now in the Hippocratic Corpus.

Where is the influence of this Epidemic text still visible? Thucydides, intellectually curious and living near to Thasos, was an exception. Otherwise, its assumed readership of fellow-doctors and its presupposition of shared background knowledge meant it was not going to be widely consulted outside its local or medical milieu. Though the earliest datable text in the Hippocratic Corpus, it already represented the features which we picked out as typical of many of the Corpus's later fifth-century texts: the emphasis on ethical principles, the relation of diseases to seasons and climate, and the exclusion of treatments or explanations which assumed the intervention of a god. This shared Hippocratic core was probably due to a shared milieu of medical discussion and teaching rather than to the Epidemic author's own direct influence. The relations of weather and sickness are an example. Several of the later texts in the Hippocratic Corpus, especially *Airs, Waters and Places* (*c.* 420 BC) and the short *Aphorisms* book 3, assume a link between the weather, the seasons and sickness, but the interconnections they make do not match those which the Epidemic doctor's year-surveys had already presented. *On Humours*, mistaken by moderns as an *Epidemics* book 8, even remarks that it is possible to predict the weather from pains and illnesses: skin conditions and pains in the joints predict rain.[1] The Epidemic doctor had never asserted such a possibility.

The text *On Prognostication*, composed perhaps *c.* 420 BC, is the one which addresses a central concern of the Epidemic doctor most fully. In antiquity it was ascribed to Hippocrates himself, although that belief happens not to be attested until *c.* 300 BC.[2] All major readers and users of *On Prognostication* then accepted that Hippocrates was indeed the author, from the Alexandrian doctors and scholars to their Roman heirs. However, its language and its numerical theory for predicting critical days, using fourths, or tetrads, are quite different from those in their predecessor, *Epidemics* 1 and 3.[3] There is another cardinal difference, evident in its very first chapter. The author discusses what a doctor must recognize in severe, possibly fatal, diseases: he states that he must consider 'their natures, how much they are beyond the power of the body and whether there is something divine in them' so as to predict their course.[4] Already in antiquity many attempts were made to define what the author meant by this 'something divine': one of the earliest suggestions, wholly implausible, was that he meant the system of critical days. A modern suggestion that he referred to the heavens and therefore the weather is no more convincing.[5] 'Something divine' refers to any divine influence which might be active in a severe disease, whether discernible in omens, dreams or other interventions. Right from the beginning, therefore, this text was at odds with the god-free nature of the Epidemic doctor's approach. The other notable exception to that approach is *On Regimen* (*c.* 400 BC), but it was addressed to non-doctors as much as to doctors: its author even recommends prayer to the gods and indicates which gods are appropriate to address in the event of favourable and unfavourable signs.[6]

Might one and the same author have written *On Prognostication* first, then the Epidemics some years later? The suggestion founders on the dating of *Epidemics* 1 and 3 some sixty years earlier in the 470s/60s BC. It is equally implausible to regard *On Prognostication* as a work of the Epidemic doctor's older age when his thinking might have changed. He would have had to change too much, his vocabulary, his theory about the calculation of critical days and his principled exclusion of divine intervention. *On Prognostication* was written by someone who shared some of the Epidemics' interests but was not the same person. Its ascription to Hippocrates is as tenuous as other such guesses in the rest of the Hippocratic Corpus.

II

The most evident impact of *Epidemics* 1 and 3 is the existence of two later clusters, *Epidemics* 2, 4 and 6 and also 5 and 7. The author of 5 and 7 was working in the 360s and 350s and is the more faithful to his earlier model, abbreviated though his work now is. The doctor-author of 2, 4 and 6 is fascinatingly different. He was at work in the last decade of the fifth century BC and I have already argued that his approach and method are strong arguments for dating 1 and 3 much earlier because their method and content diverge from his. How profound are these divergences?

His text, as we have it, is often difficult to understand. It poses complex issues of transmission and correct reading at many points, still under discussion, for which the Arabic translation of Galen's commentary on books 2 and 6 is of major value.[7] It is clear, nonetheless, that the author used elements known from *Epidemics* 1 and 3, including references to named individuals' cases and general weather conditions. However, he integrated them into a very different sort of work.

As Galen and others inferred, the text of 2, 4 and 6 was made up of notes, perhaps intended to serve as starting points for oral expositions to pupils and then eventually to be expanded and made more readily intelligible for readers. It is clear from at least one remark that the author was noting down details of a disease, that widespread cough in Perinthus, while it had not yet run its course.[8] The text differs, therefore, from *Epidemics* 1 and 3, which were written up after the conclusion of each of the cases they record. It also lacks 1 and 3's clear coherence, being notes not yet ready for publication. The fundamental point is that even if the author of 2, 4 and 6 had gone on to write his notes up smoothly, they were not notes from which a work like *Epidemics* 1 and 3 could have been constructed. Even at that preliminary stage he was not simply imitating his ultimate master.

He shares some of the interests of that master's text, but does not give them the same prominence. He is aware of concoction and of prognosis but neither is a major concern for him. He has more to say on topics which his predecessor treated much less fully. He is more explicit about kindness to patients and what it involves.[9] He is explicit about treatment, whereas books 1 and 3 were not. In book 6 especially, he states propositions without giving supporting evidence. He makes summary statements, sometimes puzzling to their reader, of what he has come to know: 'for one who is about to go mad, this signals it in advance: blood is collected into the breasts'.[10] He does not, however, claim to know everything. Whereas the author of 1 and 3 had written only once, 'this must be examined', the author of 2, 4 and 6 quite often leaves a question open or presents his uncertainty and the need for more study.[11] When he considers the particular side of the body on which pains become manifest, he raises the question of how frequent this phenomenon may be: 'is it true for all cases', he asks, 'or only when what is below rises?', citing various examples and wondering if it is as true of others as it is of effects linked to the testicles.[12] 'These things must be investigated,' he concludes, 'where and whence and why.' This little note has been called a scientific soliloquy and, indeed, it raises questions needing further research.[13] It differs altogether from the list of factors which the author of 1 and 3 at or near the end of his text stated to need consideration here and now. It also differs from 1 and 3's general use of the phrases 'what must be known' or 'what must be investigated'. In 1 and 3 they refer to items which he already knows and which his pupils must now know too or which they must note immediately.[14]

The author of 2, 4 and 6's notes show that when doing his research, he engaged in inquiries which were not at all those of 1 and 3. The most remarkable is his long discussion of the veins and inner arteries, based on his own dissection of, evidently, an animal, the only record of a dissection in the entire Hippocratic Corpus.[15] There is not a hint of any such investigation in *Epidemics* 1 and 3. Throughout 2, 4 and 6, the author is particularly interested in the channels in which blood and other humours move. He is especially interested, therefore, in deposits, or *apostaseis*, but unlike his forerunner he is interested in how they can be manipulated and directed to a particular point inside the body.[16] He is much more explicit about the 'community' between parts of the body than his predecessor ever had been.[17]

He is more explicitly interested in causes and origins, using a word for them, *aphormai*, or starting points, which is absent from 1 and 3.[18] His notes do not follow a case day by day, noting its signs until death or recovery. From most of his text prognosis would be impossible. What he records is guided by other concerns, usually unstated and, in the form in which they survive, hard for readers to decode. 'The daughter with the wife of Komes, the market supervisor: in her belly, began [to be pregnant] without a sign. At two months a phlegm-like, sometimes bilious, vomit broke out. Had a difficult birth. Was completely purged [of afterbirth]. Vomited similarly until the thirtieth day. Then her stomach was disturbed and the vomiting stopped. Diarrhoea. Her period did not occur for two years. In the winter she had piles.'[19] Prolonged observation over at least two years underlies this summary, whether it was observation by the doctor, his assistants, attendants of the patient or the patient herself: its guiding thread is hard to discern. Most of it attaches to the movement, or blockage, of fluids in or through the body, but its particular interest for the author is not specified. It is quite different from the cases mentioned in *Epidemics* 1 and 3.

Unlike 1 and 3, the author of 2, 4 and 6 never gives a separate detailed survey of the constitution of a year. So far as one is noted, it is merely integrated with a particular case. Small details of time and tenses express an interrelationship quite different from those which his Epidemic predecessor had in mind, as the author's current editor, Robert Alessi, has acutely explained: they survive correctly only in the Arabic translation of Galen's commentary and its citations of the text.[20]

In Crannon [in Thessaly] summer [cases of] anthrax; it was raining in the hot spells with violent rain continuously and this happened more with wind from the south.

There are fluid gatherings in the skin, but when caught they grow hot and cause itching.

> At that time small blisters as though from burns rose up and seemed
> like burns under the skin.[21]

Unlike his predecessor, the author distinguishes here and elsewhere
between existing features of the bodily condition and those which
occurred only with the outside stimulus of the weather.[22] The author of
1 and 3 was never so precise about a case. The author of 2, 4 and 6
continues his way of relating weather and physical conditions into other
seasons and other cases in Crannon and then states his distinctive belief
that there is a similar pattern of effect in the seasons of a year and the
times of a day. 'In autumn diseases are most acute and most deadly, in
general similar to the exacerbation in the afternoon, since the year has
a cycle ['period'] of diseases just as the day has a cycle ['period'] of disease.'[23] The author of 1 and 3 often recorded that a disease exacerbated
later in the day, but never made the point in a general statement. He also
never related it to autumn or the 'period' of a year.

The author of 2, 4 and 6's noting, approach, vocabulary, interests
and theorizing are not at all the same as 1 and 3's, their predecessor.
Yet, he was writing his notes by 406/5 (that man from Alcibiades . . .),
only two years later than the fourth and final year of *Epidemics* 1 and
3, as hitherto dated by modern scholars. Like his Epidemic predecessor, he visited Thasos. He was not a rebellious young pupil, marking
out a new path and approach at the start of his career. He wrote
with long experience, generalizing in summary form from all he
had come to know. If *Epidemics* 1 and 3 were composed as late as
412–408, he would certainly have known their author, a fellow Coan,
and even worked with him and his team. The gulf between their ways
of thinking is then extraordinarily hard to credit. Both were masters
of their craft, but now that one is dated up in the 460s, their different
outlooks and approaches can be readily accommodated. They were
separated by an interval of some sixty years.

The author of 2, 4 and 6 had a different contemporary, Thucydides
the historian, who was also active within a few miles of Thasos,
residing in his family's property near their mine at Dug-Out Forest
and writing up histories there, or so posterity believed. The doctor is
the one directly contemporary prose author who survives to be compared with the great Athenian, but he was not the role model for

Thucydides' account of the Athenians' plague. The historian had noted it and written it up soon after its occurrence in the 420s, before the author of 2, 4 and 6 started work on his notes. If Thucydides had been influenced by a medical textual model, it was the earlier text, *Epidemics* 1 and 3, a formative influence for his view of the value of history. It is fun to guess that he and the author of 2, 4 and 6 might have met on Thasos in the later stages of Thucydides' subject, the war between Athens and Sparta. Whereas Thucydides' guarded generalizations about human nature dawned on him from the events into which he inquired, the author of 2, 4 and 6 was bolder about the process: he put questions first and inquired later.

22

From Thasos to Tehran

Among Thasians on their island nobody else is known to have followed the example of their most distinguished visitor, the first Epidemic doctor, perhaps Hippocrates himself. He remained only one in a much wider range of practitioners, few of whom had any high idea of medicine as his sort of craft. Two local examples make the point. In the fourth century BC, one Hermon from Thasos attended the great shrine of the god Asclepius at Epidaurus, seeking a cure. An inscribed tablet there presented the staff's version of what happened: 'he was blind and the god healed him; after this, he did not bring the medical recompense and so the god made him blind again. So he came back and slept in the temple and the god made him healthy again.'[1] During Hermon's sleep Asclepius had appeared, it implied, and acted on his condition.

This version of Hermon's experience was both a cautionary tale, about a patient's failure to pay, and a praise of the god, about his healing twice over. It was written up as one of the many 'miracles' which aimed to impress visitors to Asclepius' shrine, but whatever happened to Hermon in real life, the temple medicine to which he turned had nothing in common with the Epidemic doctor's craft. Stories later spread, perhaps from priests and worshippers of Asclepius, that Hippocrates had copied the records of cures at Cos's temple and only then devised similar case records of his own.[2] This allegation was quite untrue. The reason is not that the earliest surviving inscriptions with these sorts of cures belong in the fourth century BC: there were probably earlier examples, copied on perishable material, none of which survives. Rather, the form and purpose from the Epidemics' named case histories were entirely different to the temple tales.[3]

In Thasos itself, the line between doctoring and religious worship also lacked Hippocratic rigour. By *c.* 400 BC the city had a fine sanctuary of the healing god Asclepius, no doubt a primary resort for sick and suffering patients, more widely used than doctors working in the Hippocratic tradition. In the first century AD, one Timagenidas, a doctor, erected a small shrine beside a cave to the god Dionysus and his attendants. His accompanying poem called on the 'leader of the maenads' and the Dionysiac rites which were to take place. The god and his group were asked to protect the doctor in his Thasian homeland, to keep him safe and free of harm and to visit from time to time.[4] Dionysus was a major god in the Thasos of the 470s, but the Epidemic doctor would surely not have stopped to dedicate such a shrine and pray in the god's honour.

The vast majority of Greek doctors elsewhere continued on their erratic way, refusing to read and preferring to propose impressive, but useless, remedies. Nonetheless, there are traces of influence from the Epidemics outside the Hippocratic Corpus. In the mid-fourth century BC, a fine grave monument in Attica commemorated the 'midwife and doctor' Phanostrate, the first female Greek doctor known to us by name.[5] Its sculpted relief shows Phanostrate seated, with four children of varying ages beneath and beside her, and below it verses describe her as 'harmful to no one, missed by all'. The words 'harmful to no one' may echo the precept laid down by the first Epidemic author for his medical pupils: 'do no harm'. Later in the century, there is an echo of the case history format too, in the supposed Royal Diaries of the last days of Alexander the Great. Their longest surviving extract purports to narrate Alexander's illness in his last ten days or so in Babylon. The text was issued as if from the secretaries of his two immediate successors, but is justly suspected of being written later for tendentious ends, to answer, implicitly, allegations that Alexander had died of poison.[6] To enhance his text's credibility at that point, the author adopted the day-by-day form of a medical report, exemplified in *Epidemics* 1 and 3.[7]

Whether or not these traces of influence derive from direct knowledge of the text itself, *Epidemics* 1 and 3 certainly continued to be studied in medical circles. One proof is *Epidemics* books 5 and 7, written (in my view by one and the same man) before *c.* 350 BC.

Again, the text has obscurities, but not because it was in note form: it became shortened by scribes during its long transmission. It is excellent evidence of *Epidemics* 1 and 3's continuing impact on doctors who were still working in its wake. Its case histories are more detailed, and at times even more lively, than those in 2, 4 and 6, but their general form, beginning with the patient's name and address, are true to the first Epidemic books' example. However, the seasons and weather are related to particular cases, rather than being presented as separate surveys of each year to form an ideal type, useful for future practitioners. It never refers to prognosis. Its focus is on treatment.

It is indeed striking that Epidemic texts were still being written more than a century after the prototype. *Epidemics* 1 and 3 continued to be studied, though not always quite as its author had hoped. In Alexandria, an edition of book 3 was produced by Baccheios in the early third century BC, followed by other commentaries on all the 'Hippocratic' books of the time, but they began to focus on the text's great puzzle, the lettering in the copy of Mnemon of Side, which they misunderstood as a sort of Greek code.[8] In the first century BC, as a recently published papyrus has shown, the distinguished doctor Asclepiades also wrote a commentary on *Epidemics* 1, but again he went off in a different direction: he tried to fill in the treatments which Hippocrates' original, he thought, had left out.[9]

Bits of *Epidemics* 2, 4 and 6 became included in the collection of brief Hippocratic *Aphorisms*, that handy text for hard-pressed doctors and students, but most strikingly, books 1 and 3 did not.[10] One reason may have been that they had less in them relating to treatment which could be excerpted usefully from its context. However, in the reign of Hadrian, commentaries on the Epidemics still continued to be written, culminating in Galen's on books 1–3 and book 6, works of the AD 170s. Yet even Galen, an avowed Hippocratean, read the texts in the light of his own contemporary interests. He is a valuable witness to older textual readings and interpretations, but he used other Hippocratic works to interpret the Epidemics' meaning.[11] As these works were not written by the same author, they are not as helpful as he assumed.

The main form of the first Epidemic text had been the case history, one which was to become a crucial tool in subsequent medical history. In antiquity it continued to be influential, but its doctor's use of it as a sober instrument for prognosis did not prevail. The fullest evidence for the change lies in twenty-one case histories which survive only in an Arabic translation, and are ascribed, surely correctly, to the important Greek doctor Rufus of Ephesus, who was active in the early second century AD.[12] Unlike the cases in *Epidemics* 1 and 3 they are written pervasively in the first person 'I'. Often they describe the author's encounter with a patient who had a named condition. Six such cases are named as cases of melancholia, a particular interest of Rufus, one of which has claims to be the most sympathetic case of it in all ancient medicine. A young man, it reveals, had provoked an increase in his black bile by brooding on geometry (*ilm al-handasa*): he had even attended 'sessions organised by the rulers', as the Arabic renders the Greek text.[13] This troubled maths student, the first attested in history, became melancholic, beginning, Rufus thinks, with 'fiery heat', in Arabic a burning of his blood (*iḥtirāq al-dam*).

Rufus does not give any general surveys of a year's constitution. Once, he remarks on a man who had pain between his ribs in spring each year, but the season is simply given as a passing comment on the condition's timing.[14] Here too, the first Epidemics' connections between the seasons of each year and its diseases and their author's concern for prognosis have given way to a simpler concern with treatment.

In his text *On Prognosis*, Galen also gives case histories as part of his argument, but once again without any preceding constitution or survey of their year's weather. Whereas Rufus had noted failed attempts by other doctors to cure a patient before he himself was called in, Galen was even more insistent: he used named cases to set off his own skill and achievement.[15] This personalized use of the case history had already, quite independently, been used by their first known Chinese exponent, Chunyu Yi, *c.* 170 BC: like Galen, he often used them to emphasize his own skill and to defend himself against critics.[16] This immodest use contrasts tellingly with the first Epidemic author's aim and tone. Admirably, he had been concerned with observation present and future, not with himself and his personal reputation.

In the sixth century AD his Epidemic books were still part of Greek medical teaching, especially in Alexandria.[17] As the terse style and language of the originals had become hard to understand, Galen's commentaries on them remained invaluable. In due course a 'Question and Answer' format was applied to them to make them more accessible and memorable for students.[18] The commentaries were then translated into Syriac by Job of Edessa in the sixth century. They served the skilled doctors of the Syriac-speaking near East, including those active at the court of the ruling Persian kings.[19]

By the late 850s, Galen's commentaries had been translated into Arabic by the great multi-lingual translator, Ḥunayn ibn Isḥāq, himself a practising physician in Baghdad. Ḥunayn even set out to translate the Greek text directly into Arabic.[20] To assist students, he also produced four abbreviated versions of the commentaries, three of which replicated the now-popular 'Question and Answer' format.[21] As a result Galen's Epidemic commentaries enjoyed a wide circulation among Muslim authors: according to their specialist, Poter Pormann, they 'were used in sex manuals, medical encyclopaedias, agricultural guides and specialist medical texts', composed throughout the Islamic world.[22]

Through these translations the Epidemics were to find their greatest Muslim admirer, Abū Bakr Muhammad ibn Zakariyyā' al-Rāzī, the celebrated doctor 'Rhazes', as he became known in Western Europe. In the early tenth century, al-Rāzī worked as a public doctor in a hospital near modern Tehran and then moved south into courtly circles in Baghdad.[23] Initially he had been taught Greek sciences in Rayy, his birthplace, but he remained an inquiring genius and included case histories which he had attended personally in his many works on medicine. The most numerous occur in his massive work, *The Comprehensive Book on Medicine (Kitāb al-ḥāwī fī l-ṭibb)*. It was written as notes in his lifetime, but was acquired from his sister by the royal vizier fourteen years after al-Rāzī's death.[24] The notes were then published in a few huge copies.

In the sixteenth book of this enormous work, al-Rāzī included thirty-three individual case histories as a group among his disorderly notes. They were intended as school exercises for students, modelled on the Hippocratic predecessors.[25] They show his openness to the notion of treating patients from all walks of life, not just magistrates and their families like those who were among the Epidemic doctor's named patients on Thasos, but a cotton merchant, a tailor, a goldsmith and even a bookseller. Seven of his chosen patients were female, but with true Epidemic reticence he named none by her own name: he too identified them by a male kinsman, if at all. However, unlike the Epidemic doctor-author, but like Rufus and Galen, he recorded his cases in the first person and included his prescriptions and treatments. He too made no attempt to relate the cases to weather and the constitution of a year. Treatment, not prediction, guided the texts he invented: unlike the Epidemic author's, almost all of his cases, he specified, were ones whose patients survived.

The Epidemic model might seem to have faded here, but in his introduction al-Rāzī explicitly remarks that he has devised unusual cases which relate to questions and narratives in Hippocrates' Epidemics.[26] He uses various secondary sources for them, but he also states that the Epidemics' cases should be read in parallel beside his own so that the two can be compared. In fact only one such case survives in his text, but it is none other than the case of Philiscos, the first case in *Epidemics* 1, who 'was residing by the wall' when he took to his bed.[27] In the vast work of this Muslim doctor, Philiscos' case, to modern eyes one of malarial blackwater fever, passed into Arabic.

In each of al-Rāzī's case histories, there is nonetheless a distinctive tone: like the Epidemic doctor, he never refers to God, prayers or divine intervention in any of the cases he records. From Thasos to Tehran, the master's realism persisted, mediated through Galen's commentaries. Thanks to it al-Rāzī, too, ignored the unmedical mumbo-jumbo of surrounding society.

In *c.* 470 BC, an Epidemic doctor had first put detailed case histories into a text, complete with names, addresses and so forth. In this degree of detail they were a Greek invention. They were intended to assist prognosis and future treatment of similar conditions. They passed into the knowledge of Islamic doctors, and were used by al-Rāzī directly because of the first Epidemic author's example.

The detail of the earliest Epidemic text and the prestige of its author, universally believed to be Hippocrates himself, had assured it an attentive audience. His silence about pagan gods and their interventions made him readily acceptable to Christian readers, but inevitably each new readership tended, like Galen, to read him through their own preoccupations.[28] In Western Europe, the texts of the Epidemic doctors attracted Latin translations, then a commentary (by a Spaniard) and during the long sixteenth-century imitations (by Guillaume de Baillou, compiling case histories of his own).[29] Their appearance in Greek in 1526 in the first printed 'Hippocratic' Corpus did not at first provoke much new interest in them. Serious interest began in the mid- to later sixteenth century, when the variety of the Hippocratic Corpus became a way to challenge Galen and his teachings.[30]

In Paris in the 1570s, self-styled Hippocratics became prominent in university medicine and championed the writing-up of patients' case histories.[31] In Italy, case histories had already begun to be written in universities a little earlier, in the 1550s. In 1573, the professor of medicine in Turin, Francesco Valleriola, stated that Hippocrates 'wrote on tablets all that he saw occurring in the sick person and narrated the complete history of the disease and what happened to the sick each day, each hour, each moment, giving specifically the name of each person'.[32] So Valleriola imitated him, explaining how 'I reworked for general use the things I wrote down', the procedure of the first Epidemic author, but specifically 'taking into considerations only those diseases that appeared to me most dangerous and of dubious treatment'.

In the sixteenth century, medical casebooks existed beside the casebooks of astrologers, texts which recorded the exact time of a consultation and then gave each patient's horoscope.[33] In the 470s BC the first Epidemic author had no such model or rival, astrology being unknown then to the Greeks. In Europe, however, astrology had been taught from the thirteenth century onwards as part of the medical curriculum: it took over the predictive aims which the first Hippocratic doctors had pursued. In medicine, case histories remained only one option among many, but in astrology they were essential: astrologers had to consult written tables and to write down their patients' times and details in order to give a prediction. Unlike healers, astrologers had to be literate.

In England, the noting and amassing of medical case histories began in connection with almanacs and simple account books, texts which itemized the fees their author received.[34] They were also used to advertised their doctor's skills, publicizing his successful remedies and acting as testimonials in the competitive arena of healing. As a database in their own right they were first established in the seventeenth century by self-styled Hippocratics, headed by Theodore de Mayerne: he was the very model of a wandering doctor, moving from Geneva to the French court in Paris and then to London and the households of King James I, Charles I and even Oliver Cromwell.[35] His daybooks ran to about a thousand cases between 1603 and 1653, 'written in beautiful script with drawings of trusses, wigs and syringes in the margins'.[36] In the 1620s, however, Francis Bacon still considered it necessary to urge English doctors to imitate Hippocrates and compile case histories even though they were already present in a few English doctors' practices: in Stratford, John Hall, Shakespeare's medical son-in-law, compiled 182 case histories in chronological sequence from 1611 to 1635.[37]

The emergence, mid-century, of experimental philosophers with new scientific interests reaffirmed the value of case histories as a resource for generalizations and future treatments and gave it renewed impetus.[38] The Hippocratic Epidemics became the model for Thomas Sydenham's case histories during the great plague years in London in the 1660s: his medical observations, in Latin, became a major textbook for the next two centuries and led to him being known as the English Hippocrates.[39] In 1726, John Floyer published the first English commentary on all forty-two of the first Epidemic doctor's case histories, and in 1734 Francis Clifton, doctor to the Prince of Wales, published an English translation of them in which he admired their rational observation. They remained an example to doctors throughout Europe, even when medicine moved to the hospital and case histories gave way to clinical notes. What had begun on Thasos c. 470 BC had a fundamental impact on medicine, persisting into the twentieth century: it lives on wherever 'topical examination' of a patient is alertly conducted.

The first Epidemic books are the ancestors of case histories of named individuals. Between the Homeric and the Hippocratic presentation of disease there had been a major conceptual shift to which they are a lasting witness. In the first book of the Iliad, after nine days of lethal plague, Achilles proposed consulting 'a seer or a priest or an interpreter of dreams, for dreams too are from Zeus' in order to find out why the god Apollo was so angry, whether for a vow or a sacrifice unpaid, in the hope that he might be willing after partaking in one 'to keep off ruinous plague from us'.[40] The Homeric plague had been extreme, but on Thasos, in season after season, the Epidemic doctor and his helpers observed details of no less fearful sicknesses, when flesh rotted away from arms and legs and burning fever drove sufferers mad. They never appealed to priests. They never recorded sacrifices or vows or cited the anger of a god.

In Hesiod's poetry, myriads of diseases, released by a woman's folly, had been presented as wandering over the earth at Zeus' direction, attacking mortals individually, often for their wicked behaviour. The Epidemic author, by contrast, observed diseases as processes inside the body, not as individualized agents attacking it from outside. Heir to the new idea of nature, of which man is a part, he never mentioned the actions of any *daimon* or any need to enlist divine help.

Aware of the new method which made their skill a rational craft, he and like-minded Greek doctors stated ethical principles by which their new craft should abide. 'Do no harm' and eventually the Hippocratic Oath are among them. Ethics were recognized to raise the standards of practice: the Hippocratic example is a precedent for raising standards in the latest craft, ethical use of the internet. The new type of medicine then raised questions about the relative standing of theory and experience, method and observation, questions which were to divide subsequent medical sects and schools in the Greek cities after Alexander and on into the Roman Empire.[41] Despite these debates, most people continued to resort to temples and sacrifices, to the shrines of Asclepius where dreams of divine healing were freely anticipated and where vows acknowledged a divine cure or hoped to elicit one. The 'empire of the seer' was breached but never overthrown.[42] In the late second century AD, when an epidemic plague swept cities of the Roman Empire, from Asia to Britain, cult was paid to the gods and goddesses according to advice dispensed by Apollo and his oracular staff in the hope of averting its onset.[43] As ever, a new shift in thinking had not swept all before it.

The significant point is that it occurred at all. In the fifth century BC, the two travelling realists still astound us, Thucydides the historian, inquiring, assessing and studying evidence from Sparta to Macedon, and the travelling doctor, active from Thessaly to Cyzicus, author of *Epidemics* 1 and 3. Both men were upper class, Hippocrates (if he is indeed the doctor-author) an exclusive Asclepiad from Cos, Thucydides from an aristocratic family with links to the aristocratic Cimon, patron of the painter Polygnotos, another genius of a nobleman in the early fifth century BC. Nonetheless, these two well-born men confronted acute suffering face to face, Thucydides as a military leader and as a victim of the plague in Athens, Hippocrates as the unflinching attendant of patients who were suffering hideously, people whom he observed day after day. Both of them, members of the upper class, aimed to benefit others, Hippocrates by helping their health, irrespective of their social background, Thucydides by helping their political decisions and understanding, the sphere of male citizens in the Greek city-states.

In his Histories, Thucydides deployed a complex, often abstract, style for his speakers, one whose contortions tested even his Greek heirs' grasp of Greek. By contrast, the Epidemic author wrote curtly and exactly. Yet he too perplexed posterity, in his case by the very concision of his style. Thucydides intended his Histories to be a 'possession for all time', not by their addition to knowledge, pure and simple, of what happened when in the war that was his subject, but by their value as a guide to what might recur in so far as human nature stayed the same.[44] He has been vindicated, as Thucydidean patterns of behaviour continue to recur in political life.

Before him the Epidemic doctor-author had approached human nature in a similar way, through cases of physical and mental suffering. Unlike Thucydides, his understanding of that physical nature was wrong on every point. Error can sometimes be creative, as the ancients exemplify, whether Euboean travellers in the time of Homer, with their fertile misunderstandings of stories they heard in the Levant, or Alexander the Great in Egypt and India, or, later, Augustine, who misunderstood scripture in a garden in Milan and then embarked on the monastic way of life which is still observed under his Rule. The errors of the Epidemic author about germs or contagion, women's bodies or even the benefits of nosebleeds were to be obstacles, not creative mistakes. His approach, however, was not. It rested on observation, humanity and caution about harmful interventions. It also rested on a belief that similar cases would recur and could then be knowledgeably treated. His Epidemics too are 'possessions for all time', the first practical texts in the invention of medical science.

Endnote 1

POLYGNOTOS AND THE STOA IN ATHENS

Polygnotos came to Athens with Cimon in the 470s, not the 460s as is still sometimes wrongly aired as a possibility (counter-evidence includes his work for the Theseion, built for Theseus's newly arrived bones in 475/4; the archaeology of the Painted Stoa; and other arguments at my pp. 168–9 above). However, the Painted Stoa, when seen by Pausanias in the AD 130s, contained a painting he presents as the Battle of Oenoe. But 'The battle of Oenoe is not easy to believe in' (D. M. Lewis, *CAH* v (1992) 117 n.77). Nonetheless, it is widely considered to date to the 450s: does its existence in the Stoa bear on the dates when Polygnotos worked there?

In his description of the Stoa, Pausanias does not name any of the artists (Paus. 1.15). He later names Panainos as the artist of one painting (Paus. 5.11); he was Pheidias's brother, but other sources and most moderns prefer to ascribe that painting mainly to Mikon, an older artist. Items in other authors show Polygnotos was the artist of another of the paintings, the *Sack of Troy* (Plut., Vit. Cim. 4.5). Who painted the Battle of Oenoe? No ancient author names an artist, a silence which contrasts with the artists named for the Stoa's other paintings. Strong arguments place the painting on a side wall inside the Stoa, whereas Polygnotos's big picture and probably at least one other were on the long middle wall (Jeffery (1965) 43 n.13; Stansbury-O'Donnell (2005) 73–88 sets out the case again very well). It is quite likely, then, that the Oenoe painting was an addition at the side, made quite a while after Polygnotos and others had worked there in the late 470s (despite Luginbill (2014) 282, the recent archaeology of the Stoa does not imply that paintings were on the side wall from the very start).

Does the painting's subject also suggest it was a later addition? Pausanias understood it to show a battle about to begin between Athenians and Spartans. It was a wholly unsuitable choice of subject for the pro-Spartan Cimon and his friends and relations who were linked to Polygnotos and the Stoa in the late 470s BC: Ballansée (1990) 91–126 proposed that it was installed in the 450s as a gesture of compromise between supporters of Cimon and 'radical democracy', but politically this proposal is wholly unconvincing, and the rest of the Stoa remained strongly Cimonian in content and origin.

Pausanias states that the battle in the picture occurred in Argive territory. In the last book of his work (Paus. 10.10.3–4), he repeats this point when presenting sculptures dedicated at Delphi and financed from the spoils of the battle: he says that the artists made them, 'as the Argives themselves say', from the victory won over the Spartans by themselves and 'Athenian helpers' in Argive territory. So the location and participants are unlikely to be only his own guess. Jeffery (1965) 52–7 made the most detailed attempt to place the battle in spring 457 in the Argolid but did not fully persuade an appreciative Meiggs (1972) 470–71. Andrewes (1975) 9–15 then argued that the terrain around Oenoe is remarkably unsuited to such a battle: 'I do not believe there is room for the battle on the ground' (15). Pritchett (1980) 46–50 revisited the area and contradicted him, but his case is not, to my mind, decisive. Thucydides also omits any reference to a battle there.

Despite Pausanias and his informants, I consider, therefore, the battle took place at another Oenoe, the one in Attic territory, located near the border with Thebes. This view has now been taken independently by others, notably by Francis and Vickers (1985) 99–113 and, in their wake, by Boardman (2005) 63–72, with prior bibliography. They connect it, albeit in differing ways, with the great battle against the Persian army at Marathon in 490 BC. If they are right, then Polygnotos might perhaps have painted an Oenoe in the late 470s, but I find their date and reconstructions wholly unconvincing. They assume Pausanias mistook both the location and one part of the troops shown: he misunderstood them as Spartans ,whereas they were actually either Persians or friendly Plataeans mustering with their Athenian allies before battle began. (In reply, Luginbill (2014) 287–8 restates why the troops shown were readily identifiable.) Boardman adds, as indeed his theory has to, that 'we have to assume that the Oenoe panel had been badly damaged' (68). Badly indeed.

Taylor (1998) 223–41, followed by Stansbury-O'Donnell (2005) 73–88, considered that the battle shown was one fought between Spartans and

Athenians in 431, an opening event in the Archidamian War. If so, it was indeed an addition to the Stoa and had nothing to do with Polygnotos. Thuc. 2.19 regards this encounter as a failed attempt by Archidamus and his Spartans to take Oenoe: it was not a real Athenian victory, and even allowing for wartime distortion it is hard to see why a defensive siege should have been added to the Stoa as a pitched battle.

As a solution I suggest that a battle at Oenoe occurred between Athenians and Spartans in summer 458, immediately after the Spartan victory at nearby Tanagra mentioned by Thucydides (1.108.1). It was a lesser affair than the closely fought main Tanagra battle, but evidently the Athenians could claim a little victory in it. At least in Diodorus' muddled account (Diod. Sic. 11.81), the battle at Tanagra seems to have extended over two days. Might a secondary engagement at Attic Oenoe have been part of the second day as the Spartiates were beginning to march off home? Some Argives were involved as allies of the Athenians, as the sculptures dedicated at Delphi from the battle-spoils attested. The sculptors of those pieces were Thebans: in the aftermath of 458 BC, they would have been working for a bitter enemy, Argos. However, they might have been at odds with their fellow Thebans (Jeffery (1965) 55 n.58 suggests political divisions in Thebes at this time), and there is no knowing what sculptors will do for a large fee and a commission for a grand public place.

A painting of a historical battle so soon after it occurred was unprecedented in Greek art. I suggest we see it in relation to another unprecedented commemoration, the big war memorial to the Argive dead at Tanagra. Recent study of yet more fragments of it have shown its scale and emphasized its site in the very centre of Athens (Papazarkadas and Sourlas (2012) 585–617). The painting of the Oenoe battle showed Athenians, Pausanias remarks, but some of the troops shown may also have been the Argive participants whom he does not mention. This tribute to a joint Athenian and Argive victory over Spartans, I hypothesize, was one more part of that Athenian willingness to please and honour their new Argive allies in this very period. In April 458 at the festival of Dionysus, Aeschylus's Oresteia had strongly emphasized that an alliance for ever must be made between the Athenians and the Argives. Macleod (1982) 124–44 adduced dramatic and thematic factors which, in his view, suited the plays' stress on the Argive alliance 'for ever', but they do not fully account for it: even the language Aeschylus uses afforces the political significance of this motif (Papazarkadas and Sourlas (2012) 604–5 and n.119, and especially de Sainte Croix (1972) 183–4). I suggest the Oenoe painting was done in 458 and that Pausanias and his sources wrongly assumed about 600 years later that it occurred at Argive Oenoe. Pausanias did not remark that there were Argives as well as Athenians in the picture, but these errors are not as gross as, supposedly, failing to recognize that Persians were being shown and that the scene was a mustering of allies before the Marathon battle thirty-two years earlier.

Luginbill (2014) 278–92 has proposed that the battle took place in the Argolid, but in 458/7, and that a (pro-Argive) painting of it replaced an earlier (anti-Argive) one in the Stoa, which had shown the Heraclids and their reception in Athens so as to escape from the Argive king Eurystheus. However, the existence of an earlier Heraclid painting is far from certain (Wycherley (1953) 26 regarded it as 'full of doubts and confusion'), and, if it was promptly replaced, it is surprising that the late scholiast on Aristophanes Plutus 385 knew its contents, if indeed it is the item it mentions. Nonetheless, Luginbill's suggestion that the Oenoe painting was a deliberately pro-Argive addition fits well with the context I have given for it.

Painted in 458/7 , the picture was not by Polygnotos. His work in the Stoa belongs to *c.* 470 BC . Then he went on to Delphi to work for the Cnidians: Stansbury-O'Donnell (2005) 81–8 has proposed a date in the late 460s for Polygnotos's work there, but he relies on a mistaken date of 466 for the Eurymedon victory (see my Chapter 13 n. 13), and I do not accept his reading of figures in Polygnotos's painting there as evidence that 'Polygnotos was ambivalent about Athenian actions generally'(85).

Endnote 2

CONFISCATIONS AND INSCRIBED ENACTMENTS
IN FIFTH-CENTURY THASOS

A ruling dated by Antiphon son of Critobulos and his two colleagues as *theo-roi* in Thasos was inscribed on the wall of its sanctuary of Apollo Pythios (Osborne and Rhodes *GHI* 177A for a recent presentation; its exact location relates to Apollo's sanctuary, for which see Picard (1921) 145, with the plan by J. Baker-Penoyre in *JHS* 29 (1909) 209, his point nr. 11). Their year of office has been dated by scholars to a year running from one November/early December to the next, some time between 412/11 and 410/9. A case for 412/11 was made by Avery (1979) 234–42, accepted by Osborne and Rhodes *GHI* (2017) 463, but it is not compelling: he argues, correctly, that the loyalties of at least one person's family affected by the ruling were pro-Athenian and democratic but considers that he must have been punished immediately, because his family were later detested by the Thirty in Athens, as indeed they were, and the grounds for this animosity must relate to the grounds behind his punishment also on Thasos. So, Avery believes, all those punished were punished in the very first months of the change to oligarchy, even before Thasos broke with the Athenians. I do not accept this argument. Confiscation of the property of these pro-Athenians in the first phase, still nominally pro-Athenian, was a very disruptive act, more likely to occur in the second openly anti-Athenian phase of oligarchy, from mid July 411 until early December, and especially likely in the third one, from December 411 to spring 407. As the ruling confiscates property, those affected by it may have fled before it occurred. Their flight might have been in the second phase of oligarchy, after the open breach with their beloved Athens, and the confiscation, dated by Antiphon, then belongs in the third phase. There, it fits with the dating advanced by French specialists in the inscribed lists. For other reasons, too, they date Antiphon's ruling to 409 and his year as *theoros*, by which it is dated, to late 410 until late 409. That dating has hitherto rested on four main arguments; it is also a lynch-pin for their dating of the columns of *theoroi* in the Thasian list (one of the issues which Osborne and Rhodes *GHI* (2017) 462–5 pass over). So I will address the four arguments in more detail; three relate to evidence outside the inscription, the fourth to a fact about its lettering.

1. The inscription attests a 'ruling of the three hundred'. They are generally interpreted to be an oligarchic body. Thasos is known to have been under an oligarchy from May/June 411 to spring 407.

2. The existence of a 'three hundred' is also attested in two rulings inscribed in Thasos about rewards for informers who reveal an uprising being plotted (Osborne and Rhodes *GHI* (2017) 176A, line 3) and then, about sixteen months later, an attempt to betray Thasos or her settlements abroad (Osborne and Rhodes *GHI* (2017) 176B, line 4). Confiscation of the property of those denounced was to follow a successful denunciation. These inscriptions were found in 1949 on one and the same stone set in a wall of Thasos's Christian basilica: most probably they had previously been displayed nearby, in the main *agora*. Informers were to be rewarded according to the value of the property of those against whom they successfully informed. These two rulings were first published by Pouilloux (1954) 139–64, who dated them between 411 and 407. Salviat (1984) 242–57 then argued in wide-ranging detail for dates of November/December 410 and March 408 respectively. Pouilloux (1954) 156 also related them to the ruling dated by Antiphon, because it records confiscations according to a resolution of the 'three hundred'. Since Pouilloux, that ruling has been widely considered to have followed either the first of the two rulings about informers (Salviat (1984) 252–4) or the second one (Pouilloux (1954) 256–7 and 161). If so, it belongs between 411 and 408 BC, and Antiphon's *theoria* is thereby dated precisely.

3. The ruling dated by Antiphon and his colleagues certainly relates to evidence found in Athens. Apemantos son of Philon, a Thasian, had his possessions confiscated under the ruling. Five sons of Apemantos were later honoured in Athens with a public inscription for being (Athenian) *proxenoi*, before 404/3 because the Thirty Tyrants destroyed it then (Osborne and Rhodes *GHI* (2017) 177B, nrs. 157, 167, 191, are other examples of the Thirty smashing inscriptions whose contents they detested).One other man whose possessions were confiscated under the

ruling dated by Antiphon was also called Apemantos: he was a Neapolitan. However, in yet another Attic inscription (I.G.II² 33), c. 385
BC and honouring Thasian exiles, a son of Apemantos is among those
honoured: his name is restored as [Amynto]r, the name of one of the
five sons of Apemantos honoured nearly twenty years earlier. So it is
safe to assume those five sons were indeed Thasians, sons of the Thasian Apemantos, not the Neapolitan. We should remember that
upper-class Athenians like Adeimantos (see my p. 147) had formerly
acquired landed estates on Thasos and knew all about keen Thasian
supporters of the Athenian democracy.

After the Thirty's fall, one of the five sons, Eurypylos, discovered
that their honorific inscription had been smashed. He was praised by
the Athenians for offering to have it re-erected at his own expense.
(Osborne and Rhodes *GHI* (2017) 177B, line 16). Eurypylos's offer
belongs soon after the Thirty's fall, probably in c. 402.

Since P. Foucart in 1904, this attestation of Apemantos's sons as loyal
pro-Athenians in the late fifth century has led scholars to place the
confiscation of the possessions of Apemantos, their father, in 411–407,
under the oligarchy which was then ruling Thasos. The ruling which
records the confiscation is dated by Antiphon: it is therefore dated to
those very years (Salviat (1984) 252–4 argues an ingenious case for 409
precisely, but his arguments are not altogether convincing).

4. The ruling dated by Antiphon is inscribed in the Ionic alphabet which,
   on present evidence, is only attested in use on Thasos from the 430s
   onwards. As it inscribes a ruling passed by an oligarchy, the oligarchy
   is therefore the one in 411–407.

None of these four arguments is conclusive. I will address them one by one.

As for 1, Thasos was under an oligarchy in the sixth and fifth century too,
at least until 463 BC. So 411–407 is not the only possible date on that point
alone.

As for 2, a 'three hundred' is also attested in the earliest surviving Thasian law about wine (Osborne-Rhodes *GHI* (2017) 103Aa, line 8). The
lettering of its inscription (Parian, written *boustrophēdon*) really does date
it up c. 470 BC. I will return to the much-discussed question of who these
'three hundred' were, but meanwhile the mention of 'the three hundred' in
the ruling dated by Antiphon does not in itself date that ruling (and his
theoric year) to a point between 411 and 408/7.

A cardinal point here has been under-emphasized. The inscription dated by Antiphon ruled that the confiscated property would be 'sacred' to Apollo Pythios. The city, then, was not to benefit directly from it or its sale. By contrast, the second of the two rulings about informers clearly presupposes a sale by the city of the property of anyone justly informed on (Osborne and Rhodes *GHI* (2017) 176B, 9–10). The informer will be given twice as big a reward if the property is worth more than 200 staters: he will get the money 'from (*ek*) the city'. Presumably it is to come from the city's sale of the property, which would also establish its value. The two procedures, one giving the property to Apollo as 'sacred', the other selling it for the city and disbursing from it, are different.

This distinction is nicely attested on Thasos in Hamon (2019A) nr. 14 (*c.* 291 BC); anyone who proposes something contrary to the decree must pay a double fine, one part to Apollo Pythios, one part to the city (line 14). This Thasian text was decreed under a democracy. It shows that declaring property or a fine to be 'sacred' to a god was not something confined to oligarchies (though Osborne and Rhodes *GHI* (2017) 464 suggest that, in the case of nr. 177A, dated by Antiphon, it was a deliberate 'avoidance of any reference to the *demos* and a denial that the new regime is enriching itself from these confiscations'). I doubt this. SEG LVII.576, a long decree about civic reconciliation in nearby Dikaia, *c.* 363/2 BC, combines the two terms 'public' and 'sacred' in one phrase: those who refuse to swear the prescribed oath (19–20) and other future offenders too (32–5) are to lose their citizen rights, and their property is to be 'public and sacred' property of the city's main god, Apollo Daphnephoros. That decree was passed by a democracy.

Among the property confiscated and declared 'sacred' to Apollo, land and houses would presumably be let, rather than sold, and the rents would go to the god's cult, as was the case with sacred lands in classical Attica. The ruling dated by Antiphon, then, does not derive directly from either of the rulings put up in the *agora* which assumed property would be sold and partly paid out to informers according to its value. Whenever those rulings were passed, they do not fix Antiphon's date.

Do the two rulings about informers really belong in *c.* 411–408? Meiggs and Lewis *GHI* (1988) 253–5 set out the arguments for and against, including the absence of any mention of the *demos* in them, and concluded by expressing shared doubts about such a dating: already F. Chamoux (1959) 351–6 had argued likewise, though not all his points stand (M. Simonton (2017) 144 n. 132 gives the subsequent bibliography, almost all of which simply adopts the date of 411–408). I share the doubts of Meiggs, Lewis and Chamoux and I consider a date somewhere between *c.* 430 and *c.* 415 to be preferable. Salviat (1984), 243–54, especially 253, argued from their dating by three *archons*, not one, that they belonged between 411 and 407, precisely in 410 and spring 408. However, I reject this conjectural dating of the return to three yearly *archons*, not one (see my Endnote 3). The rulings' reference to 'three hundred' is no obstacle to their being passed under a democracy either. The number refers in them to a court, not to the governing body. Conversely, the ruling dated by Antiphon refers to 'what pleased *the* three hundred', their *hados*. The definite article here is significant, a reference to an oligarchic governing body (emphasized clearly by Pleket (1963) 76 and n. 30). So, I wish to emphasize, is the word *hados*. It is not the word for the judgement of a court or jury, which would usually be a *krisis* or a *dikē* (Hölkeskamp (2000) 83 concurs). It occurs twice in decrees in or before the fifth century BC (Meiggs-Lewis *GHI* (1988) 2, from seventh-century Dreros on Crete, and Rhodes and Osborne *GHI* (2017) 132, from Halicarnassus *c.* 450 BC). Its related verb, *heade*, occurs six times, however, in Herodotus, always for the decision of a ruler or a governing oligarchy, twice one of 'the Spartans' (1.151.3; 3.45.1; 4.145.5; 4.153; 4.201.2; 6.106.3). Herodotus never uses it for a court or a judgement. Meiggs (1972) 576, and also Osborne and Rhodes *GHI* (2017) 463–4, tried to explain 'three hundred 'as a 'general provision for selecting 300 in certain kinds of cases' and then, 'when a 300 has been selected to make decisions they are called *the* three hundred'. The problem here is that the two rulings about informers refer to 'three hundred' but are sixteen months apart, so a three hundred surely existed by the time of the second ruling. The rulings assume that 'three hundred' exist already as a group as they are trying cases of violence.

A further problem is the script used to inscribe the two rulings: it is Parian, and as Pouilloux (1954) 139–41 well set out, Parian of a developed style. Following Pouilloux, Meiggs (1972) 575 declared that this choice of script was 'a deliberate nationalist reaction from the Attic script of recent public documents'. This reaction is generally ascribed to the oligarchs, restored to power from 411 to 407. However, this proposal is not compelling. Those who place the ruling dated by Antiphon in that same period face a problem: it is in Ionic lettering. Pouilloux (1954) 146 saw the difficulty and explained it by assuming that the 'counter current' of Ionic lettering used on walls of the sanctuary of Apollo resisted the change, whereas official inscriptions down in the city turned back meanwhile to Parian (Avery (1979) 241 replies to this. If the two rulings about informers are dated to the 420s, their Parian script is not a problem.

As for 3 , the references in Athens to Apemantos's sons are not proof of a 410/9 BC date for Antiphon either. Apemantos son of Philon could equally well have been born in *c.* 490 BC, lost his property through confiscation in 465–463 and no doubt gone into exile and fathered his sons during the 450s and early 440s. When he died, perhaps in the 430s, a loyal pro-Athenian to the very end, his sons became *proxenoi* after him and were then honoured in Athens. If so, they were in their late forties and fifties when the Thirty smashed up their honorific inscription. If [Amynto]r son of Apemantos is correctly restored in I.G.II^2 33 (to be dated *c.* 385), he was about 60–65 at the time. Hamon (2015–16) 105 n. 90 suggests that 'sans doute' another son can be added to the 'four' (actually, five) attested in Athens: Aristomenes son of Apemantos (Big Theoric List col. 6.5, name 17) whose year as a *theoros* has hitherto been dated to 377 BC. I hesitate to credit Apemantos with *six* sons, this one being unmentioned in the Athens inscription. However, if Aristomenes really is one, he can fit my higher chronology: for me he would be *theoros* in 431, born, in the 460s and then dead before the Athenians honoured his surviving brothers later in the century.

In short, the argument from the sons certainly does not fix the inscription dated by Antiphon (and his theoric year) to 410/9 BC.

There remains 4, that inscription's use of the Ionic alphabet. I have addressed this important point on p. 176, noting the very curt and abbreviated nature of its text. I propose, therefore, it was inscribed in an abbreviated form some while after its ruling's enactment, as part of a collection of related rulings, a mini-archive on that part of the wall (another text, unmentioned by Osborne and Rhodes *GHI* (2017) is visible beside it: Ch. Picard (1921) 145). As I date Antiphon's *theoria* to 464/3, near the end of the Athenian siege of Thasos, external factors may explain why the enactment was not immediately inscribed. It was a time of great turbulence for Thasos, after which the building of the neighbouring temple of Athena was abandoned, presumably directly after 463 (J.-Y. Marc, in Brunet, Coulié, Hamon et al. (2019) 44). One possible reason to inscribe the ruling later can be conjectured: the sons of those confiscated attained adulthood, in my view in the 430s or early 420s. Apollo's temple personnel might then have wished their god's right to the properties to be publicly visible in a permanent form. At that date the Ionic alphabet was used. Certainly the inscription is very brief. It is dated only by the three *theoroi*, without the usual mention of the *archons* too. All it records is the existence of a resolution of the three hundred, the fact that the properties confiscated are sacred to Apollo, and the names of the persons whose property was confiscated. The underlying resolution is likely to have been much fuller and longer (the text from Dikaia *c.* 363/2 BC, SEG LVII 576, well illustrates the complexities). For an abbreviation of an original text when eventually inscribed, we can compare the ruling of Alexander inscribed on the wall of the temple of Athena at Priene (Rhodes and Osborne *GHI* (2003) 86B). It was inscribed as part of a 'documentary archive'. The Thasos text, as Hamon (2019A) 51 n. 39 importantly observes, was later defaced and chiselled. As he dates it in *c.* 409 BC, he assumes that this erasing happened after democracy was restored in 407. If the text was inscribed in, say, the 430s, it could have been defaced earlier, between the 420s and *c.* 413, when the sons of Apemantos were politically prominent and may indeed have contrived to get their father's property returned to them.

The four arguments do not, then, fix the Antiphon ruling in 411–407. Nonetheless, it might belong there: to exclude such a date, negative arguments are needed. Its recording of the confiscation of two Neapolitans' property supports such an argument (see my p. 176). Pouilloux (1954) 156 and n. 5 saw its relevance and rightly insisted (his n. 5) that the consecration of these properties to Apollo Pythios (on Thasos's acropolis) shows that they lay in territory still directly in Thasos' control. However, he proposed either that the two Neapolitans' names were a later addition to the text (there is no evidence for that in the text, defaced as it now is) or that they might have owned properties in a part of the mainland *peraea* still controlled directly by Thasos. This proposal is very hard to credit. In 463 Thasos lost 'the mainland'. (Thuc. 1.100.3) and thereafter Neapolis and Galepsos paid tribute separately to the Athenians. In 411–407 a few other places on the mainland were still closely linked to Thasos, but, at a time of great anxiety about their possible defection, it would have been exceptionally clumsy for the Thasian oligarchy to have obliged proceeds from the sale of properties somehow still held in their territories by Neapolitans to go to Apollo Pythios on their own acropolis. Since 463, Neapolis had been independent of Thasian control, and it is very hard to believe that in 411 some of her citizens

could still be owning property in other colony-settlements attached to Thasos. I should emphasize that the word for the property being confiscated
(*chrēmata*) can certainly include land and houses: at Isoc. 16.46, Xen., Lac.
Pol. 1.9–10 and Dem. 20.40 it certainly does, and also in SEG LVII.576,
where its loss is linked to loss of citizen rights (19–20 and 32–5), land-
owning and house-ownership being one such right. The six people whose
property was to be confiscated probably left Thasos immediately and went
into exile, perhaps before being condemned to it.

For these reasons I incline to a novel dating: the ruling dated by Antiphon
belongs to the period of the Athenians' siege of Thasos in 465–463. It belongs
in a time of oligarchy (*the* Three Hundred). It is separate from the rulings
about informers (their implied procedure of confiscation is different). It confiscates property of two Neapolitans, presumably property on Thasos itself
(most implausible after 463, when Neapolitans split off and lost any citizen
rights, or *isopoliteia*, they had enjoyed in Thasos since their city's foundation). Osborne and Rhodes *GHI* (2017) 464 suggest there may be 'some
deliberate archaism' in its use of the term *hados* in a ruling of *c.* 411–408. For
me there was none .The ruling belongs, I hypothesize, in 464/3 BC.

Endnote 3

ARCHONS AND THEOROI

My case for an earlier dating of columns of names in these lists has three main props: Polygnotos son of Aglaophon and the likely date of his death; Disolympios son of Theogenes and the likely date of his birth; the ruling dated by Antiphon son of Critobulus which I have argued belongs in 464/3, not 409 BC.

These arguments are a historian's arguments, made without first-hand engagement with the stone blocks of the lists. They depend on French specialists' many years of patient engagement with those stones and their admirable reconstruction, albeit with a different chronology, which is nearing full publication. Meanwhile, I have worked with the reconstruction of the first five columns of the Big Theoric List by Salviat (1979)107–28, with the folded pages illustrating them between pp. 116 and 117. For the later columns I have used the older presentation by Pouilloux (1954) folded between pp. 263 and 264, not all of which is still valid. The fundamental discussion of the *archon* lists is Salviat (1984) 233–58. Working with Salviat's insights, Hamon (2015–16) 75 fig. 2 gives a clearer reconstruction, proposing numbered dates for each column and showing all the courses assumed to be present. This invaluable reconstruction is reproduced in an even larger format in Brunet, Coulié, Hamon et al. (2019) 78–9. Hamon's two articles (2015–16) 67–115 and (2018) 182–207 are essential contributions with a masterly grasp of the lists and their problems. Hamon (2019A) 14–25 gives a lucid summary of the lists' scale and siting, their proposed chronology, full tables of each, the blocks which are present and absent in them and the numbered dates he assigns to each one. They are still, he emphasizes, working hypotheses (Hamon (2019A) 14).

These studies advance in directions which are particularly relevant to my suggested re-dating: if accepted, they make it more difficult to sustain, and in some cases exclude it. So I give a survey of the pros and cons, at least so far as I recognize them.

1. As Pouilloux (1954) 266–7 observed, three *theoroi* five to eight years after Antiphon's year of office also appear as *archons* in the fuller of the two lists of *archons*.They are Lydos son of Lydos (Big Theoric List col. 5.25), Basinos son of Sminthios (col. 5.26) and Timyllos son of Nymphis (col. 5.28). Pouilloux also noted Boulastidas son of Labros and suggested he was the brother of Artysileōs son of Labros, who was a *theoros* eight years after Antiphon (col. 5.28), although he also saw problems in too hasty an identification (267 n. 4). He noted, too, Nymphis son of Simalion (col. 6, name 6, dating to *c.* 390 BC in his view of the theoric list). In the *archon* list as reconstructed by Salviat and now by Hamon, these people are *archons* at dates between 402 and 396. If those dates are correct, they undermine my argument that Antiphon was a *theoros* in 464/3 and that the dates in the theoric lists need to go up by fifty-four years. The interval between my suggested dates for their years as *theoroi* and Salviat and Hamon's dates for their archonships is much too long.

I do not accept Salviat's postulated date for their archonships. It rests on an ingenious hypothesis, first outlined by Pouilloux (1954) 267, which is founded on three years in the list, one certainly, one probably, and one possibly, with months of 'anarchy' (the absence of an *archon*). Like Pouilloux, Salviat links these three anarchies to external events connected with the Athenians (though they are not mentioned in the list, of course). He links the first one to 465/4 (the Thasian calendar year ran from about November to November) and the next to (I think) 464/3: the word *anarchia* does not survive in either, and although there is reference in one to a number of months, the previous word's final letter, the only one preserved, is an 's', not the ē needed for *anarchiē* (as Pouilloux (1954) 267 n. 3 also observed). The third anarchy is fully present and specified as lasting for three months: Salviat dates it to 412/11, the year in which the oligarchic coup began on Thasos. In the seven years before it, *archons* appear singly (formerly, three *archons* per year were usual). Salviat relates this change to Athenian influence, one *archon* being the eponymous *archon* of a year in Athens (though there were also eight other *archons* annually, as never on Thasos). He argues from the probable length of the interval between the third anarchy and the previous two that a single *archon* per year had been in place for fifty-two years before 411, going back to 463 when the Athenians certainly imposed a new constitution on Thasos.

Block nr. 29 of the lesser of the two *archon* lists is the one on which the archonships of Lydos, Basinos and others occur. Salviat, followed by Hamon, places it in the next but one course below the anarchy in (supposedly) 411, thereby dating these people's archonships from *c.* 402 onwards. If that is right, my suggested re-dating is wrong. However, the explanation and dating of the times of anarchy are, inevitably, hypothetical. Bouts of anarchy could have arisen for any number of reasons, internal or external, in any of the many years in which Thasian internal politics are unknown to us. Although Hamon (2019A) 15 assumes there was already a year with a single *archon* between the first and second years with *anarchia*, I do not as yet see how we can know this, as the left side of the stone, the side on which paragraph-marks indicate the separate years, is missing (however, there may be more known now than Salviat presented in 1984). The separate *archons*' names and the reference to 'months' could belong in only two years, at least as I read them.

Other possibilities remain in what is, after all, a very fragmentary list. To judge from the reconstructions published so far, block 29, with the archonships of Lydos and the others, could perhaps, to an outside eye, move over to the left, taking them earlier back in time. Another possibility is that the years with one *archon* per year, not three, may have begun much later than 462, perhaps nearer the sequence of the seven years with single *archons* which survives. If so, the interval between the second and third 'anarchy' is much less than Salviat's proposed fifty-two years, with the result that the third year with anarchy is not in fact 411, Salviat's preferred date. The chronology which he argued for in that year is anyway not one I and others, including (in my view) Thucydides 8.68, support. His attempt to link it to the two rulings about informers is also not convincing, as my Endnote 2 has argued. Alternatively, the datings of the columns could be raised by say twenty-six or twenty-seven years. That is the option I incline to favour, though it raises problems later in the list.

Athens-related events are not the only possible cue for these bouts of anarchy in the lists. My raised dating would relate the first and second ones to 492–489, years of Thasos's forced obedience to king Darius: divisive arguments over Persian relations may underlie them. The third one would then fall in 438, for internal reasons lost to us, but perhaps relating to quarrels over the practice of one *archon* per year, promptly dropped thereafter. As for the names in the lists, this is not the place to cover each particular link between them and names in the first six columns of *theoroi*, but by raising the dates of the *theoroi* by fifty-four years and the *archons* by twenty-seven, I assume, as the French do not, that a theoric magistracy was always held before an archonship (for some examples of this order, even in their alternative scheme, Hamon (2018–19) 196; others in the Big Theoric List col. 5, esp. 25–6; Hamon (2015–16) 110 n. 120). I can accommodate such overlaps as Deinostratos son of Attales (the shorter *archon* list col. 5.15 in, for me, 536 BC; Big Theoric List col. 2.24, for me 558 BC) and all the overlaps brilliantly noted by Pouilloux (1954) 267, now in the shorter *archon* list col. 6.10–15, including Telemachos son of Pithēkos and others. I put the archonship of Akēratos son of Phrasierides (its col. 3.33) in *c.* 585, but assume he lived on for a while to allow time for I.G. 12 Supplement 412, the epigram on his statue (it might even be posthumous) and his funerary memorial (I.G. 12.8.683). I can also accommodate the links presented by Hamon (2018–19) 192–3, who thereby become *theoroi* before they were *archons*. I make Antipappos son of Orges (shorter *archon* list col. 5.14) an *archon* in 538 and therefore the grandfather of the one named in Hdt.7.108 (where even the latest Oxford text should be emended from 'Antipatros' to 'Antipappos'). I put Amphērides Simaliōnos (shorter *archon* list col. 5.14) as *archon* in 538 and as *theoros* (Big Theoric List col. 2.23) in 560 BC. My dating works well for Aristokritos son of Epēratos (shorter *archon* list col. 6.13), *archon*, I propose, in 505, while Epēratos son of Aristokritos, surely his son, is now *theoros* (col. 5.17) in 467 (the hitherto proposed datings place the father as *archon* in 478, the son as *theoros* in 413/12, but the long interval is awkward, especially if, as I think, the archonship always followed the *theoria*). Nymphis son of Simaliōn becomes *archon* in 426 and *theoros* in *c.* 444, instead of being archon in 399 and *theoros* in *c.* 390 at a time when others around him were certainly holding the *theoria* before the archonship. Kallimenes son of Leontios becomes *archon* in 426 and can then be father, not grandfather (Pouilloux (1954) 267), of Leontios son of Kallimenes, a *theoros* in, for me, *c.* 414.

I recognize the hazards, but I cannot resist also noting the much-discussed Lichas son of Arkesilas, occurring in the shorter *archon* list col. 7.50 (could he really be the famous Spartan?). On my higher dating he becomes *archon* not in 397 but in 425/4, just after the Athenians had proposed a doubling of Thasos's tribute from 30 to 60 talents. The conjunction of these two personal names (known individually elsewhere than Sparta) remains striking, but nothing in the list marks him as a Spartan, and he would have had to have received honorary Thasian citizenship in order to hold an archonship. Nothing attests that.

A raising of the dates for the columns of the better-preserved list of *archons* has a more obvious advantage. It takes them back to *c.* 680 BC, even nearer to the likely date for the foundation of Thasos. Raising the dates for the *theoroi* also sets their 'first *aparchē*' further back in the seventh century. I admire, but do not accept, the hypothesis of Salviat (1979) 123–5 which links the rule of the 360 on Thasos in col. 1.5 of the Big Theoric List to the ending of a tyranny in, he suggests, 546/5 BC, with Spartan assistance. That tyranny is only attested by Plut. Mor. 859D, but, although I accept it, I consider its ending was probably much later in the century (maybe when the Spartan king Cleomenes was in Thessaly in 519 BC), and so it was not relevant to the earlier institution of the 360. As the years 411–407 and 404 onwards show, the theoric lists were remarkably undisturbed by other internal political upheavals.

2. Prosopographical links more generally. The lists certainly present Thasian upper-class families whose members can be traced as office holders through many generations: it is part of their fascination for historians. Links between names and persons, well exploited by Pouilloux (1954), have been brilliantly taken further by Hamon, no doubt with more to come (some of it is now in Hamon (2019B) 139–93 which appeared in April 2020, too late for me to include its fine points). As he persistently emphasizes, direct connections tend to be probable, not certain, and caution is needed. Throughout there are the pervasive difficulties of like-named grandfathers and grandsons, nephews, cousins and so forth. However, cumulatively the interconnections are wide ranging and impressive. I have attended closely to those in Hamon (2015–16) 100–107, especially the names he situates attractively in the reign of Alexander and those interlinked, as Salviat has also indicated, with names stamped on independently datable wine amphorae (102–3).

At several points problems arise for my proposed raising of dates in the list. The lower inscribed blocks of the theoric list col. 7 are now securely identified (96–100), but if my re-dating continues to be applied, the names there ought to belong in the 380s and 370s, not in the late 330s and 320s. However, a lease for the gardens of Heracles, acutely discussed by Hamon (2015–16) 107–10 and (2019A) nr. 42 is dated by three *theoroi*, as he well argues, who are known at a point low down in col. 7: the lease's style of lettering points to the date he prefers for them, 328 BC. I would have to override this argument from lettering.

Particularly important are rare cases where external evidence may help to date an individual in one of the lists. In Brunet, Coulié, Hamon et al. (2019), Hamon (74 and 77) proposes that Hegetorides son of Mnesistratos (Big Theoric List 3.17) is identical with the Hegetorides who made an impassioned plea to the Thasians when the Athenians were besieging the city (Polyaenus 2.33), a siege which Hamon identifies as the one in 465–3. However, it could as well be the siege ending in 407: Hegetorides the *theoros*, whom my dating would put far earlier in 531, may have been this other Hegetorides' grandfather. He might also be an ancestor of the Mnesistratos whose wife is noticed in Epid. 1.17.2, but not, as Hamon suggests, his father. If the final entry on the block at the foot of the Big Theoric List column 7 is indeed dated to 326, it is attractive to follow Hamon's further insight and identify its Aristoleōs with the Aristoleōs who is deplored by Demosthenes for having betrayed Thasos to King Philip (Hamon (2015–16) 114–16, with other possible links). If my higher re-dating were to be applied here, Aristoleōs the *theoros* would have to be a different person, an older relation of the one denounced by Demosthenes.

Especially important links are given by the few surviving decrees which are dated by a combination of *archons* and *theoroi*, establishing that they served in one and the same year. The crucial one here is Hamon (2019A) nr. 14, honouring elderly Polyaretos of Zone (nr. 15 is closely related to it). It combines a known college of *theoroi* and one of *archons*, linking them inarguably in the same year: it is currently dated to *c*. 291, a date which is also considered to suit the context of Polyaretos's deeds. If so, it counters a continuation of my updating of *theoroi* and *archons* into this later part of the lists. However, the *archons* are known only on a block of the big *archon* list, and, as only three blocks of that list survive, its placing is not certain. As for the block naming the *theoroi*, it is placed currently near the top of column 9 (Hamon (2019A) 114 n.3 citing observations by Salviat; Pouilloux (1954) 279 was unsure about its place, however). Might it perhaps move to the right, moving its neighbour there into the space, as yet blank, in the next column, column 11, over to the right? The preceding column 8 also lacks all the colleges from the middle of its sixth course on through the seventh and perhaps, too, those at the bottom of its eighth course: there might, then, be major dislocations in the sequence which is missing there. If not, Hamon (2019A) nr. 14 is indeed a counter to a continuation of my re-dating of the *theoroi* as far as that point in the list. If it is impossible there, how can it be correct in the earlier parts unless the present reconstruction of the list there is in places mistaken?

Hamon (2019A) nr. 106 is also important, the honouring of the Thasian Nossikas son of Heras by the city of Lampsacus for his actions in, surely, 322 BC. A Nossikas is attested as a *theoros* in col. 6.15, whom my re-dating would take back to 432 rather than the currently adopted 378. He might be a kinsman of the Nossikas being honoured by Lampsacus, but not a grandfather. If so, he does not refute my re-dating.

The most important challenge is Hamon (2019A) nrs. 82–3, two inscribed bases of dedications, now lost, by one and the same college of *theoroi* who are dated currently to 374 BC in the big list col. 6.20. If that date is accepted, it counters my proposed re-dating of Antiphon's year: these *theoroi*'s placing in col. 6 is just above the names (col. 6. 21–6) who are fixed for certain in the column to the right of Antiphon's because they are all on one and the same block of the little list of *theoroi* (Inv. 935, illustrated in Hamon (2015–16) 78, fig. 5). For my dating of Antiphon's year to survive, I have to raise the dates of 82 and 83 by fifty-four years. Their Leophantos son of Demalkes bears names attested in the fourth century in a wine-producing landed family, but he may simply be an earlier member of it. Nothing dates these dedications other than links between their style of lettering and the lettering of several other inscriptions, of which the most important for my purpose are Hamon (2019A) nr. 4 (the regulations about Delian Artemis's sanctuary), Hamon (2019A) nr. 37 (a dedication for a victory in a stage competition) and Hamon (2019A) nrs. 52–4 (dedications by overseers, the *epistatai*). They belong in Hamon's 'Groupe 111' (his p. 409), traceable, he suggests, to the hand of one and the same stonecutter. I therefore incline, with due caution, to an updating of these related inscriptions too. Nr. 54 is perhaps to be dissociated from the group, as it uses the more modern form of genitive singular, *-ou*, a feature of the mid to later fourth century as Hamon observes, whereas 52 and 53 do not. I do not see insuperable difficulty in raising the date of the Delion regulation (despite the link suggested by Hamon (2019A) 227 between the not uncommon name Sombrotos in it and in nr. 54) or even the stage-victory dedication. The stonecutter might have been active from *c.* 430–390, allowing these inscriptions to belong in the 390s. However, epigraphers have not even contemplated this option, so caution is necessary. I take it to save my theory, whereas they propose their datings by comparisons between the texts concerned.

3. The date at which the lists were inscribed. Hamon (2015–16) 75–83, esp. 81–2, has acutely addressed this complex question for each list,

greatly illuminating it. Here, I note only the two theoric lists, certainly inscribed at first in the fourth century, the lesser one some forty years before the big one (Hamon (2019A) 409 and 412 for the stonecutters' differing styles). The shorter list, he well notes, changes its style of lettering in its sixth column, at a point dated hitherto to 368 BC. Hamon inclines to see the early 360s, therefore, as the date at which this list was decided upon. On my higher dating, the point in column 6 would have to move up to 422. However, I would argue that the first stonecutter stopped there for other imponderable reasons – sickness, death or whatever – not that that year represented the initial length decided on for the list. After all, another hand continued to inscribe it, picking up promptly where the first cutter left off.

To sum up: in the first five columns of the Big Theoric List I am struck by problems, especially Polygnotos, Disolympios, and the date of Antiphon. Possible answers to them are not convincing: that the very famous Polygnotos lived on and worked for nearly twenty more years without any other surviving attestation; that Disolympios was born when his father was fifty or more and was named long after the year of the double victory his unique name commemorated; that the property of those Neapolitans confiscated in the ruling dated by Antiphon was movable property only, not land or houses. On the other hand, I postulate a *theoria* for Disolympios at the very early age of 18 or 19, justifying it by his immensely famous parentage. I also assume the Antiphon ruling was inscribed with another beside it some thirty years after its enactment, and hence it is in the Ionic alphabet. I then have to raise the dates of Hamon (2019A) nrs. 82 and 83 and query, perhaps unjustifiably, the placing of the second block in column 9 linked to the decree honouring elderly Polyaretos. My higher datings are certainly not in line with those built up over the decades by first-hand specialists in the inscriptions themselves, whose work is fundamental to my understanding of the problems. So I reiterate two points. Final publication of the lists, or evidence yet to be discovered, may decisively, not just 'probably', rule out my suggested re-dating. If its proposal meanwhile has helped to strengthen the eventual interpretation of the lists, I am more than content. However, my dating for the doctor-author of Epidemics 1 and 3 is a separate matter. It does not depend on a suggested re-dating of parts of the magistrates' lists. If that re-dating fails, Antiphon the patient is the grandfather of Antiphon the magistrate. The central argument of this book still stands.

Notes

Modern journals are mostly cited by the abbreviations listed in *L'Anneé Philologique*, available online at http://www.annee-philologique.com. Ancient authors are mostly cited by the abbreviations in S. Hornblower and A. Spawforth (eds.), *The Oxford Classical Dictionary* (3rd edn, Oxford, 1996), but I refer to the Epidemics as Epid., and to other works in the Hippocratic Corpus not with Hippocrates as author but by unitalicized title only, following the titles in the English translation of Jacques Jouanna, *Hippocrates* (Baltimore, 1999), pp. 372–416. I number the chapters and sections of Epid. 1 and 3 according to the edition of Jacques Jouanna (Paris, 2016), for which see below.

The outstanding general survey of ancient medicine is V. Nutton, *Ancient Medicine* (2nd edn, London, 2013). K.-H. Leven, ed., *Antike Medizin. Ein Lexikon* (Munich, 2005), is a valuable encyclopedia. On Hippocratic medicine, Jacques Jouanna's *Hippocrates* (see above), pp. 75–362, relates the medical texts to contemporary thinking very clearly. Elizabeth M. Craik, *The 'Hippocratic' Corpus: Content and Context* (London, 2015; pbk 2016), is an outstandingly helpful survey of all the 'Hippocratic' texts with valuable notes on each and has opened up this specialized field to all Anglophone scholars. The many books by G. E. R. Lloyd have central insights and illuminating comparisons with China and other cultures: his *Methods and Problems in Greek Science* (Cambridge, 1991), *In the Grip of Disease: Studies in the Greek Imagination* (Oxford, 2003) and, with Nathan Sivin, *The Way and the Word: Science and Medicine in Early China and Greece* (Yale, 2002) are particularly relevant to my book. Helen King, *Hippocrates' Woman: Reading the Female Body in Ancient Greece* (London, 1998), is a collection of brilliantly argued essays, and her *Hippocrates Now: The Father of Medicine and the Internet* (London, 2019) is a bracing evaluation of Hippocrates' fame and legacy, though it came too late for me to use.

On the Epidemic books, the English translations of 1 and 3 by W. H. S. Jones in the Loeb Library series, *Hippocrates*, vol. I (London and New York, 1923), pp. 139–288, are a starting point, though their arrangement of the contents and their Greek text and translations are not always reliable. Philip van der Eijk, 'An Episode in the Historiography of Malaria in the Ancient World', in D. Michaelides, ed., *Medicine and Healing in the Ancient Mediterranean* (Oxford, 2014), chapter 15, is a fascinating account of 'Malaria' Jones himself. Wesley D. Smith has translated Epidemics 2 and 4–7 in the Loeb Library *Hippocrates*, volume VII, but his Greek text of 2, 4 and 6 especially is open to much improvement, as its Budé edition, by Robert Alessi, will show, making due use of the Arabic evidence. The essential text, with French translation and commentary, for 1 and 3 is by Jacques Jouanna with A. Anastassiou and A. Gardasole, *Hippocrate*, vol. IV, Part 1: *Épidémies I et III* (Paris, 2016), in the Collection des Universités de France, a Budé edition (hereafter, Jouanna (2016)). For 5 and 7, the essential text, with French translation and commentary, is Jacques Jouanna, with M. R. Grmek, *Hippocrate*, vol. IV, Part 3: *Épidémies V et VII* (Paris, 2000), in the same series, a Budé edition. These two superb volumes replace earlier works on the supposed multi-authorship or rearrangement of the contents of 1 and 3, so I will refer to Jouanna (2016) instead, believing the older analyses, for all their interest, to be defunct since his edition appeared.

For English readers, the eleven volumes of the Loeb Library series, *Hippocrates*, are the obvious first resort, the first four being mainly edited by W. H. S. Jones and vols. 5–11 more recently by Paul Potter, a skilful translator and interpreter. Of the more widely read texts, the edition, with translation and commentary, of *On Ancient Medicine* by Mark Schiefsky (Leiden, 2005) is especially valuable, as is the earlier French one, *Hippocrate: l'ancienne médecine*, by the great scholar A.-J. Festugière (Paris, 1948). R. Joly, *Hippocrate: Du Régime* (Paris, 1967), is helpful on *Regimen*, and Jouanna's French editions, translations and commentaries of *Airs, eaux, lieux* (1996), *L'Art de la Médecine* (1999, with C. Magdelaine), *Prognostic* (2013) and *Le serment, les serments chrétiens, la loi* (2018), all in the Budé series (Paris), are invaluable.

Peter E. Pormann, ed., *The Cambridge Companion to Hippocrates* (Cambridge, 2018) has particularly good essays, although they reached me too late to be used systematically in my book. My bibliography is only a selection of all I have read. Jouanna (2016) gives more bibliography on

Epidemics 1 and 3, all of which I have covered, except some of the studies on manuscripts, and the excellent bibliography in C. Thumiger, *A History of the Mind and Mental Health in Classical Greek Thought* (Cambridge, 2017), adds even more than I have consulted.

On Thasos, an essential work of wide-ranging reference is Y. Grandjean and F. Salviat, *Guide de Thasos* (Athens, 2000), which I have used in the updated and even fuller edition in Greek (Athens, 2012). As a result, my chapters conform with the excellent work now presented by M. Brunet, A. Coulié, P. Hamon et al., eds., *Thasos: Heurs et Malheurs d'un Eldorado antique* (Paris, 2019), a beautifully produced summary with excellent bibliographies. On Thasos's inscriptions, P. Hamon, *Corpus des Inscriptions de Thasos III* (Athens, 2019) is an exemplary study of inscriptions from the fourth century BC onwards and a foretaste of more to come. It relates to parts of my Chapters 12 to 15 and in places implies a different chronology, to be presented fully in volumes I and II. So far, I prefer mine.

Roy Porter, *The Greatest Benefit to Mankind: A Medical History of Humanity* (London, 1997), is an outstandingly good history of the subject in general. W. H. McNeill, *Plagues and Peoples* (Harmondsworth, 1979) is an excellent study of a subject now again in the global limelight.

Introduction

1. Bickerman (1986) 196–211.
2. Folkerts (2005) 351–4, a very clear exposition.
3. Hdt. 3.80–83, which I consider to be based on a text set teasingly in Persia by its mid-5th-c. author(s), which Hdt. took to be historical when he encountered it anonymously and undated.
4. Arist., Pol. 1253B20, with Brunt (1993) 343–88, esp. 352.
5. Jouanna (1999) 210–42.
6. Ar., Eq. 191, for the noun which scans readily for comic verse: it does not have a neutral sense at Thuc. 4.21.3.
7. Xen., Eq. 1.1–3 and 11.6, surely the Simon of Ar., Eq. 242: Raulff (2017) for the Centauric pact.
8. Pl., Grg. 518B; Ath. 7.282A and 12.516C: Wilkins and Hill (1996) 144–8.
9. De Angelis (2014) 70–83.
10. Robertson (1992) 133–264.
11. Boardman (1993) 87.

12. Philipp (1990) 133–54; Jenkins (2015) 78–9 on the 1920s reconstruction, now in Munich.
13. Euripides' *Bacchae* and *Andromache* were composed first for monarchies, and his *Alcestis* for Thessalians.
14. M. Lloyd (1992) 2 and *passim*.
15. Hudson-Williams (1951) 68–73.
16. Soph., Antig. 363–4.
17. Frede (1987) 225–42; van der Eijk (2005) and esp. Jouanna (1999) 259–85 on reactions by medics against philosophy.
18. Jenkins (2015) 23.
19. Métraux (1995): rightly questioned by Osborne (2011) 40 n.62.
20. Kosak (2004).
21. Thuc. 6.14; Epid. 1.5: Jouanna (2012) 21–38 and 152–3.
22. Pl. Prt. 311B–C.

PART I: HOMER TO HIPPOCRATES
1. Homeric Healing

1. Hom., Il. 11,596–615; Hdt. 3.134.
2. Kudlien (1965) 293–9.
3. Hom., Il. 17.616–19.
4. Friedrich (2003); Griffin (1980) on 'death, pathos and objectivity' does not fully address the descriptions of wounds.
5. Van Wees (2008) 172–5 surveys similar issues in the interpretation of archaic vase paintings.
6. Hom., Il. 4.492; 5.65–9; 13.675–75; 13.650–52 with the important note of Janko (1994) 116.
7. Frölich (1879) with Nutton (2013) 37.
8. Saunders (2004) 1–14 is fundamental.
9. B. Williams (1993) 23–5 refutes it : Holmes (2010) 6–10 and 29–37 revisits it.
10. Hom., Il. 14.495 with 5.73 and the scholia, well discussed in A. S. F. Gow's note on Theoc. 25.264: Hom., Il. 5,306, with *Places in Man* 6.
11. Hom., Il., 22.325, with 24.642.
12. I differ from Irigoin (1980) 247–57, a valuable survey; *Places in Man* 6; Hom., Il. 21.337.
13. Hom., Il. 8.325–8.
14. Hom., Il. 22.396–7 with Saunders (2004) 16–17, an important insight.
15. Hom., Il. 396–7.

16. Saunders (2004) 11–12, for the list of such falls.
17. Hom., Il. 13,428–44; Saunders (1999) 348–9.
18. Hom., Il. 16.345; Saunders (2004) 361–2.
19. Hom., Il. 20.481–2.
20. Hom., Il. 13,616–17; 16,740–42; Saunders (1999) 352–4.
21. Onians (1951) 23–43, still valid though adjusted by Ireland and Steele (1995) 183–95.
22. Hom., Il. 13.345–8; Saunders (1999) 352–4.
23. Hom., Il. 5.66–7; Saunders (1999) 347–8 on 'wound–weapon mismatch'.
24. Hom., Il. 13.650–55; Saunders (1999) 352–4.
25. Loraux (1995) 88–100, with a far-fetched comparison with women's bleeding.
26. Hom., Il. 16.510–26 with 8.321–9.
27. Hom., Il. 5,415–17 and 899–900.
28. Hom., Il. 11.842–7.
29. Hom., Il. 11.814–48; 15.390–404; 19.300 and 23.280–81, on Patroclus' kindness.
30. Hom., Od. 4,220–24; Lane Fox (2008) 9 and n.20.
31. G. E. R. Lloyd (1990) 14–38; Nutton (2013) 39–40.
32. Hom., Od. 19.455–8
33. Burckhardt (1998) 129.
34. Willi (2008) 153–71, esp. 161.
35. Hom., Od. 9.399–412.
36. Hom., Il. 11.269–72
37. Hom., Od. 5,394–9.
38. Hom., Od. 11,172; 3,198–203.
39. Hom., Il. 22.30–32 with N. J. Richardson's note (1993) 109.
40. Holmes (2007) 45–84; G. E. R. Lloyd (1966) 202 on the 'complementary conceptions' of Homeric sleep.
41. Allbutt (1923) 130.
42. Hom., Il. 11.514–15.
43. Hom., Od. 17.383–5.
44. Nocita (2012) and Dana (2012) 249–66 are two good recent studies of later travelling doctors.
45. Hom., Il. 11.624–5, 639–41: Lane Fox (2008) 8.
46. Verg., Aen. 12.391–431.
47. Verg., Aen. 12.397, with Sen. Epist. 87.15.
48. Liolios, Graikou et al. (2010) 230–40.
49. Aristot., Hist. An. 612A 2–5, Cic., Nat. D. 2.126 and esp. Pease (1948) 469–74.

2. Poetic Sickness

1. Arnott (1996) 265–70; Witt (2018) 217–45.
2. Celsus, *Med.* 7 praef. 4; *Physician* 6; *On Joints* 69; *Haemorrhoids* 2 on shouting while being cauterized.
3. Xen., *Lac.*, 13.7; *Cyr.* 1.6.15–16 with Gera (1993) 65: 'it sounds like the title of a Hippocratic treatise'; *Cyr.* 3.2.12.
4. Diod. Sic. 1.82.3; Xen. *Anab.* 3.4.30 with Cawkwell (1972) 166 n.10: 'the "eight doctors" were probably slaves doing the tasks of nurses'; at 4.5.8 no doctors are mentioned in cases of snow blindness or hunger sickness ('boulimia').
5. *Physician* 14.
6. Plut., *Alex.* 8.1; 41.4; Diod. Sic. 17.103.7–8.
7. Triantaphyllou (1999) 353–64, at 363.
8. Hom., Il. 18.396–7; 1.600.
9. Lucian, *Anach.*1.
10. Hom., Il. 15.679–84.
11. *On Joints* 53.
12. Soranus, *Diseases of Women* 2.16.7 (Budé edn, 1990).
13. Pirsig, Helidonis and Velegrakis (1995) 141–2, which I owe to Tomas Alusik.
14. Dasen (1997) 5–22.
15. Dasen (1993).
16. Simpson and Pankova (2017) 161–2.
17. Hawass et al. (2010) 638–47.
18. Roberts and Manchester (2010) is an excellent survey.
19. *Iliou Persis* F3 (West).
20. Hes., FF129–33 (M – W); Bacchyl. 11; Pherecydes FGH 3 F14; Acusilaus FGH 2 F28: all analysed by Dowden (1989) 70–95.
21. Plut., *Mor.* 249 B – D.
22. Solon F4 (West) 13–22; Brock (2013) 92 with 69–76.
23. Plut., *Sol.* 15.1.
24. Solon F13 (West) 57–64.
25. Hes., Op. 95–104.
26. Edelstein (1967) 378; Frazer (1972) 235–8.
27. Hom., Il. 18.376.
28. Hom., Il. 5.749, 8.393.
29. Hes., Op. 240–41.
30. Hes., Theog. 56–62; Celotto (2017) 224–34.
31. Hes., Op. 582–8; Onians (1951) 177–8, important.

32. Sappho F31, with Hutchinson (2001) 28–9, 168–77: I take 'hōs . . . idō' in line 7 to mean 'whenever I see'.

33. Sappho F1 (Voigt).

34. Pind., Pyth. 3: Morgan (2015) is a recent survey.

35. Pind., Pyth. 3.2–8, 73–7.

36. P.Oxy. 222 cols. 1 and 19; Maehler (2004) 101, 106–7; Pind., Ol. 1; Bacchyl. 5 (ed. Maehler) 43–9.

37. Paus. 6.12.1., 8.42.9; R. R. R. Smith (2007) 124–30.

38. Pind., Pyth. 3.1–8; Hom., Il. 11.831–2; Pind., Pyth. 3.46–53, where I disagree with Farnell, *The Works of Pindar*, comm. (1932) 142 that magical amulets are meant.

39. Pind., Pyth. 3.47.

40. Hes., Theog. 811–13; Xen., Vect. 4.2.

41. Theognis 183–92 with Xen. ap. Stob. 4.29.53, who relates it, correctly, to aristocratic breeding, a corrective therefore to van Wees (2000) 63–6 and his supposed 'mafiosi'.

42. Diod. Sic. 11, 49.1–2, 51.1–2; Osborne and Rhodes *GHI* (2017) nr. 101 with commentary.

43. Arist. F 486 (Rose); Plut., Mor. 403C for the stones.

3. Travelling Doctors

1. Hom. Hym. Aphrod. 227–40.

2. Laios et al. (2012) E1–E2; D. Williams (2006) 127–32.

3. Hom., Il. 2.729–32; Willi (2008) 160–61, on the name's non-Greek, possibly Anatolian origin; Aston (2004) 18–32, esp. 22–3 nn.11–12 on Trikka, not Epidaurus.

4. Touwaide (2005) 155–73 and (2006) 558–68 are fundamental surveys, basic now to this field.

5. Hom. Hym .Dem. 208–9.

6. Schwarzmaier (2014) 155–6; Robertson (1992) 58–9.

7. Hdt. 7.181 with B.92.1–2.

8. Arr., Indica 18.5; Majno (1975) 283–4.

9. Simpson and Pankova (2017) 292.

10. Galen 18A (Kühn) 774–7; *Physician* 4 and *On Joints* 35 on showy bandaging; Salazar (2000) 52–3.

11. *On Joints* 69; Majno (1975) 152–3.

12. Liston and Preston Day (2009) 59–73.

13. Agelerakis (2006) 5–18 and (2014) 256–9.

14. Jacopi (1929) 240–45; Farmakidou (2014) 295–6; Berger (1970) 66 and 175 n.135.
15. Kind (1922) 881–90; Farmakidou (2014) 295–6, a helpful survey.
16. coreelements.uk.com/2019/02/18/what-is-dry-cupping-therapy?; Rawcliffe (1997) 63–8, with good pictures of the medieval uses.
17. Mellen (2016).
18. Agellarakis (2014) 258, citing *On Wounds in the Head*; Majno (1975) 166–9.
19. Suida, s.v. Anaximander; Barnes (1979) 19–37; in general, Irwin (1989); Batey (1999) 13–25 on nature and English garden style.
20. Osborne (2011) questions artists' depictions of the 'muscular body'; Tsingarida (1998) 59–64 and 196–8, items 1.4–1.6, for, to my eye, a different view.
21. The bases, probably by Endoios, are examples whose figures show such surface markings: Robertson 1 (1975) 226–7.
22. Robertson 1 (1975) 222 on Euphronius and 'muscularity', and 230 on Onesimos and 'movement and torsion'; Mannack (2001) on the Mannerists and muscularity too, by 500 BC.
23. Dem. 24.114 assumes gyms in Athens *c.* 600 BC: Delorme (1960) 36–7 suggests by the later 6th c.; Hdt. 6.126 for a palaestra in Sicyon c.570 BC.
24. Villard (1953) 35–43 esp. 42.
25. Philostr., Gymnastikos (Teubner, ed. Kayser) 278 lines 20–22.
26. Pl., Leg. 840A; Ael., NA 6.1
27. Craik (1995A) 387–402 and (1995B) 343–50; Jouanna (1999) 161–70.
28. Salta (2014) 331–3, an excellent overview; Samama (2003) 109 on the inscription's scansion.
29. Monterosso (2014) 327, 330.
30. Marconi (2010) 339–49.
31. Johnston (1989) 131–5; SEG 29.924.
32. Thonemann (2006) 11–43; Hdt. 4.87–9.
33. Masson (1983) = ICS no. 217; Samama (2003) 456–9 with translation; above all, Georgiadou (2010) 141–203, now the fundamental study.
34. Georgiadou (2010) 159–70 gives a balanced survey of the dating.
35. Georgiadou (2010) 172–9.
36. Masson (1983) 240; Georgiadou (2010) 184–9.
37. Masson (1983) = ICS no. 217 lines 14–17: Georgiadou (2010) 184–9.
38. Masson (1983) = ICS no. 217 lines 20–23.

4. From Italy to Susa

1. Hdt. 2.84.1.
2. Hdt. 1.197.1.
3. Geller (2004) 11–61 and (2010) and (2018) 42–54; Finkel and Geller (2007); Scurlock (2014); Asper (2015) 19–46; Burkert (1992) 75–9.
4. Hom., Od. 4.232.
5. Hdt. 2.84.1; 2.85–7 with the important findings of Marozzi (2008) 151–4.
6. Ryholt (2006) 13 and 33–46.
7. Hdt. 2.111.
8. Hdt. 2.87.
9. Totelin (2009) 153–6, noting that the words for Egyptian plants in these recipes are almost all purely Greek, and especially 179–84, an excellent survey of supposed 'Egyptian influences' in other remedies.
10. von Staden (1989) 1–26, esp.12.
11. Hdt. 4.187.2; *Airs, Waters, Places* 20.
12. Kuriyama (1999) 195–232.
13. Hdt. 3.125.1; 3.129–37 with Asheri, Lloyd, Corcella (2007) 511–61.
14. Hdt. 3.125.1.
15. Kuhrt (2007) 117–22, esp. 119 n.15 and 121 fig. 4.2.
16. Hdt. 3.132.1 and 9.16.2; Xen, An. 1.8.25.
17. Hdt. 3.131.2.
18. Hdt. 3.133.1.
19. Hdt. 3.135.
20. Hdt. 3.137.4–5, a most interesting remark; Porph., V.Pythag. 4; Phot., Biblioth. 249; Suda, s.v. Pythagoras and esp. Roubineau (2016) 171–89.
21. Ath. 10.412E – 413A; Roubineau (2016) 142–50.
22. Swerr (1961).
23. Suda, s.v. Democedes with Griffiths (1987) 38 and 48–9; Davies (2010) 31–2.
24. Hdt. 3.137.5.
25. Iamb., V.Pyth. 257,261.
26. Griffiths (1987) 37–51; Davies (2010) 19–44, esp. 39–42 and 43 n.82.
27. Davies (2010) 39, 42–4 on Atossa and 32 n.42 on the name 'Democedes'.
28. Hdt. 3.133.2; 9.109.1.
29. Hdt. 3.135.3.
30. Hdt. 3.130.1–2.

31. Hdt. 3.130.3; 3.129.2.
32. Asheri, Lloyd, Corcella (2007) 513, their note on Hdt. 3.133.1, with bibliography; Briant (2002) 265 for the abscess.
33. Grmek (1983) 288–95, at 290–92.
34. *On Fractures* 10; Epid. 5.48 for the astragalos bone.
35. Hdt. 5.47; Roubineau (2016) 60–61 lists the victors and 62–76 discusses their context.
36. Str. 6.1.12; Zenobius, Proverbs 6.27; Hesych, s.v. *Krotōnos hygiesteros*; Roubineau (2016) 71–5.
37. Iamb., V.Pyth. 82; Riedweg (2005) 77.
38. D.L. 8.12; Riedweg (2005) 10 and 36–7 on vegetarianism and 77 on harmony.
39. D–K 24 B1–39; B17 gives the names, for which see Longrigg (1993) 49; Guthrie (1962) 342 n.1 cites those against a date *c*. 500 BC, who do not persuade me. Arist. Metaph. 1.986A 29–31 synchronizes Alcmaeon with elderly Pythagoras, and I, but not others, think it is not an interpolation.
40. D–K 24 F1.
41. Iambl. VP 72; Riedweg (2005) 101–3.
42. Perilli (2001) 65–79, with Mansfeld (1975) 26–38; title is in D–K 24 A1 (citing Favorinus) and A2 (Clement and Galen).
43. D.L. 8.83.
44. Brilliantly seen by D. M. Dunlop, credited in Kirk (1956) 5–6; Lebedev (1993) 456–60 discusses it too; D–K 17 A15.
45. D–K 24 A18.
46. D–K 24 A13; Onians (1951) 118–19.
47. D–K 24 B3 and A13.
48. D–K 24 A5 with Barnes (1979) 149–50 on D–K 24 A11.
49. Onians (1951) 115–17, citing Pind., Pyth. 9.31–2; D–K 24 A5.26 and A8.
50. D–K 24 A10, citing Chalcidius.
51. G. E. R. Lloyd (1975B) 113–47 is essential, but aptly challenged by Longrigg (1993) 58–9; D–K 24 A10; Onians (1951) 76.
52. D–K 24 A1, fully discussed by Barnes (1979) 137 and Kouloumenatas (2018). I follow Lebedev (2017) 227–57.
53. Barnes (1979) 137–43; Hussey (1990) 11–38.
54. Arist. Metaph. 1.986A31 – B2.
55. D–K 24 B4.
56. Vlastos (1953) 337–66 prevails, despite Pleket (1972) 63–81 and, most recently, Mansfeld (2013) 78–95, who omits, however, Hdt. 3 142.3, a

crucial text; I do not believe that *isonomia* is a later paraphrase in D–K 24 B4.

57. D–K 24 B4, even mentioning 'tōn exōthen aitiōn'.
58. D–K 24 B4 p. 216 lines 3–4.
59. Evidence in LGPN, esp. Alcmaeon's in V.A. 22 (Colophon) and 30 (Rhodes etc.), all later than the 5th c. and 1.368 (Peirithous, on Tenos), also late.
60. Ephoros FGH 70 F23; Soph., OC 1593–4: Paus. 1.30.4.
61. Hdt. 5.37.2 and n.55 above.

5. The Asclepiads

1. Ath. 14.621D – E.
2. Hdt. 2.131.5
3. Men., *Aspis* 439–64 with Arnott (1996) 430–32 on Doric doctors: Chamoux (1953) 365–8 and plate XXII.3.
4. Berges (2006) 19–23 with the excellent map on 21.
5. Theopomp. FGH 115 F103.14; Asclepiads are first mentioned in surviving texts by Thgn. 49–438, but the lines may be later than Thgn.'s lifetime.
6. Fraser and Bean (1954) 28–30, with no. 16, lines 5, 8, 11, 12, unknown to Grensemann (1975) 2, who considered it 'nicht genau lokalisiert'.
7. Steph. Byz., s.v. Syrna.
8. Sherwin-White (1978) 61–2.158.257; Fraser 1 (1972) 343.
9. Arr. Anab. 6.11.1.
10. Tac., Ann. 12.61: Sherwin-White (1978) 149–52.
11. Sherwin-White (1978) 49 n.104.
12. SEG 16.326; Bousquet (1956) 579–91; Jouanna (1999) 34–5, 50–52: W. D. Smith (1990) challenged the interpretation of Asclepiadai, unconvincingly: as he recognized, the associations of Asclepiastai were very different. They are not relevant.
13. Lonie (1965) 1–6, (1978) 42–75 and 77–92; W. D. Smith (1973) 569–85; Thivel (1981).
14. *Regimen in Acute Diseases* 1–2, not by a Cnidian as Edelstein (1967) once proposed.
15. Galen XV (Kühn) 419.4–6; Grensemann (1975) no. 12, p. 24.
16. Vitr., *De arch.* 7, Preface 12 on Rhoecus and Theodorus writing a text on Samos's Heraion and Chersiphron and Metagenes one on Ephesus's Artemision, evidently both well before 500 BC.
17. *Ancient Medicine* 12; *Regimen* 3–4.

18. Grensemann (1975) 62–5 and 197–201.

19. Galen 10 (ed. Kühn) 474; Grensemann 1 (1975) A 20C.

20. Anon. Lond., *De Medicina* (ed. Manetti) VIII.5.

21. Grensemann (1975) and (1982) are the essential studies, supplemented by Craik (2009) 22–36, a compelling addition.

22. Tuplin (2004) 305–47 gives a full survey.

23. Hdt. 2.178.2.

24. Jouanna (2012) 6–7; Jouanna (1974) 508 n.1.

25. Grensemann (1975) 67–70 gives all texts citing the *'knidios kokkos'* and *'knēstron/kneōron'*; A25 on his p. 37 gives Rufus of Ephesus's location for it.

26. *Nature of Women* 2.32.

27. Chaouki et al. (2009) 542–6.

28. Berger (1970), updated in Hillert (1990) 70–74 and Stampolidis and Tassoulas (2014) 333–6.

29. Berger (1970) 93–7: Kappe oder Band? Men., *Aspis* 377 and the *prokomion* ('cap', I think) of the doctor, with Cordes (1994) 58 n.128.

30. Hiller (1970) 230–37; Fuchs, in Helbig 1 (4th edn, 1963), no. 875, and Berger (1970) 174 n.125. Of course, a Coan origin has also been proposed, as Stampolidis and Tassoulas (2014) 335 record, but I find it less convincing.

6. Hippocrates, Fact and Fiction

1. Ctesias F4 (Budé edn) with FF31 and 34.

2. *On Generation* 4.3; *Diseases of Women* 1.7.

3. Pl., Prt. 311B – C.

4. Pl., Phdr. 270C.

5. Pinault (1992); Jouanna (1999) 1–41 treats the fictions as historical at several points.

6. W. D. Smith (1990) 18–34; Jouanna (1999) 8 on 'undue scepticism'; Sherwin-White (1978) 260 n.23 on H. E. Siegerist's remark to the contrary.

7. Ps-Soranus Vit. Hippocr, in CMG IV.175 5–7 with Pinault (1992) 127–8; Areios of Tarsus is cited in Ps-Soranus 175.5, to be read with Pinault (1992) 9; von Staden (1999) 150 n.14 is very important on the full title of Ps-Soranus' work.

8. Pinault (1992) 24–8 and 131–4.

9. W. D. Smith (1990) is unconvincing when crediting Baccheios in Alexandria with knowledge of them.

10. W. D. Smith (1990) 2–4 and 110–25; Nelson (2005) 209–36 is important, but I do not accept his ingenious attempt to trace the text specifically to Macareus of Cos, historian and ambassador, active in the later 3rd c. BC.

11. W. D. Smith (1990) 4–6 and 108–9.

12. W. D. Smith (1990) 50–106; Brodersen (1994) 100–110; Pinault (1992) 79–93 and 145–7; I do not think that Plut., Vit. Cat. Mai. 23.4 shows that Cato already knew a letter like our nr. 3 or 4: the reference to Hippocrates there is Plut.'s own.

13. Sherwin-White (1978) 189–90 is still valid.

14. Sherwin-White (1978) 190.

15. Pherecydes FGH3 F59; Ps-Soranus, Vit. Hippocr. 1; Ps-Hippocr., *Presbeutikos* 2; Thomas (1989) 159–60 and 178 n.57 take it to have been given by the 5th-c. Pherecydes.

16. Pherekydes FGH nr. 475.

17. Parke and Boardman (1957) 276–82, at 277.

18. So, too, Hanson (1996) 159–82.

19. Helly (1993) 3–17.

20. Hom., Il. 2.676–9; Sherwin-White (1978) 18 n.36.

21. Ps-Soranus, Vit. Hipp. (CMG IV) 175.4 with Wellmann, Pauly R-E 1.2137.

22. Ps-Soranus, Vit. Hipp. (CMG IV) 176.4.

23. Ps-Soranus, Vit. Hipp. (CMG IV) 176.5; Pinault (1992) 61–73:Ogden (2017) 225–47, a very full study.

24. Ogden (2017) 23-3 on Dracontius, *Aegritudo Perdiccae* (ed. Zurlin, 1987).

25. Lucian, *Dea Syr.* 17–18 with Lightfoot (2003) 373–84; Val. Max. 5.7.3; Plut., Vit. Demetrii 38: Pinault (1992) 63–77.

26. Jouanna (1999) 40–41, citing Meyer-Steineg (1912) 1–411 and D.L. 9.42

27. Jouanna (2018B) is now fundamental to study of the Oath with bibliography at clxxxvi–cxvi and 42–9; esp. Bremmer (2002) 106–8, Edelstein (1967) 3–63, W. Jones (1924), Nutton (1997) 31–63, and von Staden (1996) 404–37 and (2007) 425–66.

28. Scrib. Largus, Epist. 3–4; Jouanna (2018B) xii.

29. P.Oxy. 31.2547, with Jouanna (2018B) lxxxviii–xciv.

30. Erotian F60 (ed. Nachmanson); Jouanna (2018B) xiii.

31. Jouanna (2018B) cxxii–cxxxvi, esp. cxxxvi n.156; Mouton and Magdelaine (2016) 217–32 present the new discovery.

32. Oath 1C line 8, ed. Jouanna (2018B) 3; my translation of 'nomōi iētrikōi' is not the more usual one, 'medical *law*'.

33. Oath 5, ed.Jouanna (2018B) 4; I share the interpretation well argued by Witt (2018) 238–42; for others, Jouanna (2018B) 32–5 with important textual details.

34. Von Staden (1996) 404–37 with Jouanna (2018B) 28–32.

35. Edelstein (1967) 55–61, esp. 60.

36. Edelstein (1967) 6 and 9–20, esp. 20: 'the Oath in its abortion-clause no less than in its prohibition of suicide echoes Pythagorean doctrines. In no other stratum of Greek opinion were such views held or proposed in the same spirit of uncompromising austerity.' But Jouanna (2018B) 3 rightly avoids translating in terms of 'suicide'.

37. Sommerstein and Torrance (2014) 370–71: Pythagoreans were not prone to swearing oaths.

38. Rosenthal (1956) 80–81 with Jouanna (2018B) cxxxv.

39. Pl., Prt. 311B–C.

40. Bremmer (2002) 106–8; Jouanna (2018B) xcii inclines to a 5th-/4th-c. date.

41. Jouanna (2018B) xxxvii–xlii, enlarging on von Staden (2007) 425–6: words like *epitelea*, *dēlēsis* and *epaurasthai* can be matched in Hdt.'s usage, as shown by Jouanna (2018B) 15 and 25,but that does not fix their use in the Oath to the 5th c.

42. Oath 3 with the texts in the Corpus relating to 'abortive pessaries' cited by Jouanna (2018B) 26–9; on 'cutting', likewise see Witt (2018) 238–42 for the apparent contradiction in other Hippocratic texts.

43. I.G. IX.1.807 lines 1–3.

44. P.Oxy. 31.2547 with 3.437 and 74.4970; Leith (2017) 39–50.

45. P.Oxy. 31.2547 with p. 63.

46. P.Oxy. 74.4970 53–4.

47. Jouanna (2018B) 158–222 is fundamental and Craik (2016) 152–5 is also important: both opt for a 5th-c. date, but I do not.

48. *Law* (ed. Jouanna) 1C; Craik (2016) 154.

49. *Law* (ed. Jouanna) 3B–4A.

50. *Law* (ed.Jouanna) 5 whose mystery language is best matched to Ar. Nu. 140 and 254ff, with Dover's notes, and Thesmo. 1150 with Austin's. However, they are highly comic passages, an important difference.

51. *Oath* (ed. Jouanna) 1C with Jouanna (2018B) 167, from whose view I differ.

52. Jouanna (2018B) 170–76 for 5th-c. parallels, including a good one in Democritus, which do not to my mind prove a late-5th-c. date for *Law* itself.

53. W. Jones, *Hippocrates*, Loeb edn II (1923) 259 suggests it is an inaugural address to students; W. Jones (1924) 58 cites a very interesting Indian parallel, also an address to medical students: Craik (2016) 155 for a 'graduation address'.

54. *Law* 3 A–C; Antiphon D–K 87 B60 with Jouanna (2018B) 175–6.

7. The Hippocratic Corpus

1. Galen XVIII A (ed. Kühn) 731.5ff; for doubts, G. E. R. Lloyd (1975A) 171–92, esp. 176–7.
2. Anon. Lond, *De Medicina* (ed. Manetti) 5.35ff with *Breaths* 4; Anon Lond., De Medicina 6.30ff, with *Breaths* 3.
3. Xenocritus in Fraser I (1972) 365 and II.540 n.248.
4. Fraser I (1972) 365–6 and II.539 n.24; 541–2 nn.252–6; Stewart (2018) 34–5, esp. nn.28–30.
5. Jouanna (1999) 63–4.
6. Jouanna (2018A) 38–62 is now the fundamental survey.
7. Littré (1839–61).
8. Jouanna (1999) 373–416 and Craik (2016) survey the texts and offer possible dates, not always agreeing.
9. Craik (2016) 208–10 well doubts the ascription of *Nature of Man* to Polybus.
10. Jouanna (1999) 412 and 375.
11. Jouanna (1996) 71–32 and 79–82
12. Craik (2007) 23–36.
13. Hesychios s.v. *sabakos* with Craik (2016) 173 on *Diseases* 1.
14. Grensemann (1975) and (1982) are the basic studies.
15. Craik (2009) 22–36 makes a compelling case.
16. *Diseases* 2.38 and 2.64; *Nature of Women* 3.2; *Diseases of Women* 1.78; *Superfoetation* 28.
17. Osborne and Rhodes, *GHI* (2017) nr. 155.
18. I discuss these problems in an article with A. Meadows, to appear in an honorary volume for E. Craik, edited by V. Nutton.
19. *Superfoetation* 28.
20. Craik (2016) xxvii–iii.
21. Jouanna and Magdelaine (1999) 319–36; Craik (2016) 30–34.
22. Witt (2018) 236–7, important on the texts about 'emergency surgery'.
23. Witt (2009), esp. 209–13.

24. *On Wounds in the Head* 9.12 and esp. 22 with Witt (2018) 233; 'it might strike modern readers that the need for trepanation is indicated fairly often, even in cases of minor injuries'. *On Joints* 3, for the Hippocratic manoeuvre.
25. *On Joints* 47.
26. *Diseases of Women* 1.70 and 3.3; *Excision of the Foetus* 1.
27. *Diseases of Women* 1.68, with men holding the woman's legs: King (2013D) 55 n.20 considers that here too the baby was already dead.
28. *Diseases of Women* 2.144; *On Joints* 42.
29. *Diseases of Women* 8: Bliquez (2014) 45–9.
30. Craik (1995B) 343–50.
31. Totelin (2009) is excellent.
32. *Affections* 15, 18, 23, 28, 40; Totelin (2009) 95–110.
33. Totelin (2009) 85; Kahn (2003) 139–61 postulated early lists and memoranda in alphabetic Greece.
34. Touwaide (2006) 558–68, an important survey.
35. Totelin (2009) 190–96 for a list.
36. Totelin (2009) 125–31, esp. 126.
37. Touwaide (2006) 564–6 is an important counter-argument.
38. Scarborough (1978) 353–85 and (1991) 138–74; Totelin (2009) 120–24.
39. *Diseases* 4.34; *Diseases of Women* 1.81 and 2.205; Totelin (2009) 151–2 and 158–61.
40. Hdt. 4.75.
41. *Nature of Women* 103; *Sterile Women* 18.3 = Loeb edn *Hippocrates* X.369.
42. *Diseases of Women* 1.75.5 = Loeb edn *Hippocrates* XI.173.
43. *Breaths* 1.
44. Ctesias F68, Budé edn, p. 219.
45. King (1998), Hanson (1990) and Dean-Jones (1994) are exceptionally helpful on all these topics, and I draw on their fine work on them.
46. Hanson (1990) 309–38, esp. 320–24.
47. *On Generation* 4 = Loeb edn *Hippocrates* X.12–13.
48. *On Generation* 4 = Loeb edn *Hippocrates* X.13.
49. *On Generation* 4 = Loeb edn *Hippocrates* X.14e–15.
50. *Regimen* 3.68 = Loeb edn Hippocrates IV.370 and 377.
51. *Nature of Women* 20.
52. *On Girls* 1 = Loeb edn *Hippocrates* IX.359–63.
53. *Nature of the Child* 20 = Loeb edn *Hippocrates* X.91–3.
54. Aphorisms 5.38: Epid. 2.6.15.
55. *Regimen* 1.28.

56. *Nature of Women* 22 = Loeb edn *Hippocrates* X.222–3; *Diseases of Women* 1.83.
57. *Sterile Women* 10 = Loeb edn *Hippocrates* X.356.
58. Lane Fox (2005) 19–50, at 24.
59. *Nature of the Child* 2 = Loeb edn *Hippocrates* X.34–7.
60. Ar. Lys. 82; Pollux 4.120.

8. The Invention of Medicine

1. *Ancient Medicine* 2.1, with Jouanna, Budé edn (1990) 38–40 on the language of 'discovery' and 50 n.1 on the 'way'.
2. *Places in Man* 46.
3. *Ancient Medicine* 5.2.
4. *Ancient Medicine* 20.1–2 uses *physis* for the 'nature of man', as Jouanna, Budé edn (1990) 208 n.5 also emphasizes. *Nutriment* 13 and 15, possibly a Hellenistic text (Jouanna (1999) 401), uses it with an even wider reference: Craik (2016) 23–6, concludes that it is likely to be early 4th-c. BC, as it may well be.
5. *Regimen* 1.2; 1.15.
6. *Ancient Medicine* 20.3.
7. Jouanna (1989A) 3–22.
8. G. E. R. Lloyd (1983) 129–35.
9. *The Sacred Disease* 4, end of chapter.
10. Willi (2003) 51–95, 96–117; Irigoin (1980) 247–57; in general, Langslow (2005) 287–302.
11. Thomas (2000) 11–16, 249–69.
12. Jouanna (2012) 39–54 is the essential study; *The Art* 1.
13. *Nature of Man* 1.
14. Hecataeus FGH1 F1; Heracleitus F1; Fowler (1996) 69 n.60.
15. Cohn-Haft (1956).
16. Hudson-Williams (1951).
17. Jouanna (2012) 44–56.
18. G. E. R. Lloyd (1979) 242–67 and (1990) 60–64 with objections already by Seaford (2004) 177–84.
19. Jouanna (2012) 44–50.
20. Jouanna (2012) 42.
21. *Regimen* 3.69; Wilkins (2005) 121–33, an excellent study.
22. *Affections* 1.
23. Nutton (2013) 86.

24. *Breaths* 1; *Airs, Waters, Places* 10 and Joly (1966) 203–4.
25. *Breaths* 6 on '*miasmata*'.
26. *Nature of Man* 9.
27. Jouanna (2012) 335–59.
28. *Nature of Man* 5; Jouanna (1999) 316.
29. Jouanna (1999) 103–11.
30. Parker (1983) 213–16.
31. *Precepts* 6; Jouanna (2012) 261–85.
32. *Physician* 1; Edelstein (1967) 87–110.
33. *Physician* 1 lines 26–9 (ed. Jones, Loeb Library *Hippocrates* vol. 11).
34. *On Fractures* 38–48; *On Joints* 62.
35. *Regimen* 1.2, 60.
36. *Regimen* 3.54.
37. *Fleshes* 19 on *hai epistamenai*; Jouanna (1999) 391–2 suggests content going back to 'fifth and sixth centuries' but I hesitate; Craik (2016) 42–8 proposes '450–400 BC', an excellent discussion .
38. *Fleshes* 19, West (1971) 365–88, excellent on uses of 'sevens', not necessarily at an early date.
39. *Fleshes* 19.
40. *Fleshes*19 again, on *hetairai* having knowledge.
41. *Fleshes* 19 on *hai epistamenai*, a source for the following sentences.
42. *Coan Prenotions* 333.
43. *Coan Prenotions* 543.

PART 2: THE DOCTOR'S ISLAND
9. The Epidemic Books

1. Plin., HN 26.3.
2. Ion FGH 392T2 and FF4–7, 9, 12–16.
3. Epid. 1.16.3; 3.3.2; 3.12.1; 7.59; Jouanna (2003) vii–viii; Graumann (2000) 35–6 with bibliography.
4. Jouanna (2003) ix and cix–xx n.152; Jouanna (2016) x.
5. G. E. R. Lloyd (2002) 32–3 and (2004) 119 n.1, to whom I owe knowledge of Chunyu Yi.
6. Sanchez and Meltzer (2012) 194–9.
7. Epid. 7.89.1.
8. Gourevitch and Gourevitch (1982) 623–4.
9. Jouanna (2003) lxxx, 1–11 and 103 line 4, where *apuretos* is Littré's emendation.

10. Epid. 3.17.4.
11. Epid. 6.8.32.
12. King (2013C) 20.
13. Epid. 6, 8, 32.
14. Brilliantly surveyed by King (2013C) 73–125.
15. Epid. 3.17.11.
16. Epid. 5.81; 7.86.
17. Epid. 5.82; 7.87.
18. King (2013A) 265–82 .
19. Epid. 3.17.6.
20. Ps-Scylax 65.4; Str. 9.5.22 and esp. Livy 44.13; Woodward (1910) 145–60; Nikolaou and Kravaritou (2012) 191–3; Anna Blomley has helped me here, dispelling the doubts of Tsiaphalias (2010) 5–26.
21. Nikolaou and Kravaritou (2012) 186; Sdrolia (2018) and Rogers (1932) 127–9, which again I owe to Anna Blomley.
22. Despite Jouanna (2000) 187n.10, I still incline to the brilliant emendation of R. Herzog at Epid. 7.4.2, reading two Coan month-names, used, therefore, by the Coan author even when abroad; Langholf (1990) 34–5 and n.109 agrees and adds more reasons.
23. Coan Prenotions 142–6, 148, 154, 508, 510, 537, 561.
24. Roselli (2008) 154–63 with SEG 58.192, a Doric epigram for a doctor.
25. Galen VII (ed. Kühn) 855; for Erotian's acceptance of all seven, Jouanna (2000) ix.
26. Galen VII (ed. Kühn) 854.11–855.9.
27. Galen VII (ed. Kühn) 890.11–891.4, where Thessalos is credited with working from Hippocrates' notes; Jouanna (2003) ix–x, esp. x n.1 on the 'five books'; I think they are 2 and 4–7.
28. Langholf (1990) 140; Deichgraeber (1971) 75, following Littré: some, including Jouanna (1999) 394, think the author is the author of 2, 4 and 6, but I disagree; Humours 7 refers to the epidemic at Epid. 6.7.1 and 6.7.10 and 20 probably draws on Epid. 6.3.24–6.4.3, but the author of Humours may be writing separately, later than Epid. 6 and using it.
29. Epid. 2.2.7, doubted wholly unconvincingly by Langholf (1977) 15–17 and esp. 107 n.5, showing he is no historian.
30. Epid. 2.1.5; 2.3–21.1; 2.3.12; 4.21; 6.2.19; 6.8.10.
31. Xen., Hell. 1.1.20, 1.3.10 and 1.5.17; Diod. Sic. 13.66: Plut., Vit. Alc. 30 and 36.3; Rhodes (2011) 95.
32. Epid. 4.45, where Smith's text and apparatus in the Loeb edn Hippocrates VII.138 with n.6 need correction.
33. Xen., Anab. 7.1.5 and 7.2.27–9.

34. Xen., Anab. 7.7.1; Lendle (1995) for a map and 457–60 for topography; Deichgraeber (1971) 75 wrongly assumes the villages were given to Medosades only in 399 BC. The gift could be as early as 410/9.
35. Epid. 4.14; 4.37; 6.3.2.
36. Epid. 4.48: May (1950) 1–10; Epid. 2.4.3 and 6.4.11 with textual variants.
37. Diod. Sic. 16.76.1–2.
38. Epid. 2.3.1; 6.7.1; Grmek (1989) 305–39.
39. Epid. 4.31; 4.56; 6.8.30 and 32; for Thasos, 6.8.29 and 32.
40. Thuc. 8.64; Xen., Hell. 1.4.9; Diod. Sic. 13.72.1; Xen., Hell. 2.2.5.
41. Aen. Tact. 18.1, 13–16, 20; Thuc. 4.135.1.
42. Chapter 12 nn.5–9 and n.34.
43. Galen VII1 (ed.Kühn) 854.11; Suda, s.v. Hippocrates IV; Hamon (2019A) 125–6, perhaps a grandson from Cos as a judge on Thasos.
44. Jouanna (2000) xxiii–xlii is essential, with observations on vocabulary which are decisive; li–lx on the relationship to Book 5.
45. Epid. 5.9; 5.32.
46. Epid. 5.1.
47. Epid. 5.3 and 5.4 with Jouanna (2003) 120 n.4; Epid. 5.11 and 5.13.
48. Ps-Scylax 33.2 and esp. Livy 42.38.10, both having 'Homolion'. I identify it with Homole (Str. 9.5.22 for the two spellings) whence comes 'Homilos'; despite Ap. Rhod. 1.594–5 and B. Helly (2013) 180, with much bibliography, I do not site it at Palaiokastro. I am grateful to Anna Blomley here too for discussions of this point.
49. Epid. 7.118.
50. Epid. 7.17.1, where I differ from Jouanna (2000) 202 n.6; for Balla, Hatzopoulos and Paschidis (2004) 796; no 'Baloion' is known, except to Steph. Byz., surely in error.
51. Epid. 7.121.1, in 356 BC, not 358, as it is promptly followed by the foundation of Philippi, in 356 BC: Loukopoulou (2004) 859–60 for the name and site, with Counillon (1998) 115–24 and the bibliography cited.
52. Plin., HN 7.124, wrongly called 'Critodemus' there.
53. Epid, 5.61 and 7.33, not 'Ainos' despite Jouanna (2003) 360 n.8, nor 'Delos', Jouanna (2000) 160 n.8; I adopt the reading of F. Robert and A. Meinecke, which Jouanna notes too.
54. Epid. 7.112–17 with 5.100–101; 7.112.2.
55. Deichgraeber (1982), to which even more can now be added.
56. Epid. 7.21, 80 and 89.
57. See n.51 above; Hammond and Griffith (1979) 246–9 and 358–9.
58. Hammond and Griffith (1979) 445–6.

59. Epid. 5.95 and 7.121.1 with Jouanna (2000) 173 n.6 on Arist., Part. An. 673 A 103, possibly a reference to Tychon's case; Ma (2008) 72–91, esp. 75 and 87 on the 'sharp end of battle'.
60. Epid. 5.60, 7,32 and perhaps 5.98, 7.29 and 5.99 and 7.30 ('Neapolis').
61. Jouanna (2000) lx–xc with Grmek: Grmek (1989) 348–55, with F. Robert.
62. Epid. 7.122.5 with Jouanna (2000) 118 n.2; I take *'porneiē achrōmos'* to mean sex with prostitutes, following Suda, s.v. *achrōmos*, well cited by Jouanna. Contrast Epid. 6.5.15 and 7.69.

10. 'On Thasos during autumn . . .'

1. Galen VII (ed.Kühn) 855.8–9; Jouanna (2003) x–xii.
2. Jouanna (2016) xii–xxvii is fundamental.
3. Langholf (1990) 169–99, on *katastasis*; Jouanna (2016) xxxiii–xlii.
4. Epid. 1.3.1.
5. Jouanna (2016) lii.
6. Jouanna (2016) cxxv–cxxvi.
7. Epid. 3.17.6.
8. Arist., Pol. 1262 A24.
9. Epid. 3.17.6.
10. Epid. 3.17.4.
11. Epid. 3.1.1 with Jouanna (2016) clvii and 301 on the lettering.
12. Galen, Comm, in Hipp. Epidem. iii 7–8 = CMG V 10.2.1, readily available in Fraser II (1972) 480 n.147 and 544 n.272.
13. Fraser I (1972) 325.
14. Fraser I (1972) 326.
15. Fraser I (1972) 367 and esp. 368.
16. Fraser I (1972) 326.
17. Fraser II (1972) 545 n.273.
18. Nollé (1983), the essential study.
19. Arr., Anab. 1.26, 4; Lane Fox (2008) 236–8 for the context.
20. Nollé (1983) 93–4 gives a list .
21. Epid. 1.1.8.
22. Jouanna (1999) 395–6 and Loeb edn *Hippocrates* vol. III.
23. Epid. 1.1.8 with Jouanna (2016) 3 n.7.
24. Epid. 1.1.6–7.
25. Coulié (1998) 445–53.
26. Epid. 1.20.2 with Jouanna (2016) cxxvii and 217.

27. Epid. 1.14.1, 2, 4; 1.15.2; 1.16.1–2.
28. Hanson (1996) 159–81.
29. Goldhill (2002) esp. 110 and 114; he does not really consider teaching.
30. Jouanna (2016) xlvi–xlviii.

11. The Thasian Context

1. Archilochus FF 21–2 (West).
2. Miller (1889) 182–3.
3. Blondé et al. (2008) 416 with bibliography. Brunet, Coulié, Hamon et al. (2019) is a fine general survey, excellently illustrated, which appeared too late to influence my text, though I will cite some points in it in notes which follow.
4. Blondé et al. (2008) 419 and Coulié (2008) support, so far, a 7th-c. date: I disagree, putting Archilochus earlier; on Demetrion, my suggestion is shared now by Holtzmann in Brunet, Coulié, Hamon et al. (2019) 94.
5. O. Picard (2011B) 1137 endorses the settlers' awareness of these metal sources, and esp. in Brunet, Coulié, Hamon et al. (2019) 61–2.
6. Tsantsanoglou (2008) 163 for the early fighting; Ornaghi (2009) 95–106.
7. I.G. XII Suppl. 412.
8. Psoma (2006) 61–86, esp. 74–6; Isaac (1986) 8–34 gives a good survey of the *peraea* as then known.
9. Loukopoulou (2004) 854–69, esp. 859 and 861; Kallet (2013A) 43–60, esp. 47 n.24 on the varying views about the identity of the settlers at Eion.
10. Vaxevanopoulos (2017) 48–58, esp. 52–3; Koukouli-Chrysanthaki (1990) 494–514.
11. Hdt. 6.46; Muller (2011) 179–92.
12. Koukouli-Chrysanthaki (1980) 309–25; Loukopoulou (2004) 861 and 864–5 on Oesyme and Galepsos.
13. Loukopoulou (2004) 860.
14. Loukopoulou (2004) 858–9.
15. Loukopoulou (2004) 872–5, esp. 875.
16. Arrington et al. (2016) 1–44, correctly located by Loukopoulou and Psoma (2008) 55–86.
17. Hdt. 6.46, who had good local informants, so I accept his figures whereas Perdrizet (1910) 1–2 simply rejected them; Pébarthe (1999) 135 n.42 for other views of the matter.

18. Thuc. I.96.2.
19. Hdt. 5.6.1.
20. Blondé, Muller and Mulliez (2002) 251–65; de Polignac (2003) 58 and figs. 3 and 3e.
21. Hansen (2006) 61.
22. Plut., Vit. Cim. 14.2.
23. Hdt. 7.168.4: Hansen (2006) 96.
24. Hansen (2006) 112–13.
25. Osborne and Rhodes *GHI* (2017) nr. 103A and B, though the dates are not entirely certain.
26. *Diseases* 3.11 with Jouanna (2014) 32; Salviat (1986) 147–54; Grandjean and Salviat (2012) 218–19 and 232–3 with bibliography; Palagia (2008) 223–34, an important study.
27. Grandjean and Salviat (2012) 285–90, 335–40, 350–52; Coulié (2002) for Chiote influence on Thasian black-figure ware.
28. O. Picard (2011A) 80–90.
29. Grandjean and Salviat (2012) 153–4 and fig. 97 with bibliography.
30. O. Picard in Grandjean and Salviat (2012) 357–61; and very clearly in Brunet, Coulié, Hamon et al. (2019) 63–4.
31. Grandjean and Salviat (2012) 152–4 and fig. 85; Evelyn Waugh, *Decline and Fall* (1928) ch. 5; Waugh had previously written an unpublished text on Silenus.
32. Hermippus in Ath. 1.29E.
33. Rolley (1965) 441–83; Grandjean (2012–13) 258–61.
34. Grandjean and Salviat (2012) 275; Salviat (1992) 261–7.
35. Maffre and Tichit (2011) 137–64 for examples.
36. So, too, Malkin (2011) 133; Pouilloux (1974) 305–16.
37. I.G. XII 276 col. 1.
38. Salviat and Servais (1964) 267–87.
39. Osborne (1987) 79; Brunet (2004) 74–86, a general overview.
40. Hdt. 6.46.2.
41. Hdt. 6.47.2.
42. Hdt. 7.118.2.
43. Hdt. 7.120.1.
44. Hdt. 8.120.
45. Hdt. 9.106.7: I assume 'those in authority' were Spartans.
46. Hdt. 1.152.2, the *rhēsis*, surely a *rhetra* .
47. Thuc. 1.100.2
48. O. Picard in Brunet, Coulié, Hamon et al. (2019) 21 proposes 'Athènes n'évince pas Thasos de son continent; elle confisque les revenus que la

cité en tirait', but Thuc. 1.100.3 disagrees, at the very least for Neapolis: like Galepsos it then pays tribute in its own right to Athens from 454/3 on; Brunet (1997) 229–42. Thuc.'s family also received that mine very near Neapolis.

49. Thuc. 8.65.2–3.
50. Xen., Hell. 1.1.12, where I take 'Thasos' to be (part of) the island, not the city; Xen., Hell. 1.1.32.
51. Xen., Hell. 1.4.9; Diod. Sic. 13.72.1.
52. Osborne and Rhodes *GHI* (2017) 524–31.
53. Epid. 1.27.1 and 1.14.3 for the year.
54. Epid. 3.1.2.
55. Hiller von Gaertringen, Pauly R-E (1934) 2.1317; Thuc. 8.64.3.
56. Hdt. 6.47.2.
57. Grandjean (2011), esp. 370–75.
58. Tyerman (2010) 33.
59. Galen, CMG V 10.1, with Jouanna (2014) 41.
60. B. Holtzmann, cited in Grandjean and Salviat (2012) 287.
61. Jouanna (2016) xiv–xxii.
62. *Nature of Man* 4 and 5.
63. Epid. 1.15.4; Epid. 4.43 with 6.2.1.
64. King (2013B) 25–53 for an up-to-date survey, esp. of fluids in women.
65. *Airs, Waters, Places* 1 and 3–7.
66. Chapter 21 nn.7–23; Epid. 2.4.1–2 is a crucial example.
67. Epid. 6.4.11.
68. Xen., Hell. 1.4.9.
69. Epid. 3.2–3.15.

12. Building Blocks of History

1. Epid. 3.17.7.
2. Epid. 3.17.10.
3. Epid. 3.17.6.
4. Epid. 3.17.9 .
5. Deichgraeber (1982) 34–5; Meinecke's proposals, in 1852, are most accessible in Littré (repr. 1971) VIII vii–xxii.
6. Deichgraeber (1982) 35 and 34 ('vielleicht Grossvater und Enkel').
7. Chryssanthaki-Nagle (2007) 3.1.354 nr. 4 and 55.
8. Epid. 3.17.3, but I disagree with Jouanna's suggestion that he was 'probablement un paysan à en juger par son régime'.

9. Chryssanthaki-Nagle (2007) 103 and 105; 109 for 'Hero-'.
10. I.G. XII 8 and Supplem. (1939); Grandjean and Salviat (2012) for a preliminary survey; Grandjean and Salviat (2006) 293–327; Osborne (2009) 103–14; Fournier, Hamon and Parissaki (2014–15) 75–93, a major survey.
11. Coulié (1998) 445–53 with bibliography; Grandjean (2011) 582–4 nr. 5 and 586.
12. Grandjean (2010) 65–80.
13. Trippé (2015–16) 43–65.
14. Tzochev (2016) 230–53 with bibliography.
15. Hamon (2015–16) 70–91 is essential and I depend on it for what follows.
16. Hamon (2018) 183–97 with 188 n.15.
17. Hamon (2015–16) 74, 76, 79, 84, 89 and Hamon (2018) 184–5.
18. Jim (2014) 13–24 omitted the Parian connection, for which see IG XII 8.358; for *theoroi* as magistrates, L. Robert (1927) 208–13 and on Thasos, I. Rutherford (2013) 128–30.
19. Hamon (2018) 192–3 for this hypothesis; see my Endnote 3.
20. Badoud (2015) *passim*, which I owe to P. J. Thonemann; Sherk (1993) 267 and 282–3; Sherk (1990) 231–95, at 292–4 on the Thasian *archons*.
21. Pl., Leg. 950–951C; I. Rutherford (2013) 336–8.
22. I. Rutherford (2013) for examples, 'at least in the Hellenistic period': I assume they could occur earlier too.
23. I.G. XII 8.263 and XII Suppl. 366 and perhaps 355.
24. Like Jim (2014) 13–24, I reject attempts to link the *aparchē* with *aparchos* or 'ruling away'. At Pind., Nem. 4.47 *aparchos* refers spatially to ruling *elsewhere*. I accept that *aparchē* in the Big Theoric List col. 1 means offerings, probably of first fruits, Apollo at Delphi being a likely recipient: Hamon (2018) 70 n.9 gives other interpretations and the bibliography.
25. Rutherford (2013) 130.
26. Hamon (2015–16) 70–73.
27. Hamon (2015–16) 73–7.
28. Hamon (2015–16) 77–80.
29. Holtzmann (1994) 29–59; Grandjean and Salviat (2012) 108–13 with bibliography, esp. Blondé, Muller and Mulliez (2000) 885–907.
30. Hamon (2015–16) presents the case very clearly, also adducing points proposed by F. Salviat.
31. Hamon (2018) 15–25; Hamon in Brunet, Coulié, Hamon et al. (2019) 76 and esp. 78–9 for the *archon*-list; my Endnote 3.

32. So also Salviat (1984) 241–12.
33. Hamon (2015–16) 83 n.27 on the spelling of the genitives.
34. Deichgraeber (1982) 15–21 and 22–8 is particularly important here.
35. Epicrates, Heracleides, Philiscos, Philon and Skymnos are examples, tabulated by Graham (2000) 325–6; their currency on Thasos can be verified in LGPN I (1987).
36. Dēithrases, Dēialkos and Cleokydes are good examples, all tabulated by Graham (2000) 325–6.
37. Hamon (2015–16) 78 with fig. 5 and 90 on the Big Theoric List col. 5.20: Jacobs (1893) 23 first noted the apparent link to Epid. 1.15.4.
38. Cyriac of Ancona (2003) Diary II.30
39. Cyriac of Ancona (2003) Diary II.36–40 and 41, evidently a fine marble sculpture of Heracles, now lost.
40. Forrest (1986) 133.
41. Bodnar and Foss, eds. (2003), Cyriac of Ancona, Diary II.43–4.
42. Osborne and Rhodes GHI (2017) nr. 177A, whose dating I reject in Endnote 2; Ch. Picard (1921) 145 for the text visible to the left.
43. Hamon (2015–16) 78 and 90 with Salviat (1984) 254, which I do not accept either, as my Endnotes 2 and 3 explain: Jouanna (2016) cxx-iii mistakes their date for Antiphon's *theoria* by one year, putting it in 409/8.
44. Epid. 1.15.4.
45. Big Theoric List cols. 4.11 and 5.30; Hamon (2019B) 179–80 now proposes we read 'Agiades' at Epid. 1.20.3, not 'Aglaides', also (later) an attested name in Thasos. He identifies him with a *theoros* in 401. If so, I regard him as a kinsman, perhaps a grandson, of the Agiades with a young daughter in, for me, *c.* 469/8.

13. Art, Sport and Office-Holding

1. I.G. XII 8.277C and D, as placed in Hamon (2015–16) 84 fig.8, following F. Salviat's arrangement.
2. I.G. XII 8.278C.
3. Robertson I (1975) 242–57 and II (1976) 658–9 give the basic sources; Lippold, Pauly R-E 21.1630–39; Pollitt (1990) 126–41, 221, 230–31 with good translations; Roscino (2010).
4. Paus. 9.4.1–2; Meiggs (1972) 277 argues for a slightly later date, relating it to 'Polygnotos's association with Athens', i.e. late 470s to early 460s.
5. Plut., Vit. Cim. 7; Thuc. 1.98.1; Hdt. 7.107.

6. Plut., Vit. Thes. 36 1–3; Plut., Vit. Cim. 8: W. G. Forrest, review in *Revue Belge* 34 (1956) 542.

7. Jeffery (1965) 43–7 is still fundamental.

8. Jeffery (1965) 45 n.20 and Plut., Thes. 36.1 for the date: Jeffery (1965) 41 n.4 on the *Peisianakteion* with Plut., Vit. Cim. 4.5

9. JHS AR (1981–2) 7–10, (2009–10) 3–4, (2010–11) 33; a reconstruction in Camp (2001) 67; Meiggs (1972) 472, citing Homer Thompson; Robertson 1 (1975) 244–5; my Endnote 1.

10. Robert Pitt, in JHS AR (2007–8) 8.

11. Jeffery (1965) 43–7.

12. Plut., Vit. Cim. 4.5; Stesimbrotos, ap. Plut., Vit. Cim. 14.5.

13. Plut., Vit. Cim. 4.7.

14. Harpokration s.v. Polygnotos, citing the orator Lycurgus, probably for the grant of citizenship.

15. Plut., Vit. Cim. 17.3.

16. Pouilloux (1960) 120–39; Tomlinson (1980) 224–8 suggests it was also a *hestiatorion*.

17. Paus. 10.25–31; Robertson 1 (1975) 247–52.

18. Manoledakis (2003) fig. 40; Stansbury-O'Donnell (1989) 203–15 and (1990) 213–35 for an earlier attempt at reconstruction.

19. Manoledakis (2003) 223–37 with bibliography.

20. At Paus. 10.30.4 I take Iaseus to be Iasos son of Triopas, with C. Robert, Pauly R-E ix.784; Triopas, honoured in Cnidos, was Apollo's son and unsuited for this painting of the Underworld.

21. Plut., Vit. Cim. 12–13, dated by Vit. Cim. 8.8, despite Meiggs (1972) 81 arguing for 466 (a proposal he abandoned in his later years).

22. Hdt. 1.174.3–6; Forrest, *CAH* iii.3 (1982) 319 for Delphi's medism.

23. Archibald, in JHS, AR (2012–13) 12 n.13.

24. Paus. 10.27.4; Anth. Pal. 9.700; Plut. Mor. 436 B (not naming the author); Page, *Epigrammata Graeca* (1975) xlviii doubted Simonides' authorship, as did Lippold, Pauly R-E 2.1634, but I see no need to do so, nor to follow the complex suggestion of Robertson 1 (1975) 24 that the epigram was originally composed for the Troy painting in the Athens stoa and only later copied on to the painting at Delphi!

25. Plin., HN 35.59, the grant being to Polygnotos personally and not to him as *proxenos* of Thasos, also an important point.

26. Plin., HN 35.123: Robertson 1 (1975) 489–90; Meiggs (1972) 277 tentatively suggested a later date.

27. Paus. 1.22.6.

28. I.G. XII 8.277C and D; placed in Hamon (2015–16) 84 fig. 8, following F. Salviat's proposal.

29. Despite Meiggs (1972) 573.

30. Ath. 12.534D; Plut., Vit. Alcib. 16; Paus. 1.22.6 for Aglaophon's paintings in 416 BC, though Gribble (2012) is cautious about accepting this. If Aglaophon was the famous Polygnotos' son he might have been born c. 475 BC when Polygnotos, on my high chronology, was c. fifty years old. Plin., HN 35.60 dates Aglaophon's zenith in c. 419–416, presumably only on the evidence of the paintings for Alcibiades. I assume Aglaophon was an Athenian citizen. The *theoros* on Thasos, hitherto dated to 447/6, cannot be this Aglaophon's son and a grandson of the famous Polygnotos. For that, Aglaophon would have to have been born c. 500 BC at the latest: he would be too old to have painted for Alcibiades in 416/15 BC. As for the Attic vase painter who signs himself as 'Polygnotos' in the 430s, he is agreed to be distinct from the great Polygnotos. Most probably he is a pseudo-Polygnotos who adopted the famous name: Robertson (1992) 210 for discussion.

31. Burckhardt (1998) 175–6.

32. Ebert (1972) 118–26 for the basic discussion.

33. Decker (1995) 134–5 for details.

34. Ebert (1972) 124.

35. P.Oxy. 222. col. 1 line 13; Paus. 6.11.4–6 and Ebert (1972) 118–26.

36. Ebert (1972) 69–70 on Euthymus, citing all the sources; Paus. 6.11.4 and 6.6.6 with Ebert (1972) 120.

37. Ebert (1972) 124.

38. As Herzog (1915) 320 already saw: so did Launey (1941) 23.

39. Dio Chrys. 31.95–6.

40. Paus. 6.11.5–8: Anderson (2000) 144–5 stresses Dio's reshaping of his story here to suit *Oration* 31's context.

41. Pouilloux (1994) 199–206; Grandjean and Salviat (2012) 97–100; O. Picard (2000) 1076–8.

42. Paus. 6.11.9.

43. Basinos Sminthiou (Big Theoric List col. 5.21), Timyllos Nymphios (col. 5.28) and Lydos Lydou (col. 5.25) are examples of *theoroi* who are also attested as *archons* in the separate *archon* list: Salviat (1984) 237 places them as *archons* at a later date than I believe to be correct.

44. My Endnote 3, for fuller discussion.

45. I.G. XII 8.263; Osborne and Rhodes *GHI* (2017) nr. 177A, but see my Endnote 2.

46. Osborne and Rhodes *GHI* (2017) nr. 176 with my Endnote 2.
47. Endnote 2.
48. Endnote 2.
49. Thuc. 1.100.3; Brunet (1997) 229–42; my Chapter 11 n. 48.
50. Osborne and Rhodes *GHI* (2017) nr. 464.
51. Pouilloux (1954) 443.
52. Ch. Picard (1921) 145.
53. See note 43 above and Endnote 3.

14. Sex and Street Life

1. Xen., Hell. 1.4.9.
2. Diod. Sic. 13.721–2.
3. Robertson (1975) 203–9. Ridgway (1970) 50 and 55 with evidence: new studies, made in 2018 and to appear, have confirmed the authenticity, as the Boston curators kindly inform me.
4. Pind., Ol. 5.13–15.
5. Epid. 1.4.2; Lazaridis (1971) Appendix 2 and 3.
6. Osborne and Rhodes *GHI* (2017) nr. 104 with pp. 20–25; Duchêne (1992) is the basic study.
7. Saliou (2003) 37–49.
8. Osborne and Rhodes *GHI* (2017) nr. 104.2–3; Grandjean and Salviat (2012) 108–14; Graham (2000) 306–11, 315.
9. Osborne and Rhodes *GHI* (2017) nr. 105.24–45; Hennig (1995) 235–82, esp. 237–43.
10. Osborne and Rhodes *GHI* (2017) nr. 104.30–33 with Graham (1998) 22–40 and Henry (2002) 217–21.
11. Osborne and Rhodes *GHI* (2017) nr. 104.41–2; Graham (2000) 311–12.
12. Osborne and Rhodes *GHI* (2017) nr. 104.7–10 (200 staters) and 46 (a *hemiekton*, a twelfth of a stater).
13. Osborne and Rhodes *GHI* (2017) nr. 104. 324 (a stater) and 28 (a *hekte*, a sixth of a stater) and 8–10 (200 staters).
14. Osborne and Rhodes *GHI* (2017) nr. 104.
15. I differ from Osborne (2009) 109–11; for urban *eukosmia*, Malay, Amendola and Rici (2018) 9–7, a fine decree from 2nd c. AD Tralles against passive homosexuals, not just the 'sexually effete' as the eds. translate: hence the ban of them from gyms and so forth where only males were present.

16. Osborne and Rhodes *GHI* (2017) nr. 104.43.
17. Wecowski (2014) 81–3, a nuanced discussion; Lewis (2002) 96, for more extreme doubts, however.
18. Osborne and Rhodes *GHI* (2017) nr. 104.43–4; Theophr., Od. 51 for the sweetened wine, albeit in the 4th c. in the *prytaneion*.
19. Schmitt-Pantel (1997) 331.
20. J.-Y. Marc in Brunet, Coulié, Hamon et al. (2019) 37–9 for topography, despite doubts earlier in Osborne and Rhodes *GHI* (2017) 25 ('outside the *agora*'); Wecowski (2014) for symposia sizes.
21. O. Picard (2011B) 1144, correcting the older view that her raised right hand indicated her to be a 'nymphe consentante'; Critias, in Ath. 11.463E for the cups.
22. Epid. 1.27.2; 3.1.5.
23. Epid. 3.17.10 and 16.
24. Brunt (1988) 92.
25. Greco (1999) 223–9; Ling (1990) 204–14.
26. Epid. 1.27.2 with Jouanna (2016) 254–5 for the textual problem.
27. Epid. 1.6.4 with Jouanna (2016) 160; Hdt. 3.29.3.
28. On *kataklinein*, Jouanna (2016) 477.
29. Epid. 3.17.1.
30. Grandjean and Salviat (2012) 116–19; Osborne and Rhodes *GHI* (2017) nr. 104.49.
31. Grandjean and Salviat (2012) 276; for the metaphor, Topper (2010) 112–14; I.G. XII Supplem. nrs. 202.382–3.
32. Grandjean and Salviat (2012) 339.
33. Epid. 1.21.3: Charneux (1987) 207–23 for *para* and the accusative meaning 'near'.
34. Grandjean and Salviat (2012) 119–21 and 306–7: Jouanna (2014) 44.
35. Epid. 3.1.5.
36. Jouanna (2016) 320–22 for the text; Jouanna (2010) 48–50.
37. Grandjean and Salviat (2012) 131–4 with pl. 68; Grandjean and Salviat (2006) 293–327, at 293–4 and 315.
38. Burkert (1985) 240–46.
39. Grandjean and Salviat (2006) 296.11 and 308–11.
40. Grandjean and Salviat (2006) 35 and 325 n.128.
41. Grandjean and Salviat (2012) 24 pl.24; Graham (2000) 303–6 on the *ochthē* .
42. Epid. 3.17.2.
43. Epid. 1.21.2; 1.27.6; 3.17.3.

44. O. Picard in Grandjean and Salviat (2012) 361–2 with fig. 281: see also 162 fig. 96.
45. Grandjean and Salviat (2012) 173–9 with plan (after p. 172) and bibliography.
46. I.G. XII Suppl. 414.
47. Epid. 1.21.2; 1.27.6; 3.17.3; Jouanna (2016) 274.
48. Epid. 3.17.15; with ch. 15 nn.6–9.
49. Epid. 3.17.11 for *leiou*. Jouanna (2016) 442–3 defends *hēlieiou*, citing the Arabic transl. of Galen at 3.17.15, fully explained by Jouanna (2014) 51–3. But I do not accept it, as no cult of the Sun is known on Thasos at this date. It is a later Hellenistic arrival.
50. Grandjean and Salviat (2012) 166–71 with figs. 99 and 100; Holtz- mann (1994) 66–79 for a late-4th-c. dating.
51. R. Parker (2000) 53–79, at 72.
52. Grandjean and Salviat (2012) 124–51; I.G. XII Suppl. nr. 409.
53. Epid. 3.1.1.
54. Epid. 1.27.5.
55. Malkin (1987) 241–90; Jouanna (2016) 270 suggests it may be 'un nom d' habitant' but I do not agree; nor do Grandjean and Salviat (2006) 38 and n.84. But Hamon (2019B) 180–82 advances another possibility, emending to Archegenes, a name attested in the upper class, as he shows.
56. Grandjean and Salviat (2006) 315–16.
57. Bruneau (1970) 413–30; Hygin., Fab. 247.
58. Epid. 3.1.8: I translate *pseudōn* here as 'lies', not 'liars'; Graham (2000) 322.
59. Arist., Pol. 7.133A30–1331B3: these markets and those in nn.61–2 below happen not to be adduced by de Sainte Croix (1972) 267–84 and 397–9 in a celebrated argument about the Athenian 'civic' *agora*.
60. Dickenson (2017) is cautious here.
61. Hdt. 1.152.3, including the word *rhēsin*.
62. Hdt. 1.153.1–2.
63. Epid. 1.20.1.
64. Jouanna (2014) 29–54 at 41–3 for a survey of the problem; however, Vagelpohl (2014) 336–7 needs to be read in full. The Arabic shows that Galen plainly scorned the textual variant adopted by Capito and others (they 'vaunt' or boast about it).
65. Ms. A with Jouanna (2016) 29 app. crit., for the alternative readings here.
66. Galen's own glossary (Kühn XIX.103) gives the meaning 'summer- house' here: his commentary, at least as given in Arabic, belittles Capito's 'threshing floor'.

67. Revermann (2017) 196, on the date of the Thorikos theatre.
68. Vit. Aeschyli 18; Garvie (2009) liii–lvii.
69. Grandjean and Salviat (2012) 134–9.

15. Patients of Quality

1. Epid. 3.17.2.
2. Conze (1860) 17.
3. Thuc. 2.52.1–2.
4. Graham (2000) 325 gives a valuable list, enhanced by Grandjean (2012–13) 225–68, on which my count is based; Hamon (2019A) 442 for important Cratistonax names later.
5. In the Big Theoric List, the four fathers in col. 5 whose names recur in Epid. 1 now 3 are Pantacles (5.22, with Hamon (2019A) 448 for, in my view, likely later kinsmen), Deialkos (5.23), Aglaides (5.28, but Hamon (2019B) 180 now reads 'Agiades') and Mnesistratos (5.30). Python (5.31) is *theoros*, but the name is not uncommon (Hamon (2019A) 450). Jouanna (2016) 465–6 indexes these names in Epid. 1 and 3.
6. Epid. 3.1.3; Jouanna (2016) 308–9 for the variant readings.
7. Epid. 3.17.15; ch.14 nn.48–9.
8. I side with Graham (2000) 326 although Jouanna (2016) 309 does not as the name lacks authority in any Ms.
9. Big Theoric List 5.23: if so, his mother might be attended in Epid. 3.17.5 for twenty-one days, aged at least fifty (the son might be *theoros* in his thirties); and she is not said when ill to have had a period. Big Theoric List 4.8 and 5.8 would be kinsmen, too early to be relevant to Epid. 3.17.5.
10. Epid. 1.16.2 and 1.20.3; Jouanna (2016) 203 cites Jacobs (1893) 24 and notes Deichgraeber (1971) 16, who suggested that the Dai-tharses in the Mss. is the author's Doric for Dēi-thrases and is a sign of the author's own Doric origin. Graham (2000) 326 agrees with Jacobs; since he wrote, a Dēithrases has become known as a magistrate in the 290s in Grandjean (2012–13) 245 line 42.
11. Epid. 1.16.2 with King (1998) 51–2 and 68–74.
12. Hanson (1992) 31–71, T 49.
13. Big Theoric List col. 4.10.
14. Epid. 1.20.3.
15. *Two Gentlemen of Verona*, Act 4.2, lines 129–30.
16. Epid. 1.27.5; 3.1.6.
17. Schaps (1977) 323–30: Dickey (1996) 243–5.

18. An example is Teleboulos in Epid. 1.16.1 and Big Theoric List col. 8.8.
19. Epid. 1.21.2 and perhaps 3.1.7.
20. Epid. 1.27.13; 3.1.10; 3.1.12; 3.17.11.
21. Epid. 3.1.4 with Jouanna (2016) 317. who opts for a female; 1.17.2.
22. Epid. 1.27.14.
23. Jouanna (2016) 295–6, an important corrective to Mendinē/Mendiē and Deichgraeber (1971) 14.
24. Fraser (2009) 160–61 for Mendaios. I realize my suggestion is unsupported in the Mss. and will therefore not appeal to strict philologists. For other ethnics in case histories, Epid. 1.20.1 and 3.17.1.
25. Epid. 3.17.2; 3.1.7; 3.1.12.
26. Trippé (2015–16) 45–9.
27. Huysecom-Haxhi (2009) 578–80; Maffre and Tichit (2011) 137–64.
28. Grandjean and Salviat (2012) 279 fig. 171; clearly shown in Ridgway (1970) pll. 62–5.
29. Ridgway (1967) 307–9 and (1970) 46 and 97 n.9; also, a later example, Grandjean and Salviat (2012) 312 fig. 212.
30. Best (1969) 5–7.
31. Despite Dugand (1979) 131–55.
32. Anastassiou and Irmer 1 (2006) 188 and 215–17; II. I (1997) 172–5 and 228–30; II.2 (2001) 122–3 and 178–9; III (2012) 179–80 and 195–6.
33. G. E. R. Lloyd (1975B) 171–92 discussed only inter-related texts supposed to be by Hippocrates: he was not discussing the claims of individual texts. At 180 n.5 he cites with approval the notion of interpolation in Epid. III, supported by Jones, Loeb edn *Hippocrates* 1.270–71: it is also raised as a possibility by Craik (2016) 75–6. Jouanna (2016) esp. xii–xxv refutes it.
34. Pl., Phdr. 270C; Jouanna (1999) 58–9, where I side with the 'minority'.
35. Galen (ed. Kühn) 131.5; G. E. R. Lloyd (1975A) 171–92, esp. 176–7.
36. Anon. Lond. *De Medicina* 5.35ff; 6.30ff.
37. Manetti (1999) 95–141.
38. Jouanna (2016) xxx–xxxiii.

PART 3: THE DOCTOR'S MIND

16. By the Bedside

1. Pleket (1993) 27–34, esp. 32 on 'the distinct upper-class flavour to Greek public physicians under the Empire'; R. Parker (1983) 210.

2. Epid. 3.17.8 with Jouanna (2016) cxiv–cxv: I do not think it is a later addition, as some have stated before Jouanna's edition made the claim most implausible.

3. Jones. Loeb edn *Hippocrates* 1 (1923) xviii summarizing Houdart (1836) 240.

4. Jones, Loeb ed. Hippocrates 1 (1923) 144.

5. Epid. 1.25.4.

6. Epid. 3.16.1.

7. Epid. 3.16.2–3.

8. Epid. 6.5.1 with Temkin (1991) 193 n.74 stressing that it does not refer to the 'self-termination of disease' but to 'the fact that all healing processes are natural processes which medicine can support but not replace'.

9. P.Oxy. 5231 and p. 45 of the commentary on it.

10. *Katechein* in Epid. 1.27.2, 1.27.8, 1.27.12; 3.17.11.

11. Fowler (1996) 69–71 on an author's 'voiceprint'; Jouanna (2016) lxix–lxxiv and lxxxii–lxxxiv for examples.

12. Epid. 3.16.1; compare 1.5.5 and 1.24.6 on 'exact' tertian fevers; for Podaleirios, Iliou Persis F4 (West); at Hdt. 7.32, Ms. A has '*akribōs*' but all editors, including N. G. Wilson in the OCT, exclude it and print '*atrekeōs*' instead, correctly.

13. Wade-Gery (1996) 1519; Epid. 3.17.3.

14. Epid. 1.27.5.

15. Epid. 1.11.2; Wootton (2007) on doctors' failures in practice to live up to this fine precept.

16. Galen, Comm. in Epid. 1.2.53: Jouanna (1997) 211–53 esp. 215–17.

17. Epid. 1.11.2.

18. Jouanna (1997) 218; Jouanna (2017) 185, 'le malade n'est pas à proprement parler un patient. Il a un rôle actif. Il doit lutter.'

19. Epid. 1.11.1.

20. Esp. *Regimen* IV; compare Epid. 1.27.6.

21. Hom., Il. 1.70.

22. Dicks (1966) 37 with Kahn (1970) 115.

23. Hdt. 5.24.1; Thuc. 1.138.3; *Regimen in Acute Diseases* 3 explicitly distinguishes medicine from divination.

24. Epid. 1.20.2 with Jouanna (2016) 217–18, an important note.

25. Epid. 1.23.1, where I disagree with Irwin (1989) 28.

26. Von Staden (2002) 23–42.

27. Epid. 3.6, to be compared with 3.17.1 (a burning fever without shivering but with delirium) and 3.17.4 (phrenitis but with delirium). For similar problems in 1.27.4, G. E. R. Lloyd (2007) 335 and in 1.11–12 G.

E. R. Lloyd (1975A) 185 n.8, answered, however, by Jouanna (2016) 187–9.

28. Withington (1920) 64–5; Celsus, De Med. 3.4.
29. Edelstein (1967) 65–85, esp. 84.
30. Epid. 1.22.1.
31. Epid. 1.26.
32. Epid. 1.24.
33. Epid. 1.27.7; Hamon (2019B) 179 notes that 'Megon' has been conjectured here, not 'Meton'.

17. Filtered Reality

1. Epid. 1.23.2.
2. Epid. 1.23.2 again; Jouanna (2016) 230–32, noting that *dianoēmata* occur nowhere else in the text. Thumiger (2017) 121–2 is good on the physical implications of a loss of voice.
3. Epid. 3.17.15.
4. Epid. 1.27.8.
5. Epid. 1.27.8.
6. Epid. 3.1.3; I.2.3.: Jouanna (2016) lxv–lxxi and 133 n.4.
7. Epid. 3.10.1.
8. Epid. 1.27.3; 3.17.7. Jouanna (2016) 410–11 for the terminology.
9. Epid. 3.7.2.
10. Epid. 1.27.1 with many other texts assessed in Jouanna (2016) lxxxv.
11. Epid. 1.23.2–4; Thumiger (2017) 32–42 surveys 'theories of mind' in other Hippocratic texts but not, significantly, Epidemics 1 and 3.
12. Jouanna (2016) lxxxii–lxxxv, an important survey.
13. *Internal Affections* 10 and 11.
14. Jones, Loeb edn *Hippocrates* 1 (1923) 144.
15. Epid. 3.14.1.
16. Epid. 1.2.1; 1.23.1; 1.27.4.
17. Epid. 1.17.2; Jouanna (2016) 198–200.
18. Epid. 1.20.2 and 23.1: Jouanna (2016) 228.
19. Epid. 3.1.10 with Jouanna (2016) 340–41 who also discusses 3.1.7.
20. Epid. 1.27.1.
21. Epid. 1.14.3 with Jouanna (2016) liii–liv.
22. Frede (1987) 128–30, an important study; *Ancient Medicine* 19 and then *Regimen* 3.70 are explicit about a cause in a way that Epid. 1 and 3 are not.

23. Epid. 3.17.12.

24. For example Epid. 3.3.1 ('from'), 1.7.3 ('dia bēchos') but *dia* is not used in this way again; at 1.2.3 it means 'during', not 'because of'.

25. Epid. 1.1; 3.1.3; 3.3.1 with 3.4.1 and 3.4.5; 3.17.11, where 'cause' is indeed the better translation.

26. Hom. Il. 19.262; F. Robert (1976) 327–42 is important here.

27. Hdt. 5.33.1; 6.133.1; Thuc. 5.53.1.

28. Jones, Loeb edn *Hippocrates* I (1923) li–liii; Langholf (1990) 79–93 is especially helpful.

29. Hdt. 5.92 already; *Nature of the Child* 12.1–6; Hanson (1995) 291–307, esp. 302–3.

30. Theophr. Od. 6.

31. Epid. 3.17.3.

32. Esp. *Regimen* 2.63.

33. Epid. 1.27.11; 3.17.2.

34. Epid. 1.16.2: 3.17.7.

35. I disagree with Langholf (1990) 80 and 84, who assumes 'bad matter' is presupposed in Epid. 1 and 3.

36. Anon. Lond. *De Medicina* (ed. Manetti) 8.5, 12.10, 16.20.

37. Epid. 1.11.1.

38. Epid. 3.4.3.

39. Epid. 1.8.4; 1.11.1; Jouanna (2016) 84 n.2, 159, 172–3.233.

40. Epid. 1.15.4.

41. Epid. 3.17.3.

42. Holladay (2002) 144 correctly emphasizes this point.

43. Nutton (1983) 1–24, an excellent survey; Parker (1983) 220; Thuc. 2.50.

44. *Airs, Waters, Places* 24; Epid. 1.2.1.

45. Epid. 1.19.1.

46. *Diseases* 4.51.

47. Epid. 1.2.3; 3.7.9 for heat and bilious excretions.

48. Epid. 1.15.4.

49. Epid. 3.14.2 and 3.17.2 at end.

50. Epid. 3.17.2, end.

51 *Nature of Man* 5.

52. Epid. 3.17.2: 11th day and end.

53. Thumiger (2013) 63–70 surveys attestations of 'melancholia'; so does Jouanna (2012), 229–58, esp. 236, but my dating of Epid. 1 and 3 affects his argument for Sophocles' primacy.

54. Grandjean and Salviat (2006) 294–327, at 311–14; R. Parker (1983) 252–4 and esp. n.105.

18. Retrospective Diagnosis

1. Graumann (2000) is an invaluable survey of modern diagnoses in the Epidemic books; on Parmeniscus, Grmek in Jouanna (2000) lxxvi–xc with Epid. 5.84 and 7.89.
2. Scheidel (1994) 151–75 and (2013) 45–59.
3. Grmek (1989) 189–92.
4. Dormandy (1999) 221–2 and *passim*, esp. 22–5.
5. Epid. 3.1.6.
6. Epid. 3.13 with Jouanna (2016) 383.
7. Epid. 3.13.1.
8. Epid. 3.14.1–2; Galen XVII B (ed. Kühn) 34.6–8: Jouanna (2016) 387.
9. Epid. 1.14.1.
10. Epid. 1.12.1 end.
11. G. E. R. Lloyd, ed. (1983) 380; contrast Grmek (1989) 336.
12. Epid. 1.27.13.
13. Epid. 3.1.2: Grmek (1989) 301.
14. Epid. 1.15.3; 1,17,1; 1.20.1, 1.27.10 end; 3.8.1; 3.17.9 end; Grmek (1989) 354–5.
15. Epid. 3.7.1–2.
16. Feigenbaum (1956) 355–7.
17. Hatemi et al. (2016) 10–22, which I owe to Prof. P. Frith and her specialized knowledge of this condition.
18. Nair and Moots (2017) 71–7.
19. Epid. 3.7.1, picked up at 3.11–12; 3.7.2 was not, I think, a *puretos*.
20. Epid. 3.17.13.
21. G. E. R. Lloyd, ed. (1983) 378–9; Grmek (1989) 302 proposes malaria and/or typhoid.
22. Epid. 3.17.8.
23. G. E. R. Lloyd, ed. (1983) 380.
24. Epid. 3.1.7; Mulhall (2019) 151–79 on the possible presence of bubonic plague but the absence of any textual attestation until the 2nd c. AD.
25. Epid. 1.27.1 end.
26. Grmek (1989) 287: *Prognostic* 2.
27. Epid. 1.6.1; 1.18.4; on phrenitis, McDonald (2009) and, briefly, Thumiger (2017) 46–7.
28. Epid. 3.17.4.
29. Epid. 7.53.
30. *Diseases* 1.30; Epid. 3.14.2.

31. *Diseases* 2.72.
32. Diseases 3.9; Acts 1.10 and 7.55.
33. Epid. 1.12.1–2; 1.14.2–4; 1.15.1–4; 1.8.2.
34. Epid. 1.12.1–2.
35. Epid. 1.20.2.
36. Goodall (1934) 525–34; G. E. R. Lloyd, ed. (1983) 380; MacArthur (1957) 146–9, who detected typhus too.
37. Epid. 3.17.1; 3.17.9.
38. Epid. 3.17.9; 3.17.7.
39. Goodall (1934) 528; Grmek (1989) 301, detecting malaria too.
40. Epid. 1.27.2.
41. Epid. 1.24.1.
42. Epid. 1.3.2; Siegel (1960) 77–98, esp. 96.
43. Goodall (1934) 528; G. E. R. Lloyd, ed. (1983) 379; Grmek (1989) 348, who diagnoses salmonellosis.
44. Epid. 1.27.10; 3.17.7.
45. Epid. 3.17.15 with Grmek (1989) 348 and 437 n.19.
46. Epid. 1.10–11 with, still, W. H. S. Jones (1907) 61–73; Sallares, Bouwman, Anderung (2004) 311–28.
47. Epid. 1.24.1–3; Sallares (2002); Oerlemans and Tacoma (2014) 219–24.
48. Epid. 1.27.1 with Grmek (1989) 284–304.
49. Grmek (1989) 301.
50. Epid. 3.17.2 with Demand (1994) 84.
51. Demand (1994) 82 n.50 and esp.84–5.
52. I owe this point to Yangos Chalazonitis and his local knowledge.
53. Agelarakis (2004) 327–49, for malaria; Endo and Eltahir (2018) 406–13 for the relevance of the winds.
54. Epid. 1.27.1, on the second day of the case.
55. Epid. 3.4.1 and 4.
56. Scheidel (1994) 151–75, esp. 152.
57. Epid. 3.15.1 with Epid. 3.4.6.
58. Oerlmans and Tacoma (2014) 222–4.
59. Epid. 3.2.5.
60. Joly (1966) is lucid and particularly important here.
61. Rovelli (2011); Leroi (2014) esp. 352–78; Clegg (2004); White (2000); and esp. Wootton (2015).
62. Leroi (2014) 365.
63. Wootton (2015) 57–109 on the 'invention of discovery'.

64. Levitin (2015) presents the deep roots of the 'new' science in ancient wisdom: Wootton (2015) esp. 560–71 and 584–6 for a powerful restatement of a real 'scientific revolution'.
65. G. E. R. Lloyd (2004) 23.

19. Philosophy, Medicine and Drama

1. Scurlock (2014) with Geller (2004) 11–61 (and 63–5 and 113–15 by others in the same volume).
2. Geller (2004) 19–21.
3. MacIntosh Turfa (2012).
4. MacIntosh Turfa (2012) 90, 92, 93, 97; in general, 164–204.
5. MacIntosh Turfa (2012) 90 and 91.
6. MacIntosh Turfa (2012) 97 and 91.
7. Theopomp. FGH 115 F 204 is a classic example; more generally, MacIntosh Turfa (2012) 217–29.
8. Epid. 1.4.1–4 with West (1978) 376–81.
9. Epid. 1.1.1 and Jouanna (2016) 116 n.3.
10. *Airs, Waters, Places* 11.2 with Dicks (1966) 33 n.8.
11. Dicks (1966) 32 insisted the 'equinoxes cannot be determined by simple observation alone', reiterated in Dicks (1972) 176–7, but Kahn (1970) 112–14 and 114. n.1 answered him convincingly; Alexander Jones has kindly confirmed the point for me, emphasizing that Babylonian texts could also date by an equinox without knowledge of the sphericity of the earth which Dicks's argument required. *On Anaximander*, D.L. 1.23 and 2.1 with Kahn (1970) 112 n.49, though rejected necessarily by Dicks (1966) 33 and n.39. Opinions still divide on this point.
12. Alcmaeon D–K 24 F4.
13. Alcmaeon D–K 24F2 with Barnes (1979) 115–20, esp. 120.
14. Empedocles D–K 31 B F 112.10–22.
15. D.L. 8.77.
16. Empedocles D–K 31 B F 108; F 100.4–5; Guthrie (1965) 211–16; Epid. 1.2.1; 3.1.6 end: Frede (1987) 229.
17. Empedocles D–K 31 A T 77
18. Empedocles D–K 31 B F.111.1–2
19. Primavesi (2008) with the review by R. Janko in *Ancient Philosophy* 30 (2010) 407–10; Nutton (2013) 47.
20. D.L. 8.70 with Guthrie (1965) 133 n.2.

21. Plin., HN 29.1.5 for Acron of Acragas; other pupils are more doubtful, despite Guthrie (1965) 216–17.
22. *Nature of Man* 1–8.
23. Hippon D–K 38A 3 and 8 with the pre-Christian A11: Guthrie (1965) 386–8 for a dating.
24. Ridgway (1970) 10; A. Stewart (2008) 377–412 and 581–612, a fine study; Adornato (2019) 557–87 for some doubts, not all of which I share: 'severe' was Winckelmann's term, so its absence from ancient epithets is irrelevant; Vout (2018) 4–11 is a timely reminder that our Tyrannicides are a composite re-creation.
25. R. R. R. Smith (2007) 83–139, esp. 112–16.
26. Theophr., H.P. 9.15.1, surely a genuine citation.
27. Aesch. F253 (ed. Sommerstein); Choeph. 281 with Garvie's note but also Sommerstein's on Eum. 785. implying it was also used of plants and crops; it is hard to know if Aesch. took such words directly from medical language, as Garvie well observes on Choeph. 185–6: Fraenkel's note on Ag. 1113 makes a good case for *eparegmos*; compare his note on Ag. 1448 and on 1480 (ichor), where the case is less compelling.
28. Aesch., Ag. 432, 792; Garvie on Choeph. 271–2.
29. Aesch., Choeph. 282 with Garvie's cautious note.
30. Aesch., Ag. 1112, but Fraenkel ad loc takes it differently: I even wonder if the 'yellow-tinged drop' here is linked in Aesch.'s thinking with urine.
31. Aesch., Choeph. 470–73 with Garvie's note ad loc.
32. Aesch., Ag. 176–81, where David Raeburn has kindly helped me.
33. Aesch., Ag. 1623 with Fraenkel's note ad loc.
34. Soph., OT 22–30.
35. Jouanna (2012) 59; Craik (2003) 45–56 well discusses Soph.'s fragments, but here too I see a secondary use of medical language, not medical thinking.
36. Soph., Philoc. 1436–7; R. Parker (1996) 184–5.
37. Epid. 1.27.9.
38. Kosak (2004); Guardasole (2000) 76–86; Craik (2001A) 81–95.
39. Holmes (2010) 237, with 236 (symptoms as '*contested sites of interpretation*') and 233 (the 'polysemy of the symptom').
40. Eur., F286 (Collard-Cropp) with their valuable note ad loc.
41. Eur., F917 (Collard-Cropp).
42. Eur., Hipp. 140–60.
43. Epid. 3.17.11 and 27.4.
44. *The Sacred Disease* 4.

45. Dodds (1929) 97–104, esp. 100.
46. Jouanna (2016) lxxxii–lxxxv on their vocabulary.
47. Eur. I.T. 282–5 and 308.
48. *The Sacred Disease* 7.17; *Breaths* 14 with Jouanna (1988) 48–9.
49. *The Sacred Disease* 9–15; I disagree here with L. P. E. Parker (2016) 116–17.
50. Eur., Or. 278.
51. Epid. 1.27.11; 3.17.7.
52. *Breaths* 14.
53. *Breaths* 14 again.
54. Eur., HF 867–70.
55. Eur., HF 932–3.
56. Epid. 1.27.2.
57. Bond (1981) 309–10 is excellent here, citing Soph., Aj. 303 too.
58. Eur., HF 942–6.
59. Dodds (1929) 100; Bond (1981) 310.
60. Eur., HF 1092–3.
61. Galen, Comm. in Epid. iii (ed. Wenkebach) 72 n.12.
62. Epid. 3.1.7 with Jouanna (2016) 73 n.7.
63. Eur., Bacc. 1122–4 with Dodds's note.
64. Devereux (1970) 35 n.6 and 42.
65. Polyaenus, Strat. 4.1.1.
66. Posedippos 44 (ed. Austin-Bastianini); Bremmer (1984) 267–86.
67. Devereux (1970) 36.
68. Epid. 3.17.2.
69. Ch. 11 n.30 and 14 n.31.
70. Holmes (2010) 228–74, on the 'polysemy of the symptom' (233–9).
71. I disagree here with Craik (2016) 75, who suggests the 'tenor' of the case histories in 1 and 3 is 'consistently formulated in terms reminiscent of the tragic stage'; Thumiger (2016) 637–64 attempts parallels with tragic actors' masks which do not persuade me, though she has much else of interest.

20. Epidemics and History

1. A. Stewart (2008), esp. 581–612, despite Adornato (2019) 557–87.
2. Dodds (1951) 36, 40 (on Aeschylus) and 49 (on Sophocles) with the detailed study of Gagné (2013), which does not, however, modify my sense of an important change.
3. Kahn (1979) 89–92, on 'linguistic density'.

4. Dodds (1951) 49 on Sophocles as 'the last great exponent of the archaic world-view'. Aen. Tact. 31.24 for another, later; 5th-c. Selinus and 5th/4th-c. Cyrene are other examples.

5. Hdt. 2.44.4; 6.47.1; 8.137–9 for his insider knowledge in Macedon; 2.51 also implies that he was initiated into the mysteries on Samothrace.

6. Thuc. 4.105.1 where *peri tauta* refers to the area round Thasos, just mentioned, not back to Amphipolis as suggested as a possibility by Hornblower (1996) 335.

7. Thuc. 4.105.1, as traditionally understood, e.g. by J. K. Davies *APF* (1971) 236–7; again Hornblower (1996) 335 cites dissenting opinions.

8. Thuc. 5.26.5: *pheugein* does not entail that he stood trial in Athens; I consider Marcellinus, Vit. Thuc. 47 to be basically correct about Thuc.'s presence at Skapte Hyle and not simply to be based on Hdt. 6.46.

9. Thomas (2000) esp. 102–35.

10. Forrest (1984) 1–11, a cardinal contribution.

11. Hdt 8.77.

12. Thomas (2000) 40, doubting the importance of 'folk wisdom', but I do not think that her examples of medical thinking on pp. 29–74 are necessarily Hippocratic. Nor do I think the Hippocratics illumine Hdt. 3.108, despite Sierra Martin (2018) 23–52.

13. Hdt .2.77.3.

14. Hdt. 4.205 is one good example: Harrison (2000) 106–7 for another; in general Demont (2018) 175–96, an excellent study which I found too late to use in my text. I do not regard Hdt. 8.113 and 117 as a significant counter-example as the origins of these afflictions were so obvious.

15. Hdt. 1.105.

16. Hdt .1.105.4.

17. *Airs, Waters, Places* 22: Jouanna (2012) 105–6.

18. *Airs, Waters, Places* 22, where the effect, I think, is erectile dysfunction, not loss of sperm.

19. Hdt. 3.33; Thomas (2000) 34–5 considers that '*ek geneēs*' means 'hereditarily', but I disagree, as does Jouanna (2005) 1–27, a very important contribution.

20. *The Sacred Disease* 1 and 22.1; Jouanna (2012) 99–100.

21. Hdt. 3.33.1, with n.19 above.

22. Dodds (1951) 179 and n.1.

23. G. E. R. Lloyd (1996) 44.

24. Thuc. 1.22.2; I.97.2; Crane (1996) 32–8 and 50–74.

25. Thuc. 1.138.3; 2.48.3; 2.65.5–6; and, implicitly, 1.22.4.

26. Thuc. 3.82–3: Weidauer (1954); Hornblower (1987) 132–5; Rechenauer (1991); Swain (1994) 303–27; Kallet (1999) 223–44 on the medical template and esp. 240.

27. Wade-Gery (1996) 1519 col. 2.

28. Thuc. 3.82.2; 8.97.2 with a significant 'ge', 'at least in my time', that is, 'later in history they may have an even better one', e.g. rule by 1000 citizens only; de Sainte Croix (1972) 6 against Thuc.'s wish to affirm *laws*.

29. Thuc. 6.14.1.

30. Jouanna (2012) 20–38 adduces Epid. 1.11.2 and discusses its relevance.

31. Thuc. 2.47.3–5; good modern discussion by Thomas (2006) 87–108 and King and Brown (2015) 449–73 and esp. Demont (2013) 73–88: on the language, Page (1953) 97–119 still prevails.

32. Diod. Sic. 14, 70–71; Lucr. 6.1138–1281 with Hutchinson (2013) 210–19, also, on other Latin poets' plagues, Procop., Pers. Wars 2.22.3, although at 22.8, unlike Thuc., he introduces God's will and at 22.35 focuses on women.

33. Thuc. 2.48.3: de Sainte Croix (1972) 30 and 32; on other lessons of history, Rutherford (1994) 53–68 and Raaflaub (2013) 3–22.

34. Thuc. 2.49.1.

35. Thuc. 2.49.2.

36. Thuc. 2. 47.3–4; 51.2–3.

37. Thuc. 2.51.3.

38. Thuc. 2.49.3.

39. Thomas (2006) 97.

40. Thuc. 2.49.2–3 and 7: Craik (2001B) 102–8 well emphasizes the head-downwards account of the disease and its fixing, *stērixeien*, but what fixes is *ponos*, as never in the Hippocratics.

41. Thuc. 2.49.6.

42. Thuc. 2.49.5.and 6.

43. Thuc. 2.49.8.

44. Thuc. 2.49.8 again, a fine observation, full of pathos.

45. The papers of Holladay and Poole from 1979 to 1988 remain fundamental, usefully collected in Holladay (2002) 123–65; yet more attempted diagnoses can be readily found on the internet, beginning with Plague of Athens on Wikipedia and in the good overview by Powell (2013), though his preferred candidate, Rickettsia, is also unconvincing (Holladay (2002) 138). Ebola was first proposed by Scarrow (1988) 4–8, amplified by Olson et al. (1996) 155–6 and various others since, but was rightly refuted by Littman, Durack et al. (2001) 674–5.

46. Plague of Athens, Wikipedia entry section 4.2.2 for bibliography.

47. Holladay (2002) 161–5 for an intriguing observation about vultures and immunity.

48. Thuc. 2.53.2–4.

49. Thuc. 2.51.4; Holladay (2002) 162: 'like sheep' is a 'simile for submission'.

50. Thuc. 2.51.6.

51. Holladay (2002) 142–5 and 159–60 are decisive here.

52. Thuc. 2.50; Plato, Apol. 32C and esp. Phaedo 67A (contrasting it with *kathareuein*) also use *anapimplesthai*, with the sense of being foully implicated or contaminated.

53. Nutton (1983) 1–24; Lucr. 6.1125–30.

54. Thuc. 2.54.2–5.

55. Ingeniously argued by Kallet (2013B) 355–82, but she does not begin to persuade me.

56. Cicero, Epist. Ad Q. Fr. 2.11.4.

57. Diod. Sic. 14.70–71.

58. Diod. Sic. 14.70.4.

59. Thuc. 1.22.4, where '*kata to anthrōpinon*' is indeed in context a 'factor making for constancy', as de Sainte Croix (1972) well argued, fully aware of Stahl (1966) and his stress in other contexts on the role of the irrational and inconsistent: Stahl (1966) 33 wrongly takes the phrase to mean here 'according to the [variable] human condition' or 'situation'.

60. De Sainte Croix (1972) 32.

61. G. E. R. Lloyd (2010) on medicine as a 'model'.

21. Hippocratic Impact

1. *Humours* 15–17, esp. 17.

2. Anastassiou and Irmer 1 (2006) 400–401.

3. Jouanna (2013) xiii–lxxxi is a masterly survey with full bibliography, esp. xli–xlv on the tetrads, l–lii on the seasons and location and esp. lxxi–lxxix; Jouanna (1989B) 60–87, esp. 61–3 on the vocabulary.

4. *Prognostic* 1.2 end.

5. Jouanna (2013) lvi–lxi gives an essential survey of opinion: I do not think that ch. 25 implies that *theion* in 1.2 refers to the heavens, including weather and so forth.

6. Jouanna (1989A) 3–22 surveys the divine in the Hippocratic corpus: *Regimen* 4.87, 89 and 90 commend prayers.

7. Alessi (2012) 71–90 is essential here with examples; Craik (2016) 68–86 for a good introduction to the Greek text's contents; Alessi (2010) 119–35 is particularly important on the author's aims and research programme.

8. Epid. 2.3.1 with Alessi (2010) 119–35, esp. 122–3

9. Epid. 6.2.6 and 6.5.3 (on coction); 6.4.7 on *charites* to be shown to patients as also, implicitly, in 2.4.4.

10. Epid. 2.6.32.

11. Epid. 1.26.3; 2.2.9; 4.25–6; 6.2.5; 6.2.20–21.

12. Epid. 6.2.5.

13. *Skepteon* at end of Epid. 6.2.5.

14. Epid. 1.11.1 and 1.23.4, 11.26.2–3 and 1.27.3 end. However, 1.17.2 is, just once, nearer to the questioning in books 2, 4 and 6.

15. Epid. 2.4.1–2.

16. Especially Epid. 2.3.8 and 6.2.7; also Holmes (2012) 49–70.

17. Especially Epid. 6.3.23.

18. Epid. 2.1.11; 6.3.12.

19. Epid. 4.24.

20. Alessi (2012) 71–90, esp. 78–82 is essential here.

21. Epid. 2.1.1.

22. Epid. 2.1.1–2; Alessi (2012) 71–90, esp. 78–82.

23. Epid. 2.1.4 and also perhaps 4.46.

22. From Thasos to Tehran

1. I.G. IV 1.122 nr. xxii; Prêtre (2009) 64 and 76; Osborne and Rhodes, *GHI* (2003) nr. 102, pp. 532–42; Dillon (1994) 239–60 on these didactic healings, quite different from the didactic aim of Epid. 1 and 3.

2. Strabo 14.2.19; Plin., HN 29.1–2; Jouanna (1999) 18–19.

3. Perilli (2009) 75–120.

4. Hamon (2019A) 166–7 on Asclepius and Thasos; Grandjean and Salviat (2012) 278–9, giving an otherwise unpublished text.

5. Daux (1972) 503–66, at 550–54.

6. Arr., Anab. 7.25–6 with Bosworth (1988) 177–9.

7. Bosworth (1988) 179, suggesting a use of 'notebooks of the royal doctors which must have documented the course of the royal illness'. But I doubt that the author had access to the originals.

8. Fraser 1 (1972) 365–7.

9. P.Oxy. 80.5231 with commentary on 44–7.

10. Roselli (1989) 182–90 for use of Epid. 2, 4 and 6 only.

11. van der Eijk (2012) 25–47, a valuable survey.
12. Ullmann (1978); Álvarez Millán (1999) 19–43, at 28–9 with nn.27–31 for some recent studies; Pormann (2008) for an edited text and translation of the separate *On Melancholy* .
13. Ullmann (1978) 72 case iii.
14. Ullmann (1978) 68 case ii.
15. Nutton (1979) for an edited text, translation and commentary; Nutton (1988) 52 for self-promotion in it.
16. G. E. R. Lloyd (2014) 67–8 and (2004) 119 n.1.
17. Manetti (2018) 327–32; Jouanna (2018) 45–6; Pormann (2018) 345–8.
18. P.Oxy. 74.4972 with intro.; 80.5239, 5241.
19. Kessel (2012) 93–123, esp. 120–23.
20. Pormann (2018) 358 translates the most revealing remarks in Ḥunayn's *Risala* about his work with a bad Greek text of Galen's Commentary on Epid. 2; Pormann (2008) 257–9 for more such details.
21. Vagelpohl (2014) 45–8.
22. Pormann (2018) 359.
23. Pines (1970–80) 323–6 for a summary; Hallum (2012) also notes al-Kaskarī, active in Baghdad hospitals in the 920s and also attentive to the Epidemic books.
24. Álvarez Millán (2000) 293–306 and (1999) 34–42; Hallum (2012) 192–8.
25. Meyerhof (1935) 321–56.
26. Meyerhof (1935) 332, but his text and translation here, Robert Alessi advises me, is not entirely correct.
27. Meyerhof (1935) 347–9.
28. Van der Eijk (2012) 25–47.
29. Nutton (1989) 434 on Baillou; Graumann (2000) 21–9.
30. Nutton (1989) 420–39; Maclean (2002) 44–5 and 76–7.
31. Pomata (2005) 105–46 surveys earlier uses of case histories and, at 132–4, the Parisians.
32. Pomata (2005) 129.
33. Kassell (2014) 595–625 is excellent here; I am much indebted to her in what follows.
34. Weber (2013) 358–402, esp. on women's medical almanacs.
35. Trevor-Roper (2006) 5–6, 66 and 269–72; Nance (2001).
36. Kassell (2014) 600 n.72.
37. F. Bacon, *The Advancement of Learning* (ed. M. Kiernan, 2000) 99; Lane (1996).
38. Cantor (2018) 362–83, at 369–75.

39. Dewhurst (1966); Anstey (2011) 457–78; Levitin (2015) 281–2 and 293–4.
40. Hom., Il. 1.62–7.
41. Nutton (2013) 142–59 and 197–221 for a survey.
42. R. Parker (1983) 210 for this fine phrase.
43. C. P. Jones (2005) 293–301.
44. Thuc. 1.22.4.

Bibliography

Adornato, G. 2019. 'Kritios and Nesiotes as Revolutionary Artists? Ancient and Archaeological Perspectives on the So-Called Severe Style Period', *AJA* 123: 557–87.

Agelarakis, A. P. 2004. 'The Clazomenean Colonization Endeavor at Abdera in Retrospect: Evidence from the Anthropological Record', in A. Moustaka et al., eds., *Klazomenai, Teos and Abdera: Metropoleis and Colony* (Thessaloniki) 327–49.

—. 2006. 'Early Evidence of Cranial Surgical Intervention in Abdera, Greece: A Nexus to "On Head Wounds" of the Hippocratic Corpus', *Mediterranean Archaeology and Archaeometry* 6: 5–18.

—. 2014. 'Woman's Cranium with Traces of Surgical Intervention', in N. C. Stampolides and G. Tasoulas, eds., *Hygieia: Health, Illness, Treatment from Homer to Galen* (Athens) 256–9.

Alessi, R. 2010. 'Research Program and Teaching Led by the Master in Hippocrates' Epidemics 2, 4 and 6', in M. Horstmanshoff, ed., *Hippocrates and Medical Education* (Leiden) 119–36.

—. 2012. 'The Arabic Version of Galen's Commentary on Hippocrates' "Epidemics", Book Two as a Source for the Hippocratic Text: First Remarks', in P. E. Pormann, ed., *Epidemics in Context: Greek Commentaries on Hippocrates in the Arabic Tradition* (Berlin) 71–90.

Allbutt, C. 1923. 'Two Books on Greek Science', *CR* 37: 129–30.

Álvarez Millán, C. 1999. 'Graeco-Roman Case Histories and Their Influence on Medieval Islamic Clinical Accounts', *Social History of Medicine* 12: 19–43.

—. 2000. 'Practice versus Theory: Tenth-Century Case Histories from the Islamic Middle East', *Social History of Medicine* 13: 293–306.

Anastassiou, A. and D. Irmer. 1997–2012. *Testimonien zum Corpus Hippocraticum*, 3 vols. (Berlin).

Anderson, G. 2000. 'Some Uses of Storytelling in Dio', in S. Swain, ed., *Dio Chrysostom: Politics, Letters and Philosophy* (Oxford) 145–52.

Andrewes, A. 1975. 'Could There Have Been a Battle at Oinoe?', in B. M. Levick, ed., *The Ancient Historian and His Materials: Essays in Honour of C. E. Stevens on his Seventieth Birthday* (Farnborough).

Anstey, P. 2011. 'The Creation of the English Hippocrates', *Medical History* 55: 457–78.

Arnott, W. G. 1996. *Alexis: The Fragments. A Commentary* (Cambridge).

Arrington, N. T. et al. 2016. 'Molyvoti, Thrace. Archaeological Project', *Hesperia* 85: 1–64.

Asheri, D., D. B. Lloyd and A. Corcella. 2007. *A Commentary on Herodotus Books I–IV*, eds. O. Murray and A. Moreno, trans. B. Graziosi (Oxford).

Asper, M. 2015. 'Medical Acculturation? Early Greek Texts and the Question of Near Eastern Literature', in B. Holmes and K. D. Fischer, eds., *The Frontiers of Ancient Science: Essays in Honor of Heinrich von Staden* (Berlin) 19–46.

Aston, E. 2004. 'Asclepius and the Legacy of Thessaly', *CQ* 34: 18–32.

Avery, H. C. 1979. 'The Three Hundred and Thasos, 411 BC', *CP* 74: 234–42.

Badoud, N. 2015. *Le Temps de Rhodes. Une chronologie des inscriptions de la cité fondée sur l'étude de ses institutions* (Munich).

Ballansée, J. 1990. 'The Battle of Oinoe in the Stoa Poikile: A Fake Jewel in the Fifth-century Athenian Crown?', *Anc. Soc.* 22: 91–126.

Barnes, J. 1979. *The Presocratic Philosophers I: Thales to Zeno* (London).

Batey, M. 1999. *Alexander Pope: The Poet and the Landscape* (London).

Berger, E., ed. 1970. *Das Basler Arztrelief* (Basel).

Berges, D. 2006. *Knidos: Beiträge zur Geschichte der archaischen Stadt* (Mainz am Rhein).

Best, J. G. 1969. *Thracian Peltasts and Their Influence on Greek Warfare* (Groningen).

Bickerman, E. J. 1986. 'Faux littéraires dans l'antiquité Classique', *Studies in Jewish and Christian History*, vol. III (Leiden) 196–211.

Bliquez, L. 2014. *The Tools of Asclepius: Surgical Instruments in Greek and Roman Times* (Leiden).

Blondé, F., A. Müller and D. Mulliez. 2000. 'Le Passage des Théores à Thasos: une énigme résolue? Questions de topographie et d'urbanisme à l'époque archaïque', *CRAI* 144: 885–907.

—. 2002. 'Évolution urbaine d'une colonie à l'époque archaïque: l'exemple de Thasos', in J.-M. Luce, ed., *Habitat et urbanisme dans le monde grec de la fin des palais mycéniens à la prise de Milet (494 av. J.-C.): Pallas. Revue d'études antiques* 58: 251–65.

— et al. 2008. 'Thasos in the Age of Archilochos: Recent Archaeological Investigations', in D. Katsonopoulou, I. Petropoulous and S. Katsarou, eds., *Archilochos and His Age* (Athens) 409–26.

Boardman, J. 1993. 'The Classical Period', in J. Boardman, ed., *The Oxford History of Classical Art* (Oxford) 83–150.

—. 2005. 'Composition and Content in Classical Murals and Vases', in J. M. Barringer and J. M. Hurwit, eds., *Periklean Athens and Its Legacy* (Austin, Texas) 63–72.

Bodnar, E. W. and C. Foss, eds. and trans. 2003. *Cyriac of Ancona: Later Travels* (Cambridge, Mass.).

Bond, G. W. 1981. *Euripides, Hercules* (Oxford).

Bosworth, A. B. 1988. *From Arrian to Alexander* (Oxford).

Bousquet, J. 1956. ' Inscriptions de Delphes', *BCH* 80: 579–91.

Bremmer, J. 1984. 'Greek Maenadism Reconsidered', *ZPE* 55: 267–86.

—. 2002. 'How Old is the Idea of Holiness (of Mind) in the Epidaurian Temple Inscription and the Hippocratic Oath?', *ZPE* 141: 106–8.

Briant, P. 2002. *From Cyrus to Alexander. A History of the Persian Empire* (Winona Lake).

Brock, R. 2013. *Greek Political Imagery from Homer to Aristotle* (London).

Brodersen, K. 1994. 'Hippokrates and Artaxerxes', *ZPE* 102: 100–110.

Bruneau, P. 1970. *Recherches sur les cultes de Délos à l'époque hellénistique et à l'époque impériale*, Bibliothèque des Écoles françaises d'Athènes et de Rome 127 (Paris) 413–30.

Brunet, M. 1997. 'Thasos et son Épire à la fin du Ve siècle et au début du IVe siècle av. J.-C.', in P. Brulé and J. Oulhen, eds., *Esclavage, guerre, économie en Grèce ancienne. Hommages à Yvon Garlan* (Rennes) 229–42.

—. 2000. 'Le territoire de Thasos, modèles et interpretation des données', in F. Kolb, ed., *Chora und Polis* (Munich) 79–86.

—. 2004. 'Le territoire de Thasos, modèles et interprétation des données', in F. Kolb, ed., *Chora und Polis* (Munich) 79–86.

Brunet, M., A. Coulié, P. Hamon et al. 2019. *Thasos. Heurs et malheurs d'un Eldorado antique* (Paris).

Brunt, P. A. 1988. *The Fall of the Roman Republic and Related Essays* (Oxford).

—. 1993. *Studies in Greek History and Thought* (Oxford).

Burckhardt, J. 1998. *The Greeks and Greek Civilisation*, trans. S. Stern and ed. O. Murray (London).

Burkert, W. 1985. *Greek Religion: Archaic and Classical* (Oxford).

—. 1992. *The Orientalizing Revolution* (Cambridge, Mass.).

Camp, J. M. 2001. *The Archaeology of Athens* (New Haven).

Cantor, D. 2018. 'Western Medicine since the Renaissance', in P. E. Pormann, ed., *The Cambridge Companion to Hippocrates* (Cambridge) 362–83.

Cawkwell, G., ed. and introd. 1972. *Xenophon: The Persian Expedition* (London).

Celotto, G. 2017. '*Eniautos* in Hesiod "Theogony" 58: One-Year Pregnancy in Archaic Greek Poetry', *Hermes* 145: 224–34.

Chamoux, F. 1953. *Cyrène sous la monarchie des Battiades* (Paris).

—. 1959. 'L'île de Thasos', *REG* 72: 351–6.

Chaouki, W., et al. 2009. 'Roots of Daphne gnidium L. inhibit cell proliferation and induce apoptosis in the human breast cancer cell line MCF-7', *Pharmazie* 64: 542–6.

Charneux, P. 1987. 'Du côté de chez Héra', *BCH* 111: 207–23.

Chryssanthaki-Nagle, K. 2007. *L'Histoire monétaire d'Abdère en Thrace (VIe s. av. J.-C. – IIe s. ap. J.-C.)* (Athens).

Clegg, R. 2004. *The First Scientist: A Life of Roger Bacon* (New York).

Cohn-Haft, L. 1956. *The Public Physicians of Ancient Greece* (Northampton, Mass.).

Conze, A. 1860. *Reise auf den Inseln des Thrakischen Meeres* (Hanover).

Cordes, P. 1994. *Iatros: das Bild des Arztes in der griechischen Literatur von Homer bis Aristoteles* (Stuttgart).

Coulié, A. 1998. 'Nouvelles inscriptions érotiques à Thasos', *BCH* 122: 445–53.

—. 2002. *La Céramique thasienne à figures noires*, Études Thasiennes 19 (Paris).

—. 2008. 'Archiloque et la colonisation de Thasos: l'apport de la céramique', in D. Katsonopoulou, I. Petropoulous and S. Katsarou, eds., *Archilochos and His Age* (Athens) 427–49.

Counillon, P. 1998. 'Datos en Thrace et le Périple du Pseudo-Scylax', *REA* 100: 115–24.

Craik, E. M. 1995A. 'Diet, *Diaita* and Dietetics', in A. Powell, ed., *The Greek World* (London) 387–402.

—. 1995B. 'Hippocratic Diaita', in J. Wilkins, D. Harvey, M. Dobson and F. D. Harvey, eds., *Food in Antiquity* (Exeter) 343–50.

—. 1998. *Hippocrates: Places in Man* (Oxford).

—. 2001A. 'Medical References in Euripides', *BICS* 45: 81–95.

—. 2001B. 'Thucydides on the Plague: Physiology of Flux and Fixation', *CQ* 51: 102–8.

—. 2003. 'Medical Language in the Sophoklean Fragments', in A. H. Sommerstein, ed., *Shards from Kolonos: Studies in Sophoclean Fragments* (Bari) 45–56.

—. 2005. 'The Hippocratic Treatise *Peri Opsios* (*De videndi acie*, On the Origin of Sight)', in P. J. van der Eijk, ed., *Hippocrates in Context* (Leiden) 191–207.

—. 2006. *Two Hippocratic Treatises: On Sight and On Anatomy* (Leiden).

—. 2009. *The Hippocratic Treatise on Glands* (Leiden).

—. 2016. *The Hippocratic Corpus. Content and Context* (London).

Crane, G. 1996. *The Blinded Eye: Thucydides and the New Written Word* (London).

Dana, M. 2012. 'Pontiques et étrangers dans les cités de la mer Noire: le rôle des citoyennetés multiples dans l'essor d'une culture régionale', in A. Heller and A.-V. Pont, eds., *Patrie d'origine et patries électives: les citoyennetés multiples dans le monde grec d'époque romaine. Actes du colloque international de Tours, 6–7 novembre 2009* (Paris) 249–66.

Dasen, V. 1993. *Dwarfs in Ancient Egypt and Greece* (Oxford).

—. 1997. 'Autour de *l'estropié* du Musée d'art et d'histoire de Genève: une représentation archaïque grecque d'hémimélie', *Gesnerus* 54: 5–22.

Daux, G. 1972. 'Stèles funéraires et épigrammes', *BCH* 96: 503–66, esp. 550–54.

Davies, M. 2010. 'From Rags to Riches: Democedes of Croton and the Credibility of Herodotus', *BICS* 53: 19–49.

De Angelis, F. 2014. 'Greek and Roman Specialized Writing on Art and Architecture', in C. Marconi, ed., *The Oxford Handbook of Greek and Roman Art and Architecture* (Oxford) 70–83.

De Sainte Croix, G. E. M. 1972. *The Origins of the Peloponnesian War* (London).

Dean-Jones, L. 1992. 'The Politics of Pleasure: Female Sexual Appetite in the Hippocratic Corpus', *Helios* 19: 72–91.

—. 1994. *Women's Bodies in Classical Greek Science* (Oxford).

—. 2010. 'Physician. A Metapaedogogical Text', in M. Horstmanshoff, ed., *Hippocrates and Education. Selected Papers Presented at the XIIth International Hippocrates Colloquium* (Leiden) 53–72.

Decker, W. 1995. *Sport in der griechischen Antike* (Munich).

Decourt, J.-C., T. H. Nielsen and B. Helly. 2004. 'Thessalia and Adjacent Regions', in M. H. Hansen and T. H. Nielsen, eds., *An Inventory of Archaic and Classical Poleis* (Oxford) 676–731.

Deichgraeber, K. 1971. *Die Epidemien und das Corpus Hippocraticum: Voruntersuchungen zu einer Geschichte der Koischen Ärzteschule*, repr. (Berlin).

—. 1982. 'Die Patienten des Hippokrates', *Akademie der Wissenschaften und der Literatur zu Mainz* 9: 8–43.

Delorme, J. 1960. *Gymnasion. Étude sur les monuments consacrés à l'éducation en Grèce (des origines à l'Empire Romain)* (Paris).

Demand, N. 1994. *Birth, Death, and Motherhood in Classical Greece* (Baltimore/London).

Demont, P. 1989. 'Les facteurs aggravants de la troisième constitution de Thasos', in G. Baader and R. Winau, eds., *Die hippokratischen Epidemien: Theorie – Praxis – Tradition*, Sudhoffs Archiv Beiheft 27 (Stuttgart) 198–204.

—. 2009. 'L'ancienneté de la médecine hippocratique: un essai de bilan', in A. Attia, G. Buisson and M. J. Geller, eds., *Advances in Mesopotamian Medicine from Hammurabi to Hippocrates* (Leiden/Boston) 129–49.

—. 2013. 'The Causes of the Athenian Plague and Thucydides', in A. Tsakmakis and E.-M. Tamiolaki, eds., *Thucydides between History and Literature* (Berlin) 73–88.

—. 2018. 'Herodotus on Health and Disease', in E. Bowie, ed., *Herodotus: Narrator, Scientist and Historian* (Boston) 175–96.

Devereux, G. 1970. 'The Psychotherapy Scene in Euripides' Bacchae', *JHS* 90: 35–48.

Dewhurst, K. 1966. *Dr. Thomas Sydenham (1624–1689): His Life and Original Writings* (London).

Dickenson, C. P. 2017. *On the Agora: The Evolution of a Public Space in Hellenistic and Roman Greece (c. 323 BC–267 AD)* (Leiden).

Dickey, E. 1996. *Greek Forms of Address from Herodotus to Lucian* (Oxford).

Dicks, D. R. 1966. 'Solstices, Equinoxes, and the Presocratics', *JHS* 86: 26–40.

—. 1972. 'More Astronomical Misconceptions', *JHS* 92: 175–7.

Dillon, M. P. E. 1994. 'The Didactic Nature of the Epidaurian Iamata', *ZPE* 101: 239–60.

Dodds, E. R. 1929. 'Euripides the Irrationalist', *CR* 43: 97–104.

—. 1951. *The Greeks and the Irrational* (California).

Dormandy, T. 1999. *The White Death: A History of Tuberculosis* (London).

Dowden, K. 1989. *Death and the Maiden* (London/New York).

Duchêne, H. 1992. *La Stèle du port. Fouilles du port 1. Recherches sur une nouvelle inscription Thasienne*, Études Thasiennes 14 (Paris) .

Dugand, J.-E. 1979. 'Les adresses de malades d'Épidémies 1 et 3 et les preuves tant archéologiques qu'épigraphiques du séjour d'Hippocrate à Thasos, capitale de l'île de ce nom', *Annales de la Faculté des lettres et sciences humaines de Nice* 35: 131–55.

Ebert, J. 1972. *Griechische Epigramme auf Sieger an gymnischen und hippischen Agonen* (Berlin).

Edelstein, L. 1967. *Ancient Medicine: Selected Papers of Ludwig Edelstein* (Baltimore).

Eijk, P. J. van der. 2005. *Medicine and Philosophy in Classical Antiquity: Doctors and Philosophers on Nature, Soul, Health and Disease* (Cambridge).

—. 2012. 'Exegesis, Explanation and Epistemology in Galen's Commentaries on Epidemics, Book One and Two', in P. E. Pormann, ed., *Epidemics in Context: Greek Commentaries on Hippocrates in the Arabic Tradition* (Berlin) 25–47.

Endo, N. and E. A. B. Eltahir. 2018. 'Prevention of Malaria Transmission around Reservoirs: An Observational and Modelling Study on the Effect of Wind Direction and Village Location', *Lancet Planet Health* 2: 406–13.

Faraone, C. A. 2006. 'Gli incantesimi esametrici ed i poemi epici nella Grecia antica', *QUCC* 84: 11–26.

—. 2009. 'Stopping Evil, Pain, Anger, and Blood: The Ancient Greek Tradition of Protective Iambic Incantations', *GRBS* 49: 227–55.

Farmakidou, E. 2014. 'Cupping Vessel, 500–450 BC', in N. C. Stampolidis and G. Tasoulas, eds., *Hygieia: Health, Illness, Treatment from Homer to Galen* (Athens) 295–6.

Feigenbaum, A. 1956. 'Description of Behçet's Syndrome in the Hippocratic Third Book of Epidemic diseases', *British Journal of Ophthalmology* 40: 355–7.

Finkel, I. L. and M. J. Geller, eds. 2007. *Disease in Babylonia* (Leiden).

Folkerts, M. 2005. 'Hippocrates of Chios', in *Brill's New Pauly: Antiquity*, vol. 1 (Leiden/Boston) 351–4.

Forrest, W. G. 1984. 'Herodotus and Athens', *Phoenix* 38: 1–11.

—. 1986. 'Epigraphy in Chios: Cyriac of Ancona to Stephanos', in J. Boardman, ed., *Chios* (Oxford).

Fournier, J. 2012. 'Les modalités de contrôle des magistrats de Thasos aux époques classique et hellénistique: réponse à Lene Rubinstein', in B. Legras and G. Thür, eds., *Symposion 2011: Études d'histoire du droit grec et hellénistique (Paris, 7–10 Septembre 2011) = Vorträge zur griechischen und hellenistischen Rechtsgeschichte (Paris, 7.–10. September 2011)* (Vienna) 355–64.

Fournier, J., P. Hamon and M. G. Parissaki. 2014–15. 'Recent Epigraphic Research in Thasos, Aegean Thrace & Samothrace (2005–2015)', *AR* 61: 75–93.

Fournier, J., P. Hamon and N. Trippé. 2011. 'Cent ans d'épigraphie à Thasos (1911–2011)', *REG* 124: 205–26.

Fowler, R. L. 1996. 'Herodotus and His Contemporaries', *JHS* 116: 62–87.

Francis, E. D. and M. Vickers. 1985. 'The Oenoe Painting in the Stoa Poikile and Herodotus' Account of Marathon', *ABSA* 80: 99–113.

Fraser, P. M. 1972. *Ptolemaic Alexandria*, 3 vols. (Oxford).

—. 2009. *Greek Ethnic Terminology* (Oxford).

Fraser, P. M. and G. E. Bean. 1954. *The Rhodian Peraea and Islands* (Oxford).

Frazer, R. M. 1972. 'Pandora's Diseases, *Erga* 102–4', *GRBS* 13: 235–8.

Frede, M. 1987. *Essays in Ancient Philosophy* (Minneapolis).

Friedrich, W.-H. 2003. *Wounding and Death in the Iliad: Homeric Techniques of Description*, trans. G. Wright (London).

Frölich, H. 1879. *Die Militärmedicin Homers* (Stuttgart).

Gagné, R. 2013. *Ancestral Fault in Ancient Greece* (Cambridge).

Garvie, A. F., ed. 2009. *Aeschylus: Persae* (Oxford).

Geller, M. J. 2004. 'West Meets East: Early Greek and Babylonian Diagnosis', in M. Horstmanshoff and M. Stol, eds., *Magic and Rationality in Ancient Near Eastern and Graeco-Roman Medicine* (Leiden) 11–61.

—. 2010. *Ancient Babylonian Medicine: Theory and Practice* (London).

—. 2018. 'A Babylonian Hippocrates: Medicine, Magic and Divination', in U. Steinert, ed., *Assyrian and Babylonian Scholarly Text Catalogues: Medicine, Magic and Divination* (Berlin) 42–54.

Georgiadou, A. 2010. 'La tablette d'Idalion réexaminée', *Cahiers du Centre d'Études Chypriotes* 40: 141–203.

Gera, D. L. 1993. *Xenophon's Cyropaedia: Style, Genre and Literary Technique* (Oxford).

Goldhill, S. 2002. *The Invention of Prose* (Oxford).

Goodall, E. W. 1934. 'On Infectious Diseases and Epidemiology in the Hippocratic Collection', *Proc. R. Soc. Med.* 27: 525–34.

Gourevitch, M. and D. Gourevitch. 1982. 'Un accès mélancolique', *L'Évolution psychiatrique* 47: 623–4.

Graham, A. J. 1998. 'The Woman at the Window: Observations on the Stele from the Harbour of Thasos', *JHS* 118: 22–40.

—. 2000. 'Thasos: The Topography of the Ancient City', *ABSA* 95: 301–27.

Grandjean, Y. 2010. 'Tessères de juge thasiennes', in A. M. Tamis, C. J. Mackie and S. G. Byrne, eds., *Philathenaios: Studies in Honour of Michael J. Osborne* (Athens) 65–80.

—. 2011. *Le Rempart de Thasos*, Études Thasiennes 22 (Paris).

—. 2012–13. 'Inscriptions de Thasos', *BCH* 136–7: 225–68.

Grandjean, Y. and F. Salviat. 2006. 'Règlements du Délion de Thasos', *BCH* 130: 293–327.

—. 2012. *Οδηγός της Θάσου*, 2nd edn (Athens).

—. 2012–13. 'Hippocrate et le sanctuaire de la Délienne à Thasos', *BCH* 136–7: 215–23.

Graumann, L. A. 2000. *Die Krankengeschichten der Epidemienbücher des Corpus Hippocraticum: medizinhistorische Bedeutung und Möglich-keiten der retrospektiven Diagnose* (Aachen).

Greco, E. 1999. 'Nomi di stradi nella città greche', in P. Orlandini and M. Castoldi, eds., *Koina. Miscellanea di studi archeologici in onore di Piero Orlandini* (Milan) 223–9.

Grensemann, H. 1970. 'Hypothesen zur ursprünglich geplanten Ordnung der hippokratischen Schriften *De fracturis* und *De articulis*', *Medizinhis-torisches Journal* 5: 217–35.

—. 1975. *Knidische Medizin. Teil 1* (Berlin).

—. 1982. *Hippokratische Gynäkologie. Die gynäkologischen Texte des Autors C nach den pseudohippokratischen Schriften* De mulieribus I, II und De sterilibus (Wiesbaden).

Gribble, D. 2012. 'Alcibiades at the Olympics: Performance, Politics and Civic Ideology', *CQ* 62: 45–71.

Griffin, J. 1980. *Homer on Life and Death* (Oxford).

Griffiths, A. H. 1987. 'Democedes of Croton. A Greek Doctor at the Court of Darius', in H. Sancisi-Weerdenburg and A. Kuhrt, eds., *Achaemenid History*, vol. II: *The Greek Sources* (Leiden) 37–51.

Grmek, M. D. 1983. 'Ancienneté de la chirurgie Hippocratique', in F. Lasserre and P. Mudry, eds., *Formes de pensée dans la Collection Hip-pocratique. Actes du IVe Colloque International Hippocratique: Lausanne, 21–26 Septembre 1981* (Geneva) 285–95.

—. 1989. *Diseases in the Ancient World* (Baltimore and London).

Guardasole, A. 2000. *Tragedia e medicina nell' Atene del V secolo a.C.* (Naples)

Guthrie, W. K. C. 1962. *A History of Greek Philosophy*, vol. I: *The Earlier Presocratics and the Pythagoreans* (Cambridge).

—. 1965. *A History of Greek Philosophy*, vol. II: *The Presocratic Tradition from Parmenides to Democritus* (Cambridge).

—. 1969. *A History of Greek Philosophy*, vol. III: *The Fifth-Century Enlightenment* (Cambridge).

Hallum, B. 2012. 'The Arabic Reception of Galen's Commentary on Hip-pocrates' "Epidemics"', in P. E. Pormann, ed., *Epidemics in Context:*

Greek Commentaries on Hippocrates in the Arabic Tradition (Berlin) 185–210.

Hammond, N. G. L. and G. T. Griffith. 1979. *A History of Macedonia*, vol. II: *550–336 BC* (Oxford).

Hamon, P. 2015–16. 'Études d'épigraphie thasienne, IV: Les magistrats thasiens du IVe s. av. J.-C. et le royaume de Macédoine', *BCH* 139–40: 67–124.

—. 2018. 'Études d'épigraphie thasienne, VI: Deux nouveaux blocs de la Grande Liste des Théores', *BCH* 142: 181–208.

—. 2019A. *Corpus des inscriptions de Thasos III: Documents publics du IVe siècle et de l'époque hellénistique*, Études Thasiennes XXVI (Athens).

—. 2019B. Étude d'épigraphie thasienne VII: en relisant les noms thasiens', *BCH* 143: 139–93.

Hankinson, R. J. 1998. *Cause and Explanation in Ancient Greek Thought* (Oxford).

Hansen, M. H. 2006. *The Shotgun Method: The Demography of the Ancient Greek City-State Culture* (Columbia, Mo.).

Hansen, M. H. and T. H. Nielsen, eds. 2004. *An Inventory of Archaic and Classical Poleis: An Investigation Conducted by the Copenhagen Polis Centre for the Danish National Research Foundation* (Oxford).

Hanson, A. E. 1990. 'The Medical Writers' Woman', in D. M. Halperin, J. J. Winkler and F. I. Zeitlin, eds., *Before Sexuality: The Construction of Erotic Experience in the Ancient Greek World* (Princeton) 309–38.

—. 1992. 'Conception, Gestation, and the Origin of Female Nature in the Corpus Hippocraticum', *Helios* 19: 31–71.

—. 1995. '*Paidopoiia*: Metaphors for Conception, Abortion and Gestation in the *Hippocratic Corpus*', in P. J. van der Eijk, H. F. J. Horstmanshoff and P. H. Schrijvers, eds., *Ancient Medicine in Its Socio-Cultural Context*, vol. I (Amsterdam/Atlanta) 291–307.

—. 1996. 'Phaenarete: Mother and Maia', in R. Wittern and P. Pellegrin, eds., *Hippokratische Medizin und antike Philosophie. Verhandlungen des VIII. Internationalen Hippokrates-Kolloquiums in Kloster Banz/ Staffelstein vom 23. bis 28. September 1993* (Hildesheim) 159–81.

—. 2010. 'Doctors' Literacy: Papyri of Medical Content', in M. Horstmanshoff, ed., *Hippocrates and Medical Education. Selected Papers Presented at the International Hippocrates Colloquium* ... (Leiden) 187–204.

Harrison, T. 2000. *Divinity and History: The Religion of Herodotus* (Oxford).

Hatemi, G. et al. 2016. 'One Year in Review 2016: Behçet's Syndrome', *Clinical and Experimental Rheumatology* 34: 10–22.

Hatzopoulos, M. and P. Paschidis. 2004. 'Makedonia', in M. H. Hansen and T. H. Nielsen, eds., *An Inventory of Archaic and Classical Poleis: An Investigation Conducted by the Copenhagen Polis Centre for the Danish National Research Foundation* (Oxford) 794–809.

Hawass, Z. et al. 2010. 'Ancestry and Pathology in King Tutankhamun's Family', *JAMA* 303–7: 638–47.

Helbig, W. 1963. *Führer durch die öffentlichen Sammlungen klassischer Altertümer in Rom* (Leipzig).

Helly. B. 1993. Ἡ Ὀδός Λάρισας-Γυρτώνης-Τέμπων, στην αναζήτηση του τάφου του Ιπποκράτη', *Θεσσαλικό Ημερολόγιο* 24: 3–17.

—. 2013. *Géographie et histoire des Magnètes de Thessalie*, vol. I: *De la plaine thessalienne aux cités de la côte égéenne (c. 750 – c. 300 av. J.-C.)* (Vareilles).

Hennig, D. 1995. 'Staatliche Ansprüche an privaten Immobilienbesitz in der klassischen und hellenistischen Polis', *Chiron* 25: 235–82.

Henry, A. 2002. 'Hookers and Lookers: Prostitution and Soliciting in Late Archaic Thasos', *ABSA* 97: 217–21.

Herzog, R. 1915. 'Zu den thasischen Theorenlisten', *Hermes* 50: 319–20.

Hiller, H. 1970. 'Zur Herkunft der Stele', in E. Berger, ed., *Das Basler Arztrelief* (Basel) 49–59.

Hillert, A. 1990. *Antike Ärztedarstellungen* (Frankfurt).

Holladay, A. J. 2002. *The Collected Papers of A. James Holladay: Athens in the Fifth Century and Other Studies in Greek History*, ed. A. J. Podlecki and F. Millar (Chicago).

Holmes, B. 2007. 'The Iliad's Economy of Pain', *TAPA* 137: 45–84.

—. 2010. *The Symptom and the Subject: The Emergence of the Physical Body in Ancient Greece* (Princeton).

—. 2012. 'Sympathy between Hippocrates and Galen: The Case of Galen's Commentary on Epidemics II', in P. E. Pormann, ed., *Epidemics in Context: Greek Commentaries on Hippocrates in the Arabic Tradition* (Berlin) 49–70.

Holtzmann, B. 1994. *La Sculpture de Thasos: Corpus des reliefs 1*, Études Thasiennes XV (Athens).

—. 2018. *La Sculpture de Thasos: Corpus des reliefs 11*, Études Thasiennes XXV (Athens).

Hornblower, S. 1987. *Thucydides* (London).

—. 1996. *A Commentary on Thucydides*, vol. II: *Books IV – V.24* (Oxford).

Horstmanshoff, H. F. J. and M. Stol. 2004. *Magic and Rationalisation in Ancient Near Eastern and Graeco-Roman Medicine* (Leiden).

Houdart, M. S. 1836. *Études historiques et critiques sur la vie et la doctrine d'Hippocrate, et sur l'état de la médecine avant lui* (Paris).

Hudson-Williams, H. L. 1951. 'Political Speeches in Athens', *CQ* 1: 68–73.

Hussey, E. 1990. 'The Beginnings of Epistemology: From Homer to Philolaus', in S. Everson, ed., *Epistemology* (Cambridge) 11–38.

Hutchinson, G. O. 2001. *Greek Lyric Poetry: A Commentary on Selected Larger Pieces* (Oxford).

—. 2013. *Greek to Latin* (Oxford).

Huysecom-Haxhi, S. 2009. *Les Figurines en terre cuite de l'Artémision de Thasos: Artisanat et piété populaire à l'époque de l'archaïsme mûr et récent*, Études Thasiennes 21 (Athens).

Ireland, S. and F. L. Steel. 1975. 'Φρένες as an Anatomical Organ in the Works of Homer', *Glotta* 53: 183–95.

Irigoin, J. 1980. 'La Formation du vocabulaire de l'anatomie en Grec: du mycénien aux principaux traités de la collection hippocratique', in M. D. Grmek, ed., *Hippocratica. Actes du colloque hippocratique de Paris (4–9 septembre 1978)*, Colloques internationaux du Centre national de la recherche scientifique no. 583 (Paris) 247–57.

Irwin, T. 1989. *Classical Thought* (Oxford).

Isaac, B. H. 1986. *The Greek Settlements in Thrace until the Macedonian Conquest* (Leiden).

Jacobs, E. 1893. *Thasiaca*, diss., Berlin.

Jacopi, G. 1929. *Scavi nella necropoli di Ialisso, 1924–1928*, Clara Rhodos III (Rhodes).

Jeffery, L. H. 1965. 'The Battle of Oinoe in the Stoa Poikile: A Problem in Greek Art and History', *ABSA* 60: 41–57.

Jenkins, I. 2015. *Defining Beauty: The Body in Ancient Greek Art* (London).

Jim, T. Suk Fong 2014. '*Aparchai* in the great list of Thasian *theōroi*', *CQ* 64: 13–24.

Johnston, A. W. 1989. 'Aeginetans Abroad', *Horos* 7: 131–5.

Joly, R. 1966. *Le Niveau de la science hippocratique* (Paris).

Jones, C. P. 2005. 'Ten Dedications "To the Gods and Goddesses" and the Antonine Plague', *JRA* 18: 293–301.

Jones W. H. S. 1907. *Malaria: A Forgotten Factor in the History of Greece and Rome* (Cambridge).

—. 1924. *The Doctor's Oath: An Essay in the History of Medicine* (Cambridge).

Jouanna, J. 1974. *Hippocrate. Pour une archéologie de l'école de Cnide* (Paris).

—. 1989A. 'Hippocrate de Cos et le sacré', *JS*: 3–22.

—. 1989B. 'Place des Épidémies dans la Collection hippocratique: le critère de la terminologie', in G. Baader and R. Winau, eds., *Die hippokratischen*

Epidemien: Theorie – Praxis – Tradition, Sudhoffs Archiv Beiheft 27 (Stuttgart) 60–87.

—. 1997. 'La lecture de l'éthique hippocratique chez Galien', in H. Flashar and J. Jouanna, eds., *Médecine et morale dans l'Antiquité*, Entretiens Fondation Hardt XLIII: 211–53.

—. 1999. *Hippocrates*, trans. M. B. DeBevoise (Baltimore).

—. 2000. *Hippocrate*, vol. IV.3: *Épidémies V et VII* (Paris).

—. 2003. *Hippocrate*, Vol. II.3: *La Maladie sacrée* (Paris).

—. 2005. 'Cause and Crisis in Historians and Medical Writers of the Classical Period', in P. J. van der Eijk, ed., *Hippocrates in Context* (Leiden) 3–27.

—. 2012. *Greek Medicine from Hippocrates to Galen – Selected Papers* (Leiden).

—. 2013. *Hippocrate*, vol. III.1: *Pronostic* (Paris).

—. 2014. 'Philologie et archéologie: la médecine hippocratique et les fouilles de Thasos', *REG* 127: 29–54.

—. 2016. *Hippocrate, Tome IV Iere Partie. Épidémies* I *et* III, with A. Anastassiou and A. Gardasole, Budé ed. (Paris).

—. 2018AB. 'Textual History', in P. E. Pormann, ed., *The Cambridge Companion to Hippocrates* (Cambridge) 38–62.

—. 2018B. *Hippocrate. Tome I 2e Partie. Le Serment. Les Serments Chrétiens. La Loi*, Budé ed. (Paris).

Jouanna, J. and C. Magdelaine, eds. 1999. *Hippocrate. L'art de la médecine* (Paris).

Kahn, C. 1970. 'On Early Greek Astronomy', *JHS* 90: 99–116.

—. 1979. *The Art and Thought of Heraclitus* (Cambridge).

—. 2003. 'Writing Philosophy: Prose and Poetry from Thales to Plato', in H. Yunis, ed., *Written Texts and the Rise of Literate Culture in Ancient Greece* (Cambridge) 139–61.

Kallet, L. 1999. 'The Diseased Body Politic, Athenian Public Finance, and the Massacre at Mykalessos (Thucydides 7.27–29)', *AJP* 120: 223–44.

—. 2013A. 'The Origins of the Athenian Economic *Arche*', *JHS* 133: 43–60.

—. 2013B. 'Thucydides, Apollo, the Plague, and the War', *AJP* 134: 355–82.

Kassell, L. 2014. 'Casebooks in Early Modern England: Medicine, Astrology and Written Records', *BHM* 88: 595–625.

Kessel, G. 2012. 'The *Syriac Epidemics* and the Problem of Its Identification', in P. E. Pormann, ed., *Epidemics in Context. Greek Commentaries on Hippocrates in the Arabic Tradition* (Berlin) 93–124.

Kind, F. E. 1922. 'Κλυστήρ', in *Paulys Real-Encyclopädie*, XI.1, 881–90.

King, H. 1998. *Hippocrates' Woman: Reading the Female Body in Ancient Greece* (London).

—. 2013A. 'Fear of Flute Girls, Fear of Falling', in W. V. Harris, ed., *Mental Disorders in the Classical World* (Leiden) 265–82.

—. 2013B. 'Female Fluids in the Hippocratic Corpus', in P. Horden and E. Hsu, eds., *The Body in Balance: Humoral Medicines in Practice* (New York/Oxford) 25–52.

—. 2013C. *The One-Sex Body on Trial: The Classical and Early Modern Evidence* (Dorchester).

—. 2013D. 'Motherhood and Health in the Hippocratic Corpus: Does Maternity Protect against Disease?', *Mètis. Anthropologie des mondes grecs anciens* 11: 51–70.

King, H. and S. Brown. 2015. 'Thucydides and the Plague', in C. Lee and N. Morley, eds., *A Handbook to the Reception of Thucydides* (Chichester) 449–73.

Kirk, G. S. 1956. 'A Passage in *De Plantis*', *CR* 6: 5–6.

Kollesch, J. 1989. 'Die diätetischen Aphorismen des sechsten Epidemienbuches und Herodikos von Selymbria', in G. Baader and R. Winau, eds., *Die hippokratischen Epidemien: Theorie – Praxis – Tradition*, Sudhoffs Archiv Beiheft 27 (Stuttgart) 191–7.

Kosak, J. C. 2004. *Heroic Measures: Hippocratic Medicine in the Making of Euripidean Tragedy* (Leiden).

Koukouli-Chrysanthaki, C. 1980. 'Οι αποικίες της Θάσου στο Β. Αιγαίο. Νεότερα ευρήματα', in *Β' Τοπικό Συμπόσιο. Η Καβάλα και η Περιοχή της* (Thessaloniki) 309–25.

—. 1990. 'Τα "μέταλλα" της θασιακής Περαίας', in T. Petridis, ed., Πόλις και χώρα στην αρχαία Μακεδονία και Θράκη. Μνήμη Δ. Λαζαρίδη (Thessaloniki) 494–514.

Kouloumenatas, S. 2018. 'Alcmaeon and His Addressees: Revisiting the Incipit', in P. Bouras-Vallianatos and S. Xenophontos, eds., *Greek Medical Literature and Its Readers: From Hippocrates to Islam and Byzantium* (London) 7–29.

Kudlien, F. 1965. 'Zum Thema "Homer und die Medizin"', *RhM* 108: 293–9.

Kuhrt, A. 2007. *The Persian Empire. A Corpus of Sources from the Achaemenid Period* (London).

Kuriyama, S. 1999. *The Expressiveness of the Body and the Divergence of Greek and Chinese Medicine* (New York).

Laios, K., M. Karamanou and G. Androutsos. 2012. 'A Unique Representation of Hypospadias in Ancient Greek Art', *Canadian Urological Association* 6: E1–E2.

Lane, J. 1996. *John Hall and His Patients: The Medical Practice of Shakespeare's Son-in-Law* (Stratford).

Lane Fox, R. 2005. 'Movers and Shakers', in A. Smith, ed., *The Philosopher and Society in Late Antiquity: Essays in Honour of Peter Brown* (Swansea) 19–50.

—. 2008. *Travelling Heroes: Greeks and Their Myths in the Epic Age of Homer* (London).

Langholf, V. 1977. *Syntaktische Untersuchungen zu Hippokrates-Texten: brachylogische Syntagmen in den individuellen Krankheits-Fallbeschreibungen der hippokratischen Schriftensammlung* (Mainz).

—. 1990. *Medical Theories in Hippocrates: Early Texts and the 'Epidemics'* (Berlin/New York).

Langslow, D. R. 2005. ' "Langues réduites au lexique"? The Languages of Latin Technical Prose', in T. Reinhardt, M. Lapidge and J. N. Adams, eds., *Aspects of the Language of Latin Prose* (Oxford), 287–302.

Launey, M. 1941. 'L'Athlète Théogène et le *Hieros Gamos* d'Héraklès Thasien', *RA* 18: 22–49.

Lazaridis, D. 1971. *Thasos and Its Peraia* (Athens).

Lebedev, A. 1993. 'Alcmaeon on Plants: A New Fragment in Nicolaus Damascenus', *PP* 48: 456–60.

—. 2017. 'Alcmeon of Croton on Human Knowledge, the Seasons of Life and Insomnia', in Chr. Vasallo, ed., *Physiologia: Topics in Pre-Socratic Philosophy and its Reception* (Trier) 227–57.

Leith, D. 2017. ' The Hippocratic Oath in Roman Oxyrhynchus', in M.-H. Marganne and A. Ricciardetto, eds., *En marge du Serment hippocratique, Papyrologica Leodiensia* 7 (Liége) 39–50.

Lendle, O. 1995. *Kommentar zu Xenophons Anabasis (Bücher 1–7)* (Darmstadt).

Leroi, A. M. 2014. *The Lagoon: How Aristotle Invented Science* (London).

Leven, K.-H., ed. 2005. *Antike Medizin. Ein Lexikon* (Munich).

Levitin, D. 2015. *Ancient Wisdom in the Age of the New Science: Histories of Philosophy in England, c. 1640–1700* (Cambridge).

Lewis, S. 2002. *The Athenian Woman: An Iconographic Handbook* (London).

Lightfoot, J. L. 2003. *Lucian: On the Syrian Goddess* (Oxford).

Ling, R. 1990. 'A Stranger in Town: Finding the Way in an Ancient City', *Greece and Rome* 37: 204–14.

Liolios, C. C., K. Graikou, E. Skaltsa and I. Chinou. 2010. 'Dittany of Crete: A Botanical and Ethnopharmacological Review', *Journal of Ethnopharmacology* 131: 229–41.

Liston, M. A. and L. Preston Day. 2009. 'It Does Take a Brain Surgeon: A Successful Trepanation from Kavousi, Crete', in L. A. Schepartz, S. C.

Fox C. Bourbou, eds., *New Directions in the Skeletal Biology of Greece*, Hesperia Supplements 43: 57–73.

Littman, R. J. 2009. 'The Plague of Athens: Epidemiology and Paleopathology', *Mount Sinai Journal of Medicine* 76: 459–67.

Littman, R. J., D. T. Durack et al. 2001. 'The Reply [to E. S. Olson et al.]', *American Journal of Medicine* 110: 674–5.

Littré, E. 1839–61. *Oeuvres complètes d'Hippocrate*, 10 vols. (Paris)

Lloyd, G. E. R. 1966. *Polarity and Analogy: Two Types of Argumentation in Early Greek Thought* (Cambridge).

—. 1975A. 'The Hippocratic Question', *CQ* 25: 171–92.

—. 1979B. 'Alcmaeon and the Early History of Dissection', Sudhoffs Archiv Beiheft 59 (Stuttgart) 113–47.

—. 1979. *Magic, Reason and Experience: Studies in the Origin and Development of Greek Science* (Cambridge).

—. 1983. *Science, Folklore and Ideology* (Cambridge).

—., ed., 1983. *Hippocratic Writings* (Harmondsworth).

—. 1990. *Demystifying Mentalities* (Cambridge).

—. 1996. *Adversaries and Authorities* (Cambridge).

—. 2002. *The Ambitions of Curiosity* (Cambridge).

—. 2004. *Ancient Worlds, Modern Reflections: Philosophical Perspectives on Greek and Chinese Science and Culture* (Oxford).

—. 2007. 'The Wife of Philinus, or the Doctors' Dilemma: Medical Signs and Cases and Non-Deductive Inference', in M. Burnyeat and D. Scott, eds., *Maieusis: Essays in Ancient Philosophy in Honour of Myles Burnyeat* (Oxford) 335–50.

—. 2009. 'Galen's Unhippocratic Case Histories', in C. Gill, T. Whitmarsh and J. Wilkins, eds., *Galen and the World of Knowledge* (Cambridge) 115–31.

—. 2010. *History and Human Nature. An Essay by G. E. R. Lloyd with Invited Responses*, Interdisciplinary Science Reviews 35 (Leeds).

—. 2014. *The Ideals of Inquiry: An Ancient History* (Oxford).

Lloyd, M. 1992. *The Agon in Euripides* (Oxford).

Longrigg, J. 1993. *Greek Rational Medicine: Philosophy and Medicine from Alcmaeon to the Alexandrians* (London).

—. 1999. 'Presocratic Philosophy and Hippocratic Dietetic Therapy', in I. Garofalo, A. Lami, D. Manetti and A. Roselli, eds., *Aspetti della terapia nel Corpus Hippocraticum. Atti del IXe Colloque International Hippocratique: Pisa, 25–29 settembre 1996* (Florence) 43–50.

Lonie, I. M. 1965. 'The Cnidian Treatises of the Corpus Hippocraticum', *CQ* 15: 1–6.

—. 1977. 'A Structural Pattern in Greek Dietetics and the Early History of Greek Medicine', *Medical History* 21: 235–60.

—. 1978. 'Cos versus Cnidos and the Historians', *History of Science* 16: 42–75 and 77–92.

—. 1985. 'The "Paris Hippocratics": Teaching and Research in Paris in the Sixteenth Century', in A. Wear, R. K. French and I. M. Lonie, eds., *The Medical Renaissance of the Sixteenth Century* (Cambridge) 155–74 and 318–26.

Loraux, N. 1995. *The Experiences of Tiresias: The Feminine and the Greek Man* (Princeton).

Loukopoulou, L. D. 2004. 'Thrace from Strymon to Nestos', in M. H. Hansen and T. H. Nielsen, eds., *An Inventory of Archaic and Classical Poleis: An Investigation Conducted by the Copenhagen Polis Centre for the Danish National Research Foundation* (Oxford) 854–69.

Loukopoulou, L. D. and S. Psoma. 2008. 'Maroneia and Stryme Revisited: Some Problems of Historical Topography', in L. D. Loukopoulou and S. Psoma, *Thrakika Zetemata I* (Athens) 55–86.

Luginbill, R. D. 2014. 'The Battle of Oinoe: The Painting in the Stoa Poikile and Thucydides' Silence', *Historia* 63: 278–92.

Ma, J. 2008. 'Chaironeia 338: Topographies of Commemoration', *JHS* 128: 72–91.

MacArthur, W. P. 1957. 'Historical Notes on Some Epidemic Diseases Associated with Jaundice', *British Medical Bulletin* 13: 146–9.

MacIntosh Turfa, J. 2012. *Divining the Etruscan World: The Brontoscopic Calendar and Religious Practice* (Cambridge).

Maclean, I. 2002. *Logic, Signs and Nature in the Renaissance* (Cambridge).

Macleod, C. W. 1982. 'Politics in the Oresteia', *JHS* 102: 124–44.

Maehler, H. 2004. *Bacchylides: A Selection* (Cambridge).

Maffre, J.-J. and A. Tichit. 2011. 'Quelles offrandes faisait-on à Artémis dans son sanctuaire de Thasos?', *Kernos* 24: 137–64.

Magdelaine, C. 2004. 'La littérature médicale aphoristique: paradoxes et limites d'un genre', in J. Jouanna and J. Leclant, eds., *Colloque: La Médecine grecque antique* (Paris) 71–94.

Majno, G. 1975. *The Healing Hand: Man and Wound in the Ancient World* (Cambridge, Mass.).

Malay, H., D. Amendola and M. Rici. 2018. 'The City of Tralleis Combats Immorality: Measures Taken against οἱ ἐν κιναιδείᾳ βιοῦντες in a New Civic Decree', *EA* 51: 91–7.

Malkin, I. 1987. *Religion and Colonization in Ancient Greece* (Leiden).

—. 2011. *A Small Greek World* (Oxford).

Manetti, D. 1999. 'Aristotle and the Role of Doxography in the Anonymus Londinensis', in P. J. van der Eijk, ed., *Ancient Histories of Medicine: Essays in Medical Doxography and Historiography* (Leiden) 95–141.

—. 2018. 'Late Antiquity', in P. E. Pormann, ed., *The Cambridge Companion to Hippocrates* (Cambridge) 316–39.

Manetti, D. and A. Roselli, eds., 1982. *Ippocrate: Epidemie, libro sesto* (Florence).

Mannack, T. 2001. *The Late Mannerists in Athenian Vase-Painting* (Oxford).

Manoledakis, M. 2003. Νέκυια. Το έργο του Πολύγνωτου στους Δελφούς (Thessaloniki).

Mansfeld, J. 1975. 'Alcmaeon: "*Physikos*" or Physician? with Some Remarks on Calcidius' "On Vision" Compared to Galen, *Plac. Hipp. Plat.* VII', in J. Mansfeld and L. M. de Rijk, eds., *Kephalaion: Studies in Greek Philosophy and Its Continuation Offered to Professor C. J. de Vogel* (Assen) 26–38.

—. 2013. 'The Body Politic: Aëtius on Alcmaeon on *Isonomia* and *Monarchia*', in V. Harte and M. Lane, eds., *Politeia in Greek and Roman Philosophy* (Cambridge) 78–95.

Marconi, C. 2010. 'Orgoglio e pregiudizio. La connoisseurship della scultura in marmo dell'Italia meridionale e della Sicilia', in G. Adornato, ed., *Scolpire il marmo: importazioni, artisti itineranti, scuole artistiche nel Mediterraneo antico. Atti del Convegno di studio tenuto a Pisa, Scuola normale superiore, 9–11 novembre 2009* (Milano) 339–59.

Marozzi, J. 2008. *The Man Who Invented History: Travels With Herodotus* (London).

Masson, O. 1983. *Les Inscriptions chypriotes syllabiques: Recueil critique et commenté*, 2nd edn (Paris).

May, J. M. F. 1950. *Ainos, Its History and Coinage 474–341 B.C.* (Oxford).

McDonald, G. C. 2009. *Concepts and Treatments of Phrenitis in Ancient Medicine*, diss. (Newcastle).

Meiggs, R. 1972. *The Athenian Empire* (Oxford).

Mellen, J. 2016. *Bore Hole*, 2nd edn (Devizes).

Métraux, G. P. R. 1995. *Sculptors and Physicians in Fifth Century Greece: A Preliminary Study* (Montreal).

Meyer-Steineg, T. 1912. 'Hippokrates-Erzählungen', *Archiv für die Geschichte der Medizin* 6: 1–11.

Meyerhof, M. 1935. 'Thirty-Three Clinical Observations by Rhazes (circa 900 A.D.)', *Isis* 23: 321–72.

Miller, E. 1889. *Le Mont Athos, Vatopédi, L'île de Thasos* (Paris).

Moisan, M., G. Maloney and D. Grenier. 1990. *Lexique du vocabulaire botanique d'Hippocrate* (Quebec).

Monterosso, G. 2014. 'Inscribed Grave Kouros. Mid-6th c. BC', in N. C. Stampolides and G. Tasoulas, eds., *Hygieia: Health, Illness, Treatment from Homer to Galen* (Athens) 327–30.

Morgan, K. A. 2015. *Pindar and the Construction of Syracusan Monarchy in the Fifth Century BC* (Oxford).

Mouton, J.-M. and C. Magdaleine. 2016. 'Le Commentaire au Serment hippocratique attribué au Galien dans un manuscrit arabe du haute Moyen Âge'. *CRAI*: 217–32.

Mueller, M. 2009. *The Iliad*, 2nd edn (London).

Mulhall, J. 2019. 'Plague before the Pandemic: The Greek Medical Evidence for Bubonic Plague before the Sixth Century', *Bulletin of the History of Medicine* 93: 151–79.

Muller, A. 2011. 'Les minerais, le marbre et le vin. Aux sources de la prospérité thasienne', *REG* 124: 179–92.

Mulliez, D. 2011. 'Histoire des fouilles de L'École française d'Athènes à Thasos', *CRAI*: 1115–33.

Nair, J. R. and R. J. Moots. 2017. 'Behçet's Disease', *Clinical Medicine* 17: 71–7.

Nance, B. 2001. *Turquet de Mayerne as Baroque Physician: The Art of Medical Portraiture* (Amsterdam).

Nelson, E. D. 2005. 'Coan Promotions and the Authorship of the Presbeutikos', in P. J. van der Eijk, ed., *Hippocrates in Context* (Leiden) 209–36.

Nikolaou, E. and S. Kravaritou, eds. 2012. *Αρχαίες πόλεις Θεσσαλίας και περίοικων περιοχών* (Larisa).

Nocita, M. 2012. *Italiotai e Italikoi: le testimonianze greche nel Mediterraneo orientale*, Hesperia 28 (Rome).

Nollé, J. 1983. 'Die "Charaktere" im 3. Epidemienbuch des Hippokrates und Mnemon von Side', *EA* 2: 85–98.

Nutton, V., ed., trans. and comm. 1979. *Galen: On Prognosis* (Berlin).

—. 1983. 'The Seeds of Disease: An Exploration of Contagion and Infection from the Greeks to the Renaissance', *Medical History* 27: 1–24.

—. 1988. 'Galen and Medical Autobiography', in V. Nutton, ed., *From Democedes to Harvey. Studies in the History of Medicine* (London) 52.

—. 1989. 'Hippocrates in the Renaissance', in G. Baader and R. Winau, eds., *Die hippokratischen Epidemien: Theorie – Praxis – Tradition*, Sudhoffs Archiv Beiheft 27 (Stuttgart) 420–39.

—. 1997. 'Hippocratic Morality and Modern Medicine', in H. Flashar and J. Jouanna, eds., *Médecine et morale dans l'Antiquité*, Entretiens Fondation Hardt XLIII: 31–63.

—. 2013. *Ancient Medicine*, 2nd edn (London).

Oerlemans, A. P. A. and L. E. Tacoma, 2014. 'Three Great Killers: Infectious Diseases and Patterns of Mortality in Imperial Rome', *AncSoc* 44: 213–41.

Ogden, D. 2017. *The Legend of Seleucus: Kingship, Narrative and Mythmaking in the Ancient World* (Cambridge).

Olson, P. E. et al. 1996. 'The Thucydides Syndrome: Ebola Déjà Vu (or Ebola Reemergent?)', *Emerging Infectious Diseases* 2: 155–6.

Onians, R. B. 1951. *The Origins of European Thought: About the Body, the Mind, the Soul, the World, Time and Fate* (Cambridge).

Ornaghi, M. 2009. *La lira, la vacca e le donne insolenti. Contesti di ricezione e promozione della figura e della poesia di Archiloco dall' arcaismo all'ellenismo* (Alessandria).

Osborne, R. 1987. *Classical Landscape with Figures: The Ancient Greek City and Its Countryside* (London).

—. 2009. 'The Politics of an Epigraphic Habit: The Case of Thasos', in L. Mitchell and L. Rubinstein, eds., *Greek History and Epigraphy: Essays in Honour of P. J. Rhodes* (Swansea) 103–14.

—. 2011. *The History Written on the Classical Greek Body* (Cambridge).

Page, D. L. 1953. 'Thucydides' Description of the Great Plague of Athens', *CQ* 3: 97–119.

Palagia, O. 2008. 'The Marble of the Penelope from Persepolis and Its Historical Implications', in S. M. R. Darbandi and A. Zournatzi, eds., *Ancient Greece and Ancient Iran. Cross-Cultural Encounters. Athens 11–13 November 2006* (Athens) 223–37.

Papadopoulou, Z. D. 2010–13. 'Παριακά', *Horos* 22–5: 403–18.

Papazarkadas, N. and D. Sourlas. 2012. 'The Funerary Monument for the Argives Who Fell at Tanagra: A New Fragment', *Hesperia* 81: 585–617.

Parke, H. W. and J. Boardman. 1957. 'The Struggle for the Tripod and the First Sacred War', *JHS* 77: 276–82.

Parker, L. P. E. 2016. *Euripides: Iphigenia in Tauris* (Oxford).

Parker, R. 1983. *Miasma: Pollution and Purification in Early Greek Religion* (Oxford).

—. 1996. *Athenian Religion: A History* (Oxford).

—. 2000. 'Theophoric Names and the History of Greek Religion', in S. Hornblower and E. Matthews, eds., *Greek Personal Names: Their Value as Evidence* (Oxford) 53–79.

Pease, A. S. 1948. 'Dictamnus', in J. Marouzeau, *Mélanges de philologie, de littérature et d'histoire anciennes: Offerts à J. Marouzeau par ses collègues et élèves étrangers* (Paris) 469–74.

Pébarthe, Ch. 1999. 'Thasos, l'empire d'Athènes et les emporia de Thrace', *ZPE* 126: 131–54.

Perdrizet, P. 1910. 'Scaptēsylē', *Klio* 10: 1–27.

Perilli, L. 2001. 'Alcmeone di Crotone tra filosofia e scienza', *QUCC* 69: 65–79.

—. 2009. 'Scrivere la medicina: la registrazione dei miracoli di Asclepio e le opere di Ippocrate', in C. W. Müller, C. Brockmann, C. W. Brunschön and O. Overwien, eds., *Antike Medizin im Schnittpunkt von Geistes- und Naturwissenschaften: internationale Fachtagung aus Anlass des 100-jährigen Bestehens des Akademienvorhabens Corpus Medicorum Graecorum/Latinorum* (Berlin) 75–120.

Philipp, H. 1990. 'Zu Polyklets Schrift "Kanon"', in P. C. Bol et al., eds., *Polyklet: der Bildhauer der griechischen Klassik* (Mainz) 135–55.

Picard, Ch. 1921. 'Fouilles de Thasos (1914 et 1920)', *BCH* 45: 86–173.

Picard, O. 2000. 'Le retour des émigrés et le monnayage de Thasos (390)', *CRAI* 144: 1057–84.

—. 2011A. 'La circulation monétaire dans le monde grec: les cas de Thasos', in T. Faucher, M.-C. Marcellesi and O. Picard, eds., *Nomisma. La Circulation monétaire dans le monde grec antique*, *BCH Suppl.* 53: 80–90.

—. 2011B. 'Un siècle de recherches archéologiques à Thasos: l'apport de la monnaie', *CRAI*: 1135–59.

Pinault, J. R. 1992. *Hippocratic Lives and Legends* (Leiden).

Pines, S. 1970–80. 'Al-Rhāzī', in C. C. Gillespie, ed., *Dictionary of Scientific Biography*, vol. XI (New York) 323–6.

Pirsig, W., E. Helidonis and G. Velegrakis. 1995. 'Medicine and Art: Facial Palsy Depicted in Archaic Greek Art on Crete', *American Journal of Otolaryngology* 16: 141–2.

Pleket, H. W. 1963. 'Thasos and the Popularity of the Athenian Empire', *Historia* 12: 70–77.

—. 1972. 'Isonomia and Cleisthenes: A Note', *Talanta* 4: 63–81.

—. 1995. 'The Social Status of Physicians in the Graeco-Roman World', *Clio Medica* 27: 27–34.

Polignac, F. de. 2003. 'Forms and Processes: Some Thoughts on the Meaning of Urbanization in Early Archaic Greece', in R. Osborne and B. W. Cunliffe, eds., *Mediterranean Urbanization 800–600 BC* (Oxford) 45–69.

Pollitt, J. J. 1990. *The Art of Ancient Greece: Sources and Documents* (Cambridge).

Pomata, G. 2005. 'Praxis Historalis: The Uses of Historia in Early Modern Medicine', in G. Pomata and N. Siraisi, eds., *Historia: Empiricism and Erudition in Early Modern Europe* (Cambridge, Mass.) 105–46.

Pormann, P. E., ed. 2008. *Rufus of Ephesus: On Melancholy* (Tübingen).

—. 2018. 'Arabo-Islamic Tradition', in P. E. Pormann, ed., *The Cambridge Companion to Hippocrates* (Cambridge), 340–61.

Potter, P. 1989. 'Epidemien 1/3: Form und Absicht der zweiundvierzig Fallbeschreibungen', in G. Baader and R. Winau, eds., *Die hippokratischen Epidemien: Theorie – Praxis – Tradition*, Sudhoffs Archiv Beiheft 27 (Stuttgart) 182–90.

Potter, P. and B. Gundert. 2005. 'Hippocrates of Cos', in *Brill's New Pauly*: 354–63.

Pouilloux, J. 1954. *Recherches sur l'histoire et les cultes de Thasos I* (Paris).

—. 1960. *Fouilles de Delphes*, vol. II: *Topographie et architecture. La région nord du sanctuaire* (Paris).

—. 1974. 'L'Héraclès thasien', *REA* 76: 305–16.

—. 1994. 'Théogénès de Thasos: quarante ans après', *BCH* 118: 199–206.

Powell, C. 2013. 'A Philological, Epidemiological and Clinical Analysis of the Plague of Athens', *Senior Honors Projects* 22. = http://collected.jcu.edu/honorspapers/22 (accessed February 2020).

Prêtre, C. 2011. 'Offrandes et dédicants dans les sanctuaires de Thasos', *REG* 124: 227–37.

Primavesi, O. 2008. *Empedokles: Physika I. Eine Rekonstruktion des zentralen Gedankengangs* (Berlin).

Pritchett, W. K. 1980. *Studies in Ancient Greek Topography, Part 3: Roads* (Berkeley).

Psoma, S. 2006. 'The "Lete" Coinage Reconsidered', in P. G. van Alfen, ed., *Agoranomia: Studies in Money and Exchange Presented to John H. Kroll* (New York) 61–86.

Raaflaub, K. A. 2013. '*Ktema es aiei*: Thucydides' Concept of "Learning through History" and Its Realisation in His Work', in A. Tsakmakis and E.-M. Tamiolaki, eds., *Thucydides between History and Literature* (Berlin) 3–22.

Radestock, S. 2015. *Prinzipien der ägyptischen Medizin: medizinische Lehrtexte der Papyri Ebers und Smith; eine wissenschaftstheoretische Annäherung* (Würzburg).

Raulff, U. 2017. *Farewell to the Horse: The Final Century of Our Relationship* (London).

Rawcliffe, C. 1995. *Medicine and Society in Later Medieval England* (Stroud) 63–8.

Rechenauer, G. 1991. *Thukydides und die hippokratische Medizin* (Hildesheim).

Revermann, M. 2017. *A Cultural History of Theatre in Antiquity* (London).

Rhodes, P. J. 2011. *Alcibiades: Athenian Playboy, General and Traitor* (Barnsley).

Ridgway, B. S. 1967. 'The Banquet Relief from Thasos', *AJA* 71: 307–9.

—. 1970. *The Severe Style in Greek Sculpture* (Princeton).

Riedweg, C. 2005. *Pythagoras: His Life, Teaching and Influence* (Ithaca, NY).

Robert, F. 1976. 'Prophasis', *REG* 89: 317–42.

—. 1979. 'Les adresses de malades dans les Épidémies II, IV et VI', in L. Bourgey, ed., *La Collection hippocratique et son rôle dans l'histoire de la médecine. Colloque de Strasbourg (23–27 octobre 1972) organisé par le Centre de recherches sur la Grèce Antique* (Leiden) 173–94.

Robert, L. 1926. 'Décret des Asclépiastes de Kolophon', *REA* 28: 7–9.

—. 1927. 'Les théores de Pergame', *REG* 40: 208–13.

—. 1964. *Les Stèles funéraires de Byzance gréco-romaine* (Paris)

Roberts, C. and K. Manchester. 2010. *The Archaeology of Disease*, 3rd edn (Stroud).

Robertson, M. 1975–6. *A History of Greek Art*, 2 vols. (Cambridge).

—. 1992. *The Art of Vase-Painting in Classical Athens* (Cambridge).

Rogers, E. 1932. *The Copper Coinage of Thessaly* (London).

Rolley, C. 1965. 'Le Sanctuaire des dieux *patrooi* et le Thesmophorion de Thasos', *BCH* 89: 441–83.

Roscino, C. 2010. *Polignoto di Taso* (Rome).

Roselli, A. 1989. 'Epidemics and Aphorisms: Notes on the History of Early Transmission of Epidemics', in G. Baader and R. Winau, eds., *Die hippokratischen Epidemien: Theorie – Praxis – Tradition*, Sudhoffs Archiv Beiheft 27 (Stuttgart) 182–90.

—. 2008. 'Medici e libri: l'epigramma per Argeo (IG II-III2 3783 = N. 8 Samama)', *Aion* 30: 154–63.

Rosenthal, F. 1956. 'An Ancient Commentary on the Hippocratic Oath', *Bulletin of the History of Medicine* 30: 52–87.

Roubineau, J.-M. 2016. *Milon de Crotone ou l'invention du sport* (Paris).

Rovelli, C. 2011. *The First Scientist: Anaximander and His Legacy* (Westholme).

Rutherford, I. 2013. *State Pilgrims and Sacred Observers in Ancient Greece* (Cambridge).

Rutherford, R. B. 1994. 'Learning from History: Categories and Case-Histories', in R. Osborne and S. Hornblower, eds., *Ritual, Finance, Politics: Athenian Democratic Accounts Presented to David Lewis* (Oxford) 53–68.

Ryholt, K. 2006. *The Petese Stories* II (Copenhagen).

Salazar, C. F. 2000. *The Treatment of War Wounds in Graeco-Roman Antiquity* (Leiden).

Saliou, C. 2003. 'Le nettoyage des rues dans l'Antiquité: fragments de discours normatifs', in P. Ballet, P. Cordier and N. Dieudonné-Glad, eds., *La Ville et ses déchets dans le monde romain: Rebuts et recyclages. Actes du colloque de Poitiers (19–21 Septembre 2002)* (Montagnac) 37–49.

Sallares, R. 2002. *Malaria and Rome: A History of Malaria in Ancient Italy* (Oxford).

Sallares, R., A. Bouwman and C. Anderung. 2004. 'The Spread of Malaria to Southern Europe in Antiquity: New Approaches to Old Problems', *Medical History* 48: 311–28.

Salta, M. 2014. 'Painted Marble Disc ca. 500 BC', in N. C. Stampolides and G. Tasoulas, eds., *Hygieia: Health, Illness, Treatment from Homer to Galen* (Athens) 330–33.

Salviat, F. 1979. 'Les colonnes initiales du catalogue des théores et les institutions thasiennes archaïques', *Thasiaca, BCH Suppl.* 5: 107–27.

—. 1984. 'Les archontes de Thasos', in Πρακτικα τοῦ ἡ συνεδρίου ελληνικής και λατινικής επιγραφικής, vol. I (Athens) 233–58.

—. 1986. 'Le vin de Thasos. Amphores, vin et sources écrites', in J.-Y. Empereur and Y. Garlan, eds., *Recherches sur les amphores grecques. Actes du Colloque international organisé par le Centre national de la recherche scientifique, l'Université de Rennes II et l'École française d'Athènes (Athènes, 10–12 septembre 1984)*, Bulletin de Correspondance Hellénique Supplément 13 (Athens) 145–95.

—. 1992. 'Calendrier de Paros et calendrier de Thasos. Boēdromia, Badromia et la solidarité des armes', in M.-M. Mactoux and E. Geny, eds., *Melanges P. Lévêque, 6. Annales Littéraires de l'Université de Besançon* 429 (1992) 261–7.

Salviat, F. and J. Servais. 1964. 'Stèle indicatrice thasienne trouvée au sanctuaire d'Aliki', *BCH* 88: 267–87.

Samama, E. 2003. *Les Médecins dans le monde grec: Sources épigraphiques sur la naissance d'un corps médical* (Geneva).

Sanchez, M. S. and E. S. Meltzer. 2012. *The Edwin Smith Papyrus. Updated Translation of the Trauma Treatise and Modern Medical Commentaries* (Atlanta).

Saunders, K. B. 1999. 'The Wounds in Iliad 13–16', *CQ* 49: 345–63.

—. 2004. 'Frölich's Table of Homeric Wounds', *CQ* 54: 1–17.

Scarborough, J. 1978. 'Theophrastus on Herbals and Herbal Remedies', *Journal of the History of Biology* 11: 353–85.

—. 1991. 'The Pharmacology of Sacred Plants, Herbs, and Roots', in C. A. Faraone and D. Obbink, eds., *Magika Hiera: Ancient Greek Magic and Religion* (Oxford) 138–74.

Scarrow, G. D. 1988. 'The Athenian Plague: A Possible Diagnosis', *AHB* 2: 4–8.

Schaps, D. 1977. 'The Woman Least Mentioned: Etiquette and Women's Names', *CQ* 27: 323–30.

Scheidel, W. 1994. 'Libitina's Bitter Gains: Seasonal Mortality and Endemic Disease in the Ancient City of Rome', *AncSoc* 25: 151–75.

—. 2013. 'Disease and Death', in P. Erdkamp, ed., *The Cambridge Companion to Ancient Rome* (Cambridge) 45–59.

Schmitt-Pantel, P. 1997. *La Cité au banquet* (Rome).

Schofield, M. 2012. 'Pythagoreanism: Emerging from the Presocratic Fog. Metaphysics A 5', in C. Steel and O. Primavesi, eds., *Aristotle's Metaphysics Alpha: Symposium Aristotelicum* (Oxford) 141–66.

Schwarzmaier, A. 2014. 'Attic Red-Figure Kylix ca. 500 BC', in N. C. Stampolides and G. Tassoulas, eds., *Hygieia: Health, Illness, Treatment from Homer to Galen* (Athens) 155–6.

Scurlock, J. A. 2014. *Sourcebook for Ancient Mesopotamian Medicine* (Atlanta).

Sdrolia, S. 2018. Η Ιστορία της Αμπελοκαλιέργειας στο Δήμο Αγιάς' = http://www.dimosagias.gr/component/k2/item/154.html (accessed February 2019).

Seaford, R. 2004. *Money and the Early Greek Mind: Homer, Philosophy, Tragedy* (Cambridge).

Sherk, R. K. 1990. 'The Eponymous Officials of Greek Cities: Mainland Greece and the Adjacent Islands', *ZPE* 84: 231–95.

—. 1993. 'The Eponymous Officials of Greek Cities V: The Register: Part VI: Sicily', *ZPE* 96: 267–95.

Sherwin-White, S. M. 1978. *Ancient Cos* (Göttingen).

Siegel, E. 1960. 'Epidemics and Infectious Diseases at the Time of Hippocrates. Their Relation to Modern Accounts', *Gesnerus* 17: 77–98.

Sierra Martin, C. 2018. 'Hares and Lions: A Hippocratic Reading and Commentary on Herodotus III.108', *Ágora. Estudios Clássicos em Debate* 20: 33–52.

Simonton, M. 2017. *Classical Greek Oligarchy: A Political History* (Princeton).

Simpson, St. J. and S. Pankova. 2017. *Scythians: Warriors of Ancient Siberia* (London).

Siraisi, N. 1994. 'Cardano, Hippocrates and Criticism of Galen', in E. Kessler, ed., *Girolamo Cardano: Philosoph, Naturforscher, Arzt* (Wiesbaden) 131–55.

Smith, R. R. R. 2007. 'Athletes and the Early Greek Statue Habit', in S. Hornblower and C. Morgan, eds., *Pindar's Poetry, Patrons and Festivals* (Oxford) 83–139.

Smith, W. D. 1973. 'Galen on Coans versus Cnidians', *Bulletin of the History of Medicine* 47: 569–85.

—. 1990. *Hippocrates: Pseudepigraphic Writings* (Leiden).

Sommerstein, A. H. and I. C. Torrance, eds., 2014. *Oaths and Swearing in Ancient Greece* (Berlin).

Staden, H. von, ed., trans. and comm. 1989. *Herophilus: The Art of Medicine in Early Alexandria: Edition, Translation and Essays* (Cambridge).

—. 1996. ' "In a pure and holy way": Personal and Professional Conduct in the Hippocratic Oath?', *Journal of the History of Medicine and Allied Sciences* 51: 404–37.

—. 1999. 'Rupture and Continuity: Hellenistic Reflections on the History of Medicine', in P. J. van der Eijk, ed., *Ancient Histories of Medicine: Essays in Medical Doxography and Historiography in Classical Antiquity* (Leiden) 143–87.

—. 2002. 'ὡς ἐπὶ τὸ πολύ: "Hippocrates" between Generalization and Individualization', in A. Thivel and A. Zucker, eds., *Le Normal et la pathologique dans la Collection hippocratique,* vol. I (Nice) 23–42.

—. 2006. 'Interpreting "Hippokrates" in the 3rd and 2nd centuries BC', in C. W. Müller, C. Brockmann and C. W. Brunschön, eds., *Ärzte und ihre Interpreten: medizinische Fachtexte der Antike als Forschungsgegenstand der Klassischen Philologie* (Leipzig) 15–47.

—. 2007. 'The *Oath*, the Oaths, and the Hippocratic Corpus', in V. Boudon-Millot, A. Guardasole and C. Magdelaine, eds., *La Science médicale antique: Nouveaux regards. Études réunies en l'honneur de Jacques Jouanna* (Paris) 425–66.

Stahl, H. P. 1966. *Thukydides. Die Stellung des Menschen im geschichtlichen Prozess* (Munich).

Stampolidis, N. C. and Y. Tassoulas. 2014. *Hygieia: Health, Illness, Treatment from Homer to Galen* (Athens).

Stansbury-O'Donnell, M. D. 1989. 'Polygnotos's Iliupersis: A New Reconstruction', *AJA* 93: 203–15.

—. 1990. 'Polygnotos's Nekyia: A Reconstruction and Analysis', *AJA* 94: 213–35.

—. 2005. 'The Painting Program in the Stoa Poikile', in J. M. Barringer and J. M. Hurwit, eds., *Periklean Athens and Its legacy: Problems and Perspectives* (Austin) 73–88.

Stewart, A. 2008. 'The Persian and Carthaginian Invasions of 480 B.C.E. and the Beginning of the Classical Style: Part 1, 2 and 3', *AJA* 112: 377–412 and 581–615.

Stewart, K. A. 2018. *Galen's Theory of Black Bile* (Leiden).

Swain, S. 1994. 'Man and Medicine in Thucydides', *Arethusa* 27: 303–27.

Swerr, A. 1961. *Arzt der Tyrannen: das Leben des grössten praktischen Arztes der Antike* (Munich).

Taylor, J. G. 1988. 'Oinoe and the Painted Stoa: Ancient and Modern Misunderstandings', *AJP* 119: 223–43.

Temkin, O. 1991. *Hippocrates in a World of Pagans and Christians* (Baltimore).

Thivel, A. 1981. *Cnide et Cos? Essai sur les doctrines médicales dans la Collection Hippocratique* (Paris).

Thomas, R. 1989. *Oral Tradition and Written Record in Classical Athens* (Cambridge).

—. 2000. *Herodotus in Context* (Cambridge).

—. 2006. 'Intellectual Milieu and the Plague', in A. Rengakos and A. Tsakmakis, eds., *Brill Companion to Thucydides*, vol. I (Leiden) 87–108.

Thonemann, P. 2006. 'Neilomandros. A Contribution to the Study of Greek Personal Names', *Chiron* 36: 11–43.

Thumiger, C. 2013. 'The Early Greek Medical Vocabulary of Insanity', in W. V. Harris, ed., *Mental Disorders in the Classical World* (Leiden).

—. 2016. 'The Tragic Prosopon and the Hippocratic Facies', *Maia*: 637–64.

—. 2017. *A History of the Mind and Mental Health in Classical Greek Medical Thought* (Cambridge).

Tomlinson, R. A. 1980. 'Two Notes on Possible Hestiatoria', *ABSA* 75: 221–8.

Topper, K. 2010. 'Maidens, Fillies and the Death of Medusa on a Seventh-Century Pithos', *JHS* 130: 109–19.

Totelin, L. M. V. 2009. *Hippocratic Recipes: Oral and Written Transmission of Pharmacology Knowledge in Fifth- and Fourth-Century Greece* (Leiden).

—. 2016. 'Technologies of Knowledge: Pharmacology, Botany, and Medical Recipes', in *Oxford Handbooks Online* (Oxford).

Touwaide, A. 2005. 'Healers and Physicians in Ancient and Medieval Cultures', in Z. Yaniv and U. Bachrach, eds., *Handbook of Medicinal Plants* (New York) 155–73.

—. 2006. 'Medicinal Plants', in *Brill's New Pauly* 558–68.

Trevor-Roper, H. 2006. *Europe's Physician: The Various Life of Sir Theodore de Mayerne* (New Haven/London).

Triantaphyllou, S. 1999. 'An Early Iron Age Cemetery in Ancient Pydna, Pieria: What Do the Bones Tell Us?', *ABSA* 93: 353–64.

Triebel-Schubert, C. 1984. 'Der Bergriff der Isonomie bei Alkmaion', *Klio* 66: 40–50.

Trippé, N. 2015–16. 'Une lettre d'époque classique à Thasos', *BCH* 139–40: 43–65.

Tsantsanoglou, K. 2008. 'Archilochos Fighting in Thasos. Frr. 93a and 94 from the Sosthenes Inscription', in D. Katsonopoulou, I. Petropoulous and S. Katsarou, eds., *Archilochos and His Age* (Athens) 163–80.

Tsiafalias, A. and B. Helly. 2010. 'Το Μυστήριο της Αρχαίας Μελίβοιας', in A. Tsiafalias et al., eds., *Αναζητώντας την Αρχαία Μελίβοια* (Meliboia) 5–26.

Tsingarida, A. 1998. 'À la découverte du corps humain', in A. Verbanck-Piérard, ed., *Au temps d'Hippocrate: Médecine et société en Grèce antique* (Mariemont) 53–64.

Tuplin, C. 2004. 'Doctoring the Persians: Ctesias of Cnidus, Physician and Historian', *Klio* 86: 305–47.

Tyerman, C. 2010. *New College* (London).

Tzochev, C. 2016. 'Markets, Amphora Trade and Wine Industry: The Case of Thasos', in E. M. Harris, D. M. Lewis and M. Woolmer, eds., *The Ancient Greek Economy: Markets, Households and City-States* (Cambridge) 230–53.

Ullmann, M., ed. and trans. 1978. *Rufus von Ephesos, Krankenjournale* (Wiesbaden).

Vagelpohl, U. 2014. *Galeni in Hippocratis Epidemiarum librum I commentariorum I–III Versionem Arabicam* (Berlin).

Vaxevanopoulos, M. 2017. *Recording and Study of Ancient Mining Activity on Mount Pangaion, E. Macedonia, Greece*, diss., Thessaloniki.

Villard, F. 1953. 'Fragments d'une amphore d'Euphronios au Musée du Louvre', *Monuments et mémoires* 47: 35–46.

Vlastos, G. 1953. 'Isonomia', *AJP* 74: 337–66.

Vout, C. 2018. *Classical Art: A Life History from Antiquity to the Present* (Princeton).

Wade-Gery, H. T. 1996. 'Thucydides', in S. Hornblower and A. Spawforth, eds., *The Oxford Classical Dictionary*, 3rd edn (Oxford) 1516–21.

Weber, A. S. 2013. 'Early Modern Medical Almanacs in Historical Context', *English Literary Renaissance* 33: 358–402.

Wecowski, M. 2014. *The Rise of the Greek Aristocratic Banquet* (Oxford).

Wees, H. van. 2000. 'Megara's Mafiosi: Timocracy and Violence in Theognis', in R. Brock and S. Hodkinson, eds., *Alternatives to Athens* (Oxford) 52–67.

—. 2008. 'Violence', *JHS* 128: 172–5.

Weidauer, K. 1954. *Thukydides und die hippokratischen Schriften: der Einfluss der Medizin auf Zielsetzung und Darstellungsweise des Geschichtswerks* (Heidelberg).

West, M. L. 1971. 'The Cosmology of "Hippocrates", De Hebdomadibus', *CQ* 21: 365–88.

—, ed. 1978. *Hesiod: Works and Days* (Oxford).

White, M. 2000. *Leonardo: The First Scientist* (London).

Wilkins, J. 2005. 'The Social and Intellectual Context of *Regimen II*', in P. J. van der Eijk, ed., *Hippocrates in Context* (Leiden) 121–33.

Wilkins, J. and S. Hill. 1996. 'Mithaikos and Other Greek Cooks', in H. Walker, ed., *Cooks and Other People. Proceedings of the Oxford Symposium on Food and Cookery 1995* (Totnes) 144–8.

Willi, A. 2003. *The Languages of Aristophanes. Aspects of Linguistic Variation in Classical Attic Greek* (Oxford).

—. 2008. 'νόσος and ὁσίη: Etymological and Sociocultural Observations on the Concepts of Disease and Divine (Dis)favour in Ancient Greece', *JHS* 128: 153–71.

Williams, B. 1993. *Shame and Necessity* (Berkeley).

Williams, D. 2006. 'The Chian Pottery from Naukratis', in A. Villing and E. Schlotzhauer, eds., *Naukratis: Greek Diversity in Egypt* (London) 127–32.

Withington, E. 1920. 'The Meaning of *Krisis* as a Medical Term', *CR* 34: 64–5.

Witt, M. 2009. *Weichteil- und Viszeralchirurgie bei Hippokrates. Ein Rekonstruktionsversuch der verlorenen Schrift* De vulneribus et telis (Berlin).

— 2018. 'Surgery', in P. E. Pormann, ed., *The Cambridge Companion to Hippocrates* (Cambridge) 217–45.

Woodward, A. M. 1910. 'Inscriptions from Thessaly', *Liv. Ann.* 3: 145–60.

Wootton, D. 2007. *Bad Medicine: Doctors Doing Harm since Hippocrates* (Oxford).

—. 2015. *The Invention of Science: A New History of the Scientific Revolution* (London).

Index

Robin Lane Fox is an emeritus fellow at New College, Oxford. He's the author of many books on ancient and classical history, includinɢ *Augustine*, *Alexander the Great*, and *The Classical World*, which was named one of the Top Ten Nonfiction Books of 2006 by the *Washington Post* Book World. He lives in Oxford, England.